T0192549

Analog Synthesizers: Understanding, Performing, Buying

Making its first huge impact in the 1960s through the inventions of Bob Moog, the analog synthesizer sound, riding a wave of later developments in digital and software synthesis, has now become more popular than ever.

Analog Synthesizers charts the technology, instruments, designers, and musicians associated with its three major historical phases: invention in the 1960s–1970s and the music of Walter Carlos, Pink Floyd, Gary Numan, Genesis, Kraftwerk, The Human League, Tangerine Dream, and Jean-Michel Jarre; re-birth in the 1980s–1990s through techno and dance music and jazz fusion; and software synthesis. Now updated, this new edition also includes sections on the explosion from 2000 to the present day in affordable, mass market Eurorack format and other analog instruments, which has helped make the analog synthesizer sound hugely popular once again, particularly in the fields of TV and movie music.

Major artists interviewed in depth include:

- Hans Zimmer (Golden Globe and Academy Award nominee and winner, "Gladiator" and "The Lion King")
- Mike Oldfield (Grammy Award winner, "Tubular Bells")
- Isao Tomita (Grammy Award nominee, "Snowflakes Are Dancing")
- Rick Wakeman (Grammy Award nominee, Yes)
- Tony Banks (Grammy, Ivor Novello and Brit Awards, Genesis)
- Nick Rhodes (Grammy Award winner, Duran Duran)

and from the worlds of TV and movie music:

- Kyle Dixon and Michael Stein (Primetime Emmy Award, "Stranger Things")
- Paul Haslinger (BMI Film and TV Music Awards, "Underworld")
- Suzanne Ciani (Grammy Award nominee, "Neverland")
- Adam Lastiwka ("Travelers")

The book opens with a grounding in the physics of sound, instrument layout, sound creation, purchasing, and instrument repair, which will help entry level musicians as well as seasoned professionals appreciate and master the secrets of analog sound synthesis. *Analog Synthesizers* has a companion website featuring hundreds of examples of analog sound created using dozens of classic and modern instruments.

Mark Jenkins has written about electronic music for *Melody Maker*, *International Musician*, *Keyboard Player* (UK), *Keyboard* (USA), and many other publications. He has performed and recorded solo and with members of Tangerine Dream, Can, Gong, White Noise, and Van Der Graaf Generator in the UK, USA, Europe, Brazil, Russia, and China, at venues including the Queen Elizabeth Hall, the London Planetarium, the Carnegie Science Center Pittsburgh, the Vanderbilt Planetarium, and the *Teatro Nacional* in Brazil.

Analog Synthesizers: Understanding, Performing, Buying

From the Legacy of Moog to Software Synthesis

Second Edition

Mark Jenkins

Routledge
Taylor & Francis Group

NEW YORK AND LONDON

Second edition published 2020
by Routledge
52 Vanderbilt Avenue, New York, NY 10017

and by Routledge
2 Park Square, Milton Park, Abingdon, Oxon, OX14 4RN

Routledge is an imprint of the Taylor & Francis Group, an informa business

First edition published by Focal Press 2007

Library of Congress Cataloging-in-Publication Data
Names: Jenkins, Mark, 1960– author.
Title: Analog synthesizers : understanding, performing, buying from the legacy of
    Moog to software synthesis / Mark Jenkins.
Description: Second edition. | New York, NY : Routledge, 2019. | Includes
    bibliographical references and index.
Identifiers: LCCN 2018057713 (print) | LCCN 2018059729 (ebook) |
    ISBN 9780429844386 (pdf) | ISBN 9780429844362 (mobi) |
    ISBN 9780429844379 (epub) | ISBN 9781138319387 (hardback : alk. paper) |
    ISBN 9781138319363 (pbk. : alk. paper) | ISBN 9780429453991 (ebook)
Subjects: LCSH: Synthesizer (Musical instrument)
Classification: LCC ML1092 (ebook) | LCC ML1092 .J47 2019 (print) |
    DDC 786.7/4—dc23
LC record available at https://lccn.loc.gov/2018057713

ISBN: 978-1-138-31938-7 (hbk)
ISBN: 978-1-138-31936-3 (pbk)
ISBN: 978-0-429-45399-1 (ebk)

Typeset in Garamond
by Apex CoVantage, LLC

Visit the companion website: www.routledge.com/cw/jenkins

Contents

Contents

Acknowledgements

SECOND EDITION 2018

Thanks to Claire Margerison and Lara Zoble at Focal Press; Simon Lowther and Stuart Steele at MSL Pro Distribution for Studio Electronics and Studiologic; Alex Theakston at Source Distribution for Arturia; Gus and Jerome at Wunjo Keyboards, London for invaluable market insights; Eric Lukac-Kuruc at Klavis for some valuable comments on the explanation of circuit design; and all the new interviewees, in particular, Suzanne Ciani for her insights into Buchla.

Dedicated to the memories of Keith Emerson and Edgar Froese.

FIRST EDITION 2007

Thanks to Matt Russell; Catharine Steers and David Bowers at Focal Press; my hosts at the Electro-Music festivals in the USA, Howard Moscowitz and Greg Waltzer; all the interviewees, in particular, Bob Moog, Jean-Michel Jarre, Mike Oldfield, Tony Banks, Nick Rhodes, Jeff Wayne, and Irmin Schmidt; all the contributors to the *Music for the 3rd Millennium* CDs, including Keith Emerson, Isao Tomita, Rick Wakeman, David Vorhaus/ White Noise, Patrick Moraz, Larry Fast, Karl Bartos, Richard Pinhas, Steve Baltes and Harald Grosskopf, Martyn Ware and Vince Clarke, Johannes Schmoelling, Michael Rother and Dieter Moebius, Michael Stearns, Geoff Downes, Paul Haslinger, and Dave Greenslade.

Dedicated to those who have gone to meet the great synthesist in the sky: Dr Robert A. Moog, Peter Bardens, Tim Souster, Neil Ardley, Michael Karoli, Ruediger Lorenz, Michael Garrison, Richard Burmer.

About the author

Born in Wales in 1960 and influenced by the early albums of Mike Oldfield and Tangerine Dream, Mark Jenkins independently developed the tape loop recording system also used by Terry Riley and Brian Eno, and started to record in 1978 using an electronic organ and modified frequency generators from his school physics department. His first analog instruments were a kit-built synthesizer and sequencer from *ETI* and *Practical Electronics* magazines.

Gaining an Honours Degree in English Literature, Philosophy, Psychology and History of Science at the University of Leicester, he used the university's music and video studios to create a series of synthesizer and video graphic concerts. Becoming Music Editor of *Electronics and Music Maker* magazine, he reviewed and used all the major instrument releases of the time, including the first commercially available MIDI synthesizer, the Sequential Prophet 600, and met many of the electronic music scene's most influential artists. From 1983 he was a committee member, and later organiser, of the UK Electronica "Future Age Music" festivals.

As Technical Editor of *Electronic Soundmaker* magazine, he used PPG, Prophet and other keyboards for recordings and for concerts at the Festival of Mind, Body and Spirit at London Olympia. In 1985, becoming Technical Editor of the weekly national music paper, *Melody Maker*, he launched AMP Music, which remains the UK's longest-established label for synthesizer, experimental and progressive music.

Releases for various artists on the label were accompanied by concerts at the London Planetarium and elsewhere, and Mark continued to write for music, video and computing magazines, including *International Musician*, *Music Technology*, *Studio Sound*, *Music Week*, *Future Music*, *Sound On Sound*, *Keyboard Player*, *Keyboard* (USA), *Virtual Instruments* (USA), *iCreate*, and *Music Mart*.

Composing steadily for computer games, film and theatre productions, books and virtual reality systems, he performed around the UK and at the Theatre National in Brazil in 1992, starting a schedule of one CD release per year. Mark's first CD, *Space Dreams*, was chosen for sale by the London Science Museum, becoming one of the UK's most popular synthesizer releases ever, while reviews of further albums compared his music favorably to that of Jean-Michel Jarre, Jan Hammer, Mike Oldfield, and Tangerine Dream.

On the AMP Music label, Mark released two albums plus his remix of a CD single for Keith Emerson of Emerson, Lake & Palmer, completing an almost clean sweep of the original innovators of European synthesizer and progressive music with other releases, including albums from White Noise, Tangerine Dream, Daevid Allen of Gong, and Richard Pinhas of Heldon.

Mark Jenkins in his studio in 1993: Korg SQ10 sequencer, ARP Odyssey, OSC OSCar, Moog Multimoog & Sonic 6; EMS Polysynthi, Roland Vocoder Plus, Elka Rhapsody, Yamaha CS80, Moog Liberation; Simmons SDS6 sequencer, Sequential Prophet 600, EDP Wasp, Linn 9000, Oberheim SEM, PPG Wave 2.2, Sequential 700 Programmer, Oberheim MiniSequencer, Korg PS3100 and Casio AZ1; Orla DMK7 master keyboard, Roland SH101, Roland MC202, EMS Synthi Aks, Minimoog, Roland VK09 organ, Roland EM101 module, Moog Polymoog.

Mark Jenkins has recorded and performed with members of Tangerine Dream, Can, Gong, Heldon, White Noise, and Van Der Graaf Generator. He has given multimedia concerts in the USA at the Carnegie Science Center Planetarium, Pittsburgh, the Franklin Institute Fels Planetarium, the Cheltenham Arts Center, Philadelphia, the Ocean County College Planetarium, New Jersey, in the UK at the London Astoria Theatre, the South Bank Purcell Room, and the Queen Elizabeth Hall, and at many other venues.

His most recent releases are tracks on the three CD volumes of *Music for the 3rd Millennium*, on which he also included rare or unreleased music from Keith Emerson, Rick Wakeman, and many others; the multi-synthesizer studio CDs *Sequencer Loops* and *Sequencer Loops 2*; the all-virtual synthesis CD *If The World Were Turned On Its Head, We Would Walk Among The Stars*; the double CD *Live in the USA*; and CD versions of Terry Riley's *A Rainbow In Curved Air*, and Mike Oldfield's *Tubular Bells*, all available through Amazon.co.uk or direct from his own website.

WEBSITES

www.markjenkinsmusic.com
www.youtube.com/markjenkinsmusic

Introduction: what's so great about analog?

How do we usually describe modern high-tech musical instruments? Terms such as "digital", "sampled" or "virtual" are commonplace, though tell us very little about what the instrument actually sounds like. But when we hear the word "analog", we get an instant idea of the type of noise we can expect – a certain sense of power, simplicity, and richness, which dates back to the earliest days of electronic music.

Of course, those times aren't so very far in the past. We're talking about a field which has a modern history of only 60-odd years – so that a few musicians still very active in the field today have been creative throughout the whole period of its development, not something which can be said of the guitar, the folk music group or the symphony orchestra.

Of course, electronic musical instruments, as part of a general high-technology trend which also includes computers in general, have developed at a pace previously unknown in any musical genre. It is amusing to be able to refer as "antique" to an instrument which may have only been launched in the 1960s, remained state of the art for just three or four years, lost almost all value in the 1970s and 1980s, and is now commanding prices, adjusted for inflation, even higher than when it was launched.

But that's exactly what has happened to many instruments in the analog field, and one major reason for the publication of this book is to put into perspective the real value and musical usefulness of these original instruments, as compared to the perceived collector's value, if any, which they may have developed in more recent years.

A second reason is to act as a tuition guide to the techniques and possibilities of analog, which are often now unclear to the latest generations of musicians. In the 1960s it was possible to play for hours with a new instrument and sometimes (particularly when the handbook, if one even existed, had become mysteriously separated from the machine) completely fail to get any kind of musical sound out of it. But the general concepts of analog did become pretty familiar after a while; it was widely realised that most machines were laid out in more or less the same way, and the instruments were mastered by a whole generation of musicians who could then concentrate on making them sound more expressive and genuinely musical.

In the 1970s and 1980s, new generations of instrument designs using new techniques and technologies were introduced, some incorporating aspects of analog sound, but some diverging from it completely. Some of these digital synthesis technologies seemed even more obscure than had analog originally (to be fair, in many senses they could be more powerful and flexible too) and in the effort to master them, many musicians were forced to neglect the simpler routines of analog sound creation. To the extent that, by the mid-1980s, a whole generation of musicians had appeared who experienced as much difficulty mastering analog sound synthesis as those of 20 years before.

So what were these half-forgotten analog sounds, and why are they still worth the effort of mastering them? In the late nineteenth and first half of the twentieth century, various attempts had been made to create electronic musical instruments, mostly keyboard equipped, but these were generally large, heavy and expensive designs. The degree of control they provided over the eventual sound was extremely limited and avant-garde composers, particularly those working in Germany in the 1950s, started to look for smaller electronic components that could be adapted to act as musical sound sources.

The problem, though, with these ad hoc instruments – often adapted from electronic test equipment, signalling equipment or radio components – was one of control, resulting in the creation of a style of music that incorporated some interesting sounds, but that would inevitably tend towards the unconventional and atonal.

The offer made by analog sound synthesis, starting from the early 1960s, was to combine the easy control of the earliest electronic keyboard instruments with the wider sonic possibilities of the avant-gardists' self-built equipment. This would bring whole new areas of possibility to sound creation without the necessity of abandoning traditional performance and compositional styles – something which appeared highly desirable once the first flush of rock and roll sensibility had passed in the early 1960s.

Although many others had been working towards the same end, the man generally associated with the invention of analog sound synthesis was Dr Robert (Bob) A. Moog, whose early 1960s patents led to his successful creation of a modular analog synthesizer system. Despite the fact that Moog himself was a non-musician, this turned out to be no dry technical achievement; the sound of the Moog synthesizer, beyond offering the relatively easy control that had been desired for some years, was actually surprisingly rich, powerful, and flexible.

The impact of the Moog sound (the name becoming generic for the synthesizer for a while) is well documented, and to some extent has remained the standard for which to aim, despite the passing of Bob Moog himself. The basic tone is rich, strong, and powerful; the Moog "twang" is distinctive (no instrument having previously offered such marked tonal changes through the course of a single note) and the cutting quality of the Moog leadline sound has yet to be excelled.

The impact of the analog synthesizer was more or less instant. Used as much to imitate conventional instruments as to generate new sounds, within only a few years it was raising questions from musicians' unions about possible unemployment amongst their members. Within ten years it was a staple element of all types of popular and experimental music, and it certainly was replacing orchestral musicians in many applications.

Perhaps it was through attempting to compete in the orchestral music arena that analog brought about its own downfall, since the distinctive richness of analog sound is of a quite different sort to that of conventional acoustic orchestral instruments. Whatever the reasons, in the 1980s more "realistic" additions and alternatives to analog synthesis began to appear,

Mark Jenkins with Dr Robert Moog in 2004.

and by the late 1980s there were virtually no analog instruments still in production.

The story of the subsequent "analog revival" is an exciting one, driven by the new, cutting-edge styles of dance music, which were paradoxically calling for the use of sounds that had first been heard some 20 years earlier. As a result, a huge second-user market for original analog instruments came into existence and (after a considerable delay caused by market uncertainty) a few analog instruments went back into production too.

After a few years, when the analog revival was really established as more than a passing enthusiasm, some of the larger Japanese manufacturers came onto the scene, though typically their contributions involved the use of only the most up-to-date technology available. "Virtual analog" instruments initially from Clavia in Sweden sought to emulate the abilities of the classic designs, while offering stability of tuning, enhanced computer control and other facilities denied to the first generations of instruments.

Whether these latest imitative analog instruments match or exceed the abilities and audio quality of those from the 1960s and 1970s is still a subject of massive debate. Certainly, the prices being asked for the classic instruments show no sign of decreasing, except in the case of the handful of models that at some point had become really seriously overvalued.

In this book, the technical explanation of the nature of analog sound creation is followed by the story of its birth and of its subsequent development by various designers, manufacturers and performers.

The individual components of analog sound creation are then examined in detail, with step-by-step examples of sound creation techniques. Then, the modern imitative analog instruments and software instruments are examined, again with detailed information on programming and playing them. This is followed by a new Chapter 8 on the most recent developments in hardware, software, and particularly the booming Eurorack modular field. The book is completed with appendices listing the major instrument lines currently available, hints on values and purchasing, other sources of information, and a discography of readily available recordings that give good examples of analog sound synthesis, including some modern releases on vinyl.

The website accompanying the book, www.routledge.com/cw/jenkins, gives many audio examples of analog sound creation basics, as well as more advanced techniques, and of the abilities of the individual instruments associated with classical and "virtual" analog sound synthesis, and includes video demos, interviews and other material.

The history and techniques of analog sound synthesis make a fascinating study in terms of the development of modern music, the practical techniques available to sound creators today, and of the design and technology of a popular consumer product spanning the last 60 years or so. As such, it is a subject that has long deserved detailed examination.

1

What is analog?

Before starting to look at the creative aspects of analog sound synthesis, it will be a huge help to develop a basic understanding of the principles of physics governing the whole subject of sound – perhaps because of all the methods of electronic sound creation, analog synthesis is probably the closest to those basic scientific principles. If that means a few hours thinking about what sound really is, how it can most easily be created electronically and how it is interpreted by the human ear, then that time spent will be more than paid back through a deeper, more intuitive approach to the handling of analog synthesizers, modules, and effects. So let's briefly take a look at some very basic physics before starting to look at analog instruments themselves.

SOUND

Sounds detected by the human ear only exist because of the medium of the air (sounds can also be transmitted through solid objects or liquids, but for our purposes we're discussing sounds created electronically and ultimately reproduced by some kind of conventional speaker system). A sound is a repeated pressure wave – a more or less regular change in the pressure of air arriving at the human ear, specifically at the eardrum.

When a sound strikes the ear, it arrives in the form of a rapid series of changes in the air pressure at that point in space. The most obvious way to create such pressure waves is to move an object contained in the same air medium somewhere nearby. You could do this just by striking two pieces of wood together, but in the case of electronic sound creation, the moving object is usually the cone of a speaker, which moves because it is attached to a magnet encircled by a coil of wire through which an electrical signal is passed – this is the same principle as an electric motor, but designed to create backward and forward rather than circular motion. The electrical signal, and so the sound reproduced, will have three major parameters: frequency, amplitude, and wave shape. Each is explained in the following paragraphs.

600 BC Greek philosopher Thales finds that rubbing amber *(electron)* makes it attract small objects

What is analog?

(a, b) Representation of a sound with increasing frequency (pitch). (c, d) Representation of a sound with increasing amplitude (volume). (e, f) Representation of simple organ and piano volume envelopes, and a complex volume envelope showing attack and decay times, sustain level and release time.

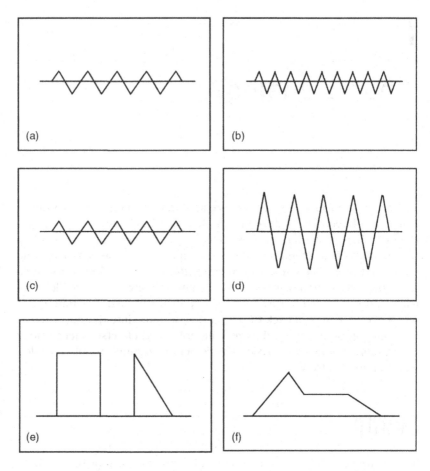

(a)

(b)

(c)

(d)

(e)

(f)

FREQUENCY

1500 William Gilbert extends Thales' electrostatics discoveries to include sulphur and glass

If one area of increased air pressure hits the ear each second, we call this a one cycle per second (CPS) or one hertz (1 Hz) sound, and the value in Hz of the sound is referred to as the frequency, or, in musical terms, the note or pitch of the sound. In fact, a 1 Hz sound is not audible to the human ear; sounds begin to become audible at around 20 Hz. A very low hum, such as those sometimes generated by electrical equipment, will be at 50 or 60 Hz. Low-pitched instruments, such as bass drums and bass guitars, produce sounds predominantly around 100–200 Hz; 440 Hz is often used as a tuning standard (in musical terms it is an "A" and so is also referred to as A440). Human voices and stringed instruments tend to produce pitches up to around 4000 Hz (4 kHz or 4 k); high-pitched whistles and other instruments will be producing sound up to around 12 kHz, and the highest pitches audible to the human ear are at around 16 kHz (young people manage somewhat higher). Some electronic equipment is designed to handle pitches even higher than this because there is a belief that higher pitched sounds can have a psychological effect, although not consciously audible. Whether this is true or not, it is assumed that equipment that can handle pitches higher than normally necessary will be more easily capable of handling pitches within the normal range of human hearing.

There is a further musical way of referring to pitch: classical church organs create sound using resonant pipes, the length of which were measured in feet. The longest pipes (perhaps 32 ft in length) give the deepest pitch, so on many analog synthesizers the pitches produced are referred to in terms of "footage", as 64', 32', 16', 8', 4', 2' and 1' – a range which covers six octaves or more (pitch increasing by one octave each time the footage is halved). Octaves are the musical intervals most clearly seen in the repeated pattern of the piano-style keyboard, and though the term derives from the inclusion of eight white keys per octave, in Western music there are five black keys per octave to be taken into account as well.

Electronic circuits can very easily be designed to create variations in output across the whole range of human hearing and beyond, and we will briefly look at what sort of electronic circuits are used in the analog synthesizer in particular. When a speaker reproduces a sound at a given pitch, it has to be able to vibrate at that pitch, which if it is well constructed, it will readily be able to do. Very small speakers are unable to vibrate at very slow speeds, though, and so the lowest pitches they can reproduce may well be relatively high; we would refer to these speakers as "lacking in bass". Very large speakers can vibrate slowly and reproduce bass frequencies well, but may have difficulty in vibrating very quickly; the highest pitches they can reproduce may be relatively low and we would refer to them as "lacking in treble". For this reason, speakers of different sizes are usually found in combination, to efficiently cover all the frequency ranges within the range of human hearing: usually, a small speaker (or "tweeter") for high frequencies and a large speaker (or "woofer") for low frequencies. Sometimes other speakers are included, designed to handle the middle frequencies ("mid-range" speakers), or the very lowest frequencies ("sub-woofers") and the very highest frequencies ("super-tweeters").

AMPLITUDE

The next parameter of any electrical signal being converted into a sound is amplitude; in other words, the size of the variation in the electrical signal level, and therefore the size of the variation in the air pressure level created, or its volume. A very great change in air pressure, from very high to very low and back again, when repeated, is a loud sound; a very small change in air pressure, from high to low and back again, when repeated, is a quiet sound. Again, in electronic music we are concerned with sounds reproduced by a speaker, and the loudness literally becomes a factor of the amount of air the speaker can move. The cone in a large speaker will actually move several centimetres when in

1729 Stephen Grey discovers properties of electrical conductors and insulators

1790 Death of Benjamin Franklin, who experimented with Leyden jars, an early form of battery

1837 First practical electric telegraph developed in the UK by Cooke and Wheatstone

action, and can create pressure waves comprising very large quantities of air. A very small speaker may be able to vibrate at just the same speed, but since the cone is smaller, this vibration moves much smaller quantities of air and the sound is quieter.

Trying to handle a very loud sound with a very small speaker can simply tear the speaker apart (it can do the same to your ears, in which the eardrum or tympanic membrane is acting like a tiny microphone, the reverse action of a speaker cone); this is even assuming that the electrical coil in the speaker can handle the very wide variation in electrical voltage represented by a very loud sound, which may simply melt the coil through excessive heating effects.

By the way, the "loudness" button found on some hi-fi systems has a related but less obvious function: it slightly boosts the highest and lowest frequencies when listening at lower volumes, to compensate for the fact that the human ear is slightly less sensitive to both of these at low listening volumes.

WAVE SHAPE

1867 Death of Michael Faraday, who made important discoveries in the science of electricity

The third major parameter of an electrical signal being converted into a sound is its wave shape. Two sounds of exactly the same pitch and exactly the same volume can sound quite different from one another. What exactly is happening here? The answer lies in the way in which the air pressure varies over time in each cycle. A regular variation that builds up towards a maximum value with its rate of increase slowing as it does so, reaches a peak, dies down to a minimum value, and then begins to build up again can be plotted against time to create a smooth repeated curve, which is referred to as a sine wave. This is the most basic type of wave, the one usually heard from an electronic tuning reference device; it is not particularly harsh or "cutting" and is often compared to the sound made by a flute or whistle.

But it is easily possible to make an electronic circuit vary its output in a quite different way; the signal can build at a constant rate until it reaches a maximum value, then immediately begin to decrease at exactly the same rate. When plotted against time this creates a series of triangular shapes, and so is referred to as a triangle wave. This sound is noticeably different to a sine wave – somewhat sharper and more cutting, and more comparable perhaps to the sound of a bassoon.

A third way of varying a signal is for it to move almost instantaneously from a low level to a maximum level, then fall back gradually to the low level before repeating the cycle at a constant rate. Because of the series of shapes made when this is plotted against time, this is referred to as a sawtooth wave; the opposite, building up at a constant rate then dropping almost instantaneously to the lowest level, can be referred to as an inverse sawtooth wave. Like the triangle wave, both waves sound more cutting than a sine wave, but a little more nasal.

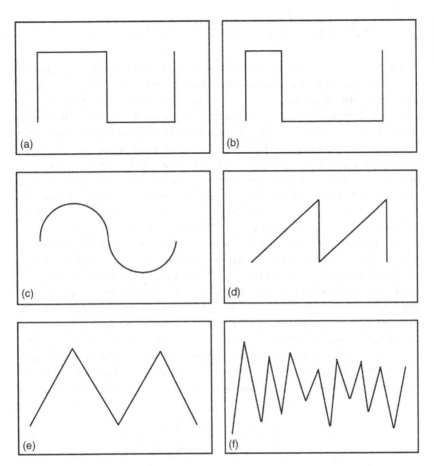

Waveforms: (a) square; (b) pulse; (c) sine; (d) sawtooth; (e) triangle; (f) noise.

A fourth way of varying the signal is to bring it up to a maximum level almost instantaneously, hold it at the high level for a time, then drop it almost instantaneously to the lowest level, hold it at that level for the same amount of time, then repeat the process. Plotted against time, this shows a series of square shapes and so is known as a square wave, and sounds stronger and more cutting than any of the previous shapes. Of course, the time for which the wave holds at its maximum value does not have to be exactly the same as the time for which it holds at its minimum value; the ratio between these two times is known as the mark/space ratio or pulse width, and can be expressed as a percentage, 20% or even 10% for "thin pulse" waves, which sound progressively thinner and weaker as the ratio decreases.

One interesting technique is applicable only to the square wave and does not readily apply to other waveforms. If an electronic circuit is designed so that the pulse width can be altered, it is possible to make the sound vary from thin and weak to strong and rich at will. Doing this under the control of another circuit is referred to as pulse width modulation (PWM), a common technique for making sounds more interesting and for introducing some apparent "movement" within a sound.

1874 Elisha Gray patents "singing telegraph", including an elementary keyboard

HARMONICS AND OVERTONES

1876 Scots-born Alexander Graham Bell patents telephone while at Boston University

There is another way to look at the construction of wave shapes, involving the consideration of "harmonics". When we consider a simple wave such as a sine wave we know its "fundamental" pitch or frequency, as discussed previously. Another wave at twice that frequency sounds naturally "in tune" with the first; we refer to it as the second harmonic and it's musically as well as mathematically related, since it's an octave above (another wave at half the frequency is one octave below the fundamental, and so on). A wave at three times the frequency of the fundamental can be referred to as the third harmonic, four times as the fourth harmonic and so on, and these frequencies are usually generated by musical instruments to some extent, though more quietly than the fundamental.

Interestingly enough, all other waves can be created from a combination of sine waves of different frequencies or harmonics superimposed on one another. To create a square wave from only sine waves, simply generate the basic frequency (the "fundamental") plus all its odd-numbered harmonics – the third, fifth, seventh, and so on. The sine wave quickly transforms into a rounded-off square, and as more and more oddly numbered harmonics are added, becomes a more or less perfect square. The same can be done, using different combinations of harmonics, to generate the triangle, sawtooth and all other waves, and so the tone of a musical instrument or sound depends largely on its content in terms of harmonics.

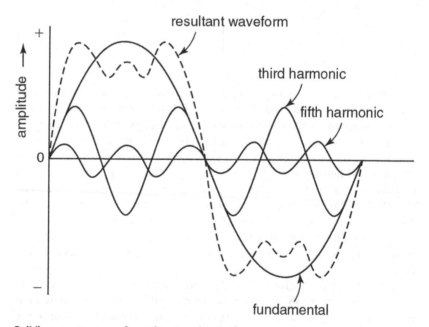

Building a square wave from sine wave harmonics.

Although some advanced digital synthesizers actually work this way – this so-called "harmonic" or "additive" synthesis method was found on the Kawai K5, Korg DSS1 and a few others – in using analog synthesizers, the common wave shapes have already been created for you, and the four common simple waveforms – sine, triangle, sawtooth, and square (and its variation the thin pulse) – are all we usually need to deal with. But, of course, sounds in the real world are much more complex than this. Even musical instruments that appear to have a fairly simple tone, such as an oboe, are actually producing something much more complex than any of these basic wave shapes. In what ways are sounds in the real world more complex than these simple electronically created sounds?

The fact is that a pure pitch at a certain frequency, such as can be created by a simple electronic circuit, will almost never be heard in the real world. Suppose a flute (constructed from a combination of wood and metal parts, and "brought to life" by the breath of a human player) is playing a note around 440 Hz. The loudest element of this note will indeed be at 440 Hz, but there will also be quieter elements at twice this frequency and at half this frequency, at three and four times the frequency, and at other intervals that will be mathematically but not necessarily musically related. The basic tone of the flute is similar to a sine wave, but these "overtones" generated at different frequencies tend to obscure this fact, and a printed display of the actual waveform put out by a flute can easily be seen to be much more complex than a simple sine wave. As for other instruments, such as an oboe, for which even the basic waveform is perhaps a more complex version of a sawtooth wave, the added overtones make the picture even more complicated. Both these are relatively simple instruments compared to something like an acoustic piano, which has a complex basic waveform as well as different sets of overtones created by the strings, body, and metal framework, in addition to interference effects created by the fact that strings are arranged in pairs each tuned to slightly differing pitches. So the overall tone or "timbre" of any acoustic instrument is a more complex wave than any of those created by the analog electronic circuits at which we will be looking, and if you combine several musical instruments together, as on a recording of a complete orchestra, the result is a waveform that is extremely complex and that becomes vastly different from one moment to the next.

1879 Death of James Clerk Maxwell, who unified laws of electricity and magnetism

NOISE

What, then, is the difference between organised sound, or music, and completely random noise? The waveform of a piece of music looks chaotic, so why does it sound tuneful coming out of the speaker? Many analog synthesizers do include a source of truly random noise, the "white noise" generator; white noise is a hissing sound like that coming from a radio tuned between stations, and is often used as a basis for creating sounds of the sea or wind. These applications give a clue to its nature, like the sea or wind, a white noise generator is creating tiny sounds at completely random frequencies and times, and the result has no easily discernible pitch. There are various different ways to create white noise electronically, and in fact it often creates itself and becomes a problem to eliminate completely from some circuit designs, but that is another matter.

1891 Stoney introduces the term electron for the theoretical smallest unit of electricity

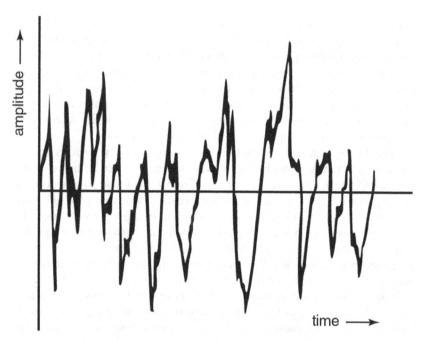

The frequency display for random (white) noise.

But all organised sounds or musical output, however complex they may appear, will comprise many elements which repeat in an organised way for however short a time, so while they may appear random, they show much more organisation than a white noise source.

PHASE

A final consideration, particularly relevant to stereo sound reproduction, is the question of phase. Consider a simple sine wave of a particular frequency; it repeatedly reaches the peak of its cycle, then starts to die away again. Another sine wave of exactly the same frequency can also be generated, starting at a very slightly later time, so that when the first is reaching the peak of its cycle, the second is halfway down. The two signals may be of the same frequency, loudness and waveform, but we say that one is out of phase with the other. If the two are added together, they cancel each other to a greater or lesser extent; if they are completely out of phase (we say they are 180 degrees out), they may cancel each other completely, so no sound is heard at all.

This is the principle of the phase shifter or "phaser", an (originally analog) sound effects unit that splits the incoming signal, adds a very short delay to one part so it is out of phase with the original, then mixes the two back together. Making the delay length change gradually alters the degree of cancellation and produces the distinctive swirling, "phasing" effect. But phase cancellation is something to be avoided in music reproduction; if, for instance, one speaker of a stereo pair is used with its connections reversed, one half of the sound will be reproduced 180 degrees out of phase with the other, and while the effects on a complex piece of music may not at first be obvious, you can be sure that the music is not being heard as it was intended.

The conclusion to be made from all these facts is that analog synthesis, in its most basic form, generally creates sounds that are much more simple than those heard all the time in the natural world. With some artistic imagination and technical ability, they can be altered and combined to give good imitations of some simple acoustic instruments, or to create new and less easily describable sounds that have just as much interest as some acoustic instruments. But the techniques of analog sound creation using simple wave shapes are never likely to match the more complex sounds of the real world, which is why other methods of electronic sound creation, such as digital sound synthesis and digital sound sampling, recording and manipulation, were also developed.

SYNTHESIZER COMPONENTS

It is time to take a look at what types of electronic circuitry actually create the sounds heard from analog synthesizers. This is clearly the subject of a whole (more technical) publication in itself, since the subject of electronic circuit design is beyond the scope of this book. But it is enough to say that although developed relatively recently – just over 50 years ago – the basic circuits in a conventional analog synthesizer can be extremely simple.

The first element is the one that actually makes an audible sound, the oscillator. As its name suggests, this is simply an electronic circuit that puts out an electrical signal which oscillates, or varies in value, in a regular and repeated way. One such circuit is known as the multivibrator. It can comprise just one or two cheap transistors (which are effectively just electrically controlled switches), a couple of resistors (which, as their name suggests, resist the passage of electrical current to a greater or lesser extent) and a handful of other components. When a steady power supply (of perhaps a few volts, easily obtained from a small battery) is applied to one end of the circuit, the other end of the circuit produces a voltage that varies rapidly in value. By selecting the value of the resistors and other components (or by using resistors that can be varied in value, otherwise known as potentiometers, which are quite simply what lie behind almost all synthesizer front-panel controls), the frequency, amplitude and wave shape coming from the circuit can be varied. Convert this to sound through a speaker and you have a simple (though not very tuneful) synthesizer, with (in musical terms) variable pitch, loudness, and tone. For reasons that we'll examine in detail elsewhere, analog oscillators are usually designed so that these various parameters can be precisely changed by varying the voltage (a measure of electrical potency rather than current flow) of an electrical signal applied to them, and so they are usually known as voltage-controlled oscillators (VCOs).

1896 Thaddeus Cahill develops the Telharmonium to give concerts over the US telephone system

The history of analog synthesizer design is the history of the electronics engineers who created rather more complex and increasingly musical-sounding circuit designs, and the musicians (or, in some disastrous cases, accountants) who decided exactly how they should be controlled and packaged. We have already looked at the oscillator, but of course there are many other circuits to be found on an analog synthesizer. Perhaps the next most important of these is the filter, a circuit that responds differently to electrical signals of different frequencies reaching it. Trumpet players have been achieving the effect of one type of filter for hundreds of years by simply placing their hands over the open end of the trumpet; the human hand tends to stop the higher harmonic elements of the trumpet's sound from escaping, but does not do so much to cut out the lower harmonic elements. The sound becomes softer, smoother and duller, and this is what we call "low pass filtering", because low harmonic elements are passed through and can still be heard, while high harmonic elements are prevented from passing through and can no longer be heard.

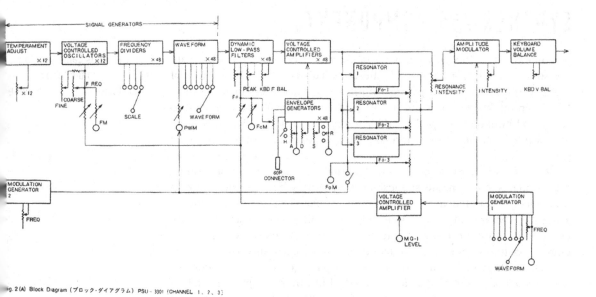

Fig. 2 (A) Block Diagram (ブロック・ダイアグラム) PSU – 3301 (CHANNEL 1, 2, 3)

A synthesizer creates sound by connecting various electronic circuits, here, in a Korg PS3300.

A graphic equaliser on a hi-fi system is also acting as a low pass filter if you reduce the level of the top three or four bands and leave the others centred; the result, again, is a smoother and duller tone with less of a "cutting" quality. On a synthesizer, the low pass filter is an electronic circuit that allows through the slow, low-pitched frequencies that are presented to it, but prevents the passage of fast, high-pitched frequencies. The exact frequency at which the filter starts to prevent signals from passing through is referred to as the "cut-off frequency" and can usually be varied; the amount by which signals above the cut-off frequency are reduced in volume, or the power of the filter, is measured in decibels per octave (dB/oct). A 10dB/oct filter (as commonly found in graphic equalisers) will make a frequency one octave above the cut-off point, subjectively half as loud as one exactly at the cut-off point. Some synthesizer filters operate at 12 dB/oct; more powerful ones operate at 24 dB/oct.

You may ask what effect this type of filtering would have on a sine wave-like signal like a flute sound at a particular frequency, and the answer is very little, unless the filter is actually set to cut out frequencies lower than the sine wave itself, because the sine wave-like sound has very few higher harmonics to be affected. But for other types of more complex waveform that have some higher harmonic content, applying a low pass filter audibly changes their tone, and if a sound consists of a good combination of different wave shapes, filtering can change the effect of the sound substantially. If a good-quality filter is set to a very low cut-off frequency, the incoming sound can disappear completely.

An early tape-based analog electronic music studio featuring EMS and Roland synthesizers and banks of graphic equalisers.

Some filters also slightly increase the level of the signal specifically around the cut-off point, this effect is referred to as emphasis, resonance or "Q". If this setting can be varied by the user, setting a high resonance level will start to add a distinct tone to the sound, which will vary just as the filter cut-off position is varied. Using a high "Q" setting has become a staple technique of analog sound synthesis, leading to sounds described as "wet" or "squelchy", and closing down a filter with a very high "Q" setting can be a very dramatic effect. As the resonance setting becomes very strong, a whistling sound is heard at a frequency exactly around the cut-off point and, on some filter designs, going to the highest possible resonance makes the filter begin to act as an oscillator, creating its own (sometimes very loud) whistling sound. This can be useful as a sound creation technique in itself, but has to be handled very carefully as it can unexpectedly create very strong, very low or high-pitched signals that can easily damage speakers. Some synthesizer filters are intentionally designed so that their maximum "Q" setting falls just short of making the filter oscillate.

Filters can also be applied to incoming sounds much more complex than simple combinations of oscillators. If a filter is used to treat a whole piece of music, setting a high cut-off point will make no audible difference to the music. Lowering the cut-off point to perhaps 12 kHz will start to reduce the audibility of high-pitched elements such as whistles and high-hat cymbals. As the cut-off point is lowered, string sounds and human voices will become duller and smoother, and finally very little will be left except bass drums and bass guitars. Close the filter down completely and the whole piece may become inaudible. For reasons that we'll examine elsewhere, filters are usually designed so that at least their cut-off frequency, if not other parameters such as their resonance as well, can be changed by varying the voltage of an electrical signal applied to them, and so they are usually known as voltage-controlled filters (VCFs).

With slight modifications to the electrical circuit, other designs of filter are available, including: "high pass", which acts in the opposite way to a low pass filter, allowing high frequencies through while cutting low frequencies; "band pass", which is a combination of the two, cutting high and low frequencies but leaving mid frequencies unaltered (a graphic equaliser is a series of band pass filters centred around different frequency points, while a parametric equaliser is a single band pass filter with its central frequency variable); and "band-reject" or "notch" filters, which act in the opposite way, reducing the volume of certain frequencies while leaving higher and lower frequencies unaltered. A "comb" filter, used in some effects unit designs, is simply a series of band pass or band-reject filters set to different frequencies but all acting together in parallel. It's possible to build a multi-mode filter that can be switched from low pass to high pass, band pass or other responses.

The third most important circuit found on a synthesizer is the envelope shaper. This is a circuit that varies the level of a signal over a short period of time, and is most obviously used to vary the volume (loudness) of the final sound by applying its output to the amplifier circuit, which is usually a voltage-controlled amplifier circuit (VCA). We have already discussed the pitch, timbre and loudness of musical instrument sounds, but not the way in which the volume varies during the course of an individual note. The easiest example here is the sound of a simple electric organ: hold a key down and a note sounds instantly, release it and the note stops instantly. But this is not the case, for instance, on a guitar; if a string is struck, it certainly begins to make a sound instantly, but the sound does not stop instantly, rather fading away gradually to nothing. On other instruments, such as the violin, the sound does not start instantly either, it may begin quietly, then build up to its loudest volume, then fade away again.

This variation in the volume of a sound is referred to as its volume "envelope", and an electronic circuit that can control this is an "envelope shaper" or "envelope generator". In an envelope-shaping circuit with variable values, the time taken for the sound to build up to its full volume is known as "attack"; the time taken to die away to nothing after the note is no longer being played is known as "release". Other parameters can also usually be set; the time taken for the sound to die down to the "sustain" level, even though the note is still being held, is known as "decay", and the volume level at which the sound continues to play while it is still being held (if any) is known as the "sustain" level. From the abbreviations A, D, S and R for attack, decay, sustain and release, envelope shapers are sometimes know as ADSRs, but note that while attack, decay and release are expressed as lengths of time, sustain is a level represented as a percentage of the maximum available volume achieved just at the end of the decay period. Envelope shapers can be simpler, perhaps just attack-release with optional sustain (AR switchable to ASR), or alternatively more complex, offering further stages of decay or release at different points, for instance, attack-decay-sustain-decay-release, or ADSDR.

Although envelope shapers are most importantly applied to volume, they can also be used to control other parameters. The most usual one is the filter cut-off point, so that the filter cut-off frequency, and so the overall tone of the sound, varies during the course of each and every note. Sometimes a separate second envelope shaper is provided for this purpose, and sometimes the job also has to be done by the volume envelope shaper. Either way, this is a very common technique in analog sound synthesis, creating one of the most distinctive types of analog sound available, and will be looked at in much more detail elsewhere.

It is important also to understand that, to some extent, the distinction between a sound and an electrical signal becomes blurred within the analog synthesizer. If you take the electrical signal from an oscillator and apply it directly to a speaker, you will hear a sound. But if you take the same electrical signal and apply it instead to the part of the filter circuit that controls its cut-off frequency, you will hear the cut-off frequency changing more or less rapidly as a function of the frequency of the oscillator. If you apply the output of an envelope generator to a voltage-controlled amplifier, you will hear a particular volume envelope, but if you apply it to the cut-off control input of a filter, you will hear a change in tone through the course of a note. In other words, a varying electrical voltage can either become a sound in itself or a way of controlling some aspect of a different sound altogether. On a well-designed analog synthesizer, any voltage can be sent anywhere without the possibility of causing damage to an inappropriate circuit, though on some, such as the Buchla designs, audio signals are often made distinct from control signals with different types of cables or connections.

Early compact analog synths such as the Micromoog were perfect instruments for beginners.

CIRCUIT DESIGN

While the earliest analog synthesizers used small individual electronic components, such as transistors, resistors, and capacitors ("discrete" components) in their circuit designs, these were quickly superseded by ICs (integrated circuits, or "chips"), which in their simplest form comprise just a handful of miniaturised transistors, resistors, and other components assembled into a small, easily handled block. Some of the earliest chips comprised a whole oscillator design in a single component, and it was not long before filters, envelope shapers, amplifiers and even entire synthesizer circuits in the form of single ICs became available through companies such as Solid State Micro Technology (SSM) and Curtis Electro Music (CEM). Although the functional difference between discrete component and IC designs matters very little to the musician, there have been endless arguments about perceived slight differences in sound quality between synthesizers using discrete components and those using ICs, and even between those using different types or versions of an IC. Only one thing is for certain: while discrete components are relatively easily replaced if they fail, some older chips are now no longer obtainable, and so some models of analog synthesizer using ICs can now be extremely difficult to repair.

SOUND DESIGN

We have looked at how the laws of physics define the nature of a sound, and how the relevant parameters are produced and controlled using electronic circuits such as oscillators, filters, and envelope shapers. Understanding what these circuits are contributing to the sound is fundamental to any imaginative use of analog instruments, and certainly to any attempt to imitate particular types of realistic instrumental sounds using analog technology. As will be seen in subsequent chapters, a good combination of basic knowledge of the physics of sound, plus some artistic imagination, needs to go into any attempt to create interesting sounds using analog techniques. For instance, imagine the process of attempting to reproduce a fairly complex acoustic instrument sound, such as that of a plucked string instrument like a bouzouki, using analog techniques. The timbre of the string is quite rich and complex, with many high and low harmonics, so a simple wave shape such as a sine wave will not be of very much help. A combination of richer waves such as sawtooth and thin pulse waves at various multiples of the basic frequency will be of more help, and will add in the overtones that give the instrument its rich sound.

But other factors add to the original sound, such as the resonance from the wooden body of the instrument, which could be simulated with a sub-octave of a smoother sine wave. The overall loudness envelope is like that of a piano, starting instantaneously but dying away slowly, and is easy enough to set. But if the instrument is to sound as if it is being played with a plectrum, the sound of the plectrum also has to be added. This will be

The Technosaurus Selector B was a powerful small modular analog synthesizer.

very short, percussive and at no particular pitch, so a different envelope controlling another unpitched sound source, such as white noise, has to be added in parallel to the first.This sort of complex task is a typical one in advanced analog synthesis, and more examples can be found elsewhere.

Given a basic understanding of the principles of physics behind the creation of sound, and of the type of electronic circuit used in the analog synthesizer to generate and modify these sounds, we are now ready to look in Chapters 3 and 4 at the designers, manufacturers, and artists involved in the field, but firstly in Chapter 2 at how the wonders of analog sound synthesis can actually be applied to making music (or at least musical sounds), and hopefully to having a little fun.

2

Aspects of analog sound

In Chapter 1 we looked at the principles of physics that underlie the electronic creation of sound, and the simple circuits used to do this. In this chapter we will look more systematically at the facilities available on analog synthesizers, and give some ideas about how to apply them practically and imaginatively.

As we will see in Chapter 3, the first analog synthesizers were modular – they were divided into individual circuits, each one carrying out a different task – which had to be connected or "patched" together in order to create a complete sound. This may well have been because Robert Moog solved the problems of designing a voltage-controlled oscillator, a voltage-controlled filter, and a voltage-controlled amplifier one at a time, but whether this is true or not, the fact is that the modular synthesizer has advantages and disadvantages. A disadvantage is that it will not make a sound until some substantial work has been done to set it up; an advantage is that it is extremely flexible and rewards experimentation and imaginative use.

For practical reasons there was soon a lot of pressure towards the introduction of a non-modular synthesizer, and from the Minimoog onwards the majority of synthesizer designs were not modular, a trend which has now markedly reversed. But even in the early days, the legacy of modular design persisted, even if individual elements of the synthesizer were internally connected and did not need to be patched together, the front panels of instruments still tended to be marked up as if the instrument comprised separate modules. In fact, the Minimoog more or less defined the standard arrangement of facilities on analog synthesizers: one or more oscillators to create a basic sound, leading to a filter to alter its tone, then an envelope shaper to control the loudness and filter setting during the course of a note, and perhaps with white noise or an external sound source mixed in.

But as early as the time of the Minimoog design, some of the fine details of modular synthesizer design were being obscured. On a truly modular synthesizer, volume levels are controlled by voltage-controlled amplifiers, and to alter a level during the course of a note, these have to be "patched" to an envelope generator. On the Minimoog and later designs, it is assumed that one of the envelope shapers will always be used to control output volume, so the voltage-controlled amplifier for audio output is permanently connected to an envelope generator, not separately accessible as such (see Appendix C for a photo of the more modular Minimoog Model B prototype).

1920 Leon Termen develops the gesture-responsive Theremin in Russia

1928 Keyboard-equipped Ondes Martenot invented in France, influenced by Theremin

6

Other modules are also conflated on integrated instrument designs. Some low-frequency oscillators (LFOs) used for modifying pitch or filter cut-off levels will offer a "random" setting as well as the more conventional sine, sawtooth or square wave shapes. But this obscures the fact that random voltages in strictly analog systems were generally created using three separate circuits: a white noise or random voltage source; a sample-and-hold circuit, which sets itself to the level of the white noise source in a fraction of a second and then holds at that level; and the LFO itself, which controls the frequency with which a new sample is taken (of course, the term sample-and-hold is not to be confused with the more modern definition of "sound sampling", though it does have some aspects in common). This sort of design can obscure rather than clarify the facilities available on a particular instrument, for instance, some instruments which clearly must have an internal random or white noise source do not make this available as an independent audio signal.

For these reasons it's best to study the details of analog synthesizer design using a fully modular system, applying the lessons learned to simpler integrated instruments later on. Rather than appearing to favour a current manufacturer, in this chapter many of the examples are taken from the long discontinued Selector system by Technosaurus, a Swiss manufacturer which began to introduce its designs in 1995. The Selector resembled the earliest Moog modular systems using full-sized, quarter-inch jack sockets, which helped make it both more reliable and easier to patch into other professional audio systems than smaller modular systems made elsewhere in Europe and the UK, using mini-jack patching, or US-made designs such as PAiA and Buchla using banana sockets. Also, the Selector placed its interface sockets in groups away from the related controls, so its layouts are very clear. Alongside this powerful but now discontinued system we'll compare the modern Studio Electronics Boomstar SEM, a small MIDI-equipped monophonic module released in 2018, which offers similar facilities in a compact integrated format not requiring patch cables.

1929 Laurens Hammond launches tonewheel organ in the USA

1934 Birth of Robert Moog in Flushing, NY

1939 Second generation of Hammond organs now using valve technology

VOLTAGE-CONTROLLED OSCILLATOR

The voltage-controlled oscillator (VCO) is the basic building block and most frequently used sound source in any synthesizer application, and its output will exhibit at the very least the three most obvious parameters described in Chapter 1, frequency, wave shape, and amplitude, or, in more musical terms, pitch, tone, and loudness.

On most synthesizer designs the VCO would more correctly be described as a voltage-controlled audio oscillator because it is designed to create electrical signals, and so eventually sounds more or less within the range of human hearing. The low-frequency oscillator (LFO), described later, uses a very similar circuit, specified simply to give an output with a much lower speed of variation, and so generally directed towards controlling other synthesizer functions rather than itself creating a sound. On some synthesizers, such as the Minimoog, one or more oscillators are switchable between LFO and VCO applications, but when only audio

1947 Selmer Clavioline invented and licensed for manufacture in several countries

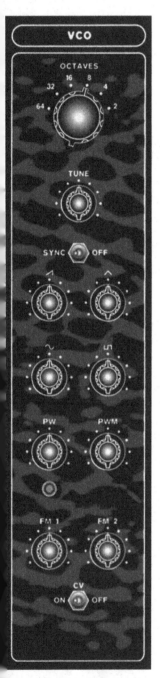

The Technosaurus Selector VCO module alongside the SE Boomstar, SEM version.

applications are being considered, there are various ways of adjusting a VCO to span only the range of pitches used in normal musical applications.

As we discussed in Chapter 1, the range of human hearing, and so the range of sounds that we consider musical, spans from somewhere below 50 Hz to somewhere above 16000 Hz. As a doubling in frequency represents an increase in pitch of one octave, this represents a range of something over seven octaves – exactly as found on the full-length, 88-note piano keyboard. This is quite a range to be covered with any accuracy by any single variable control, and so most synthesizer designs provide a selector for the rough frequency range, often marked in "footages" (a term derived from the length of the different pipes found on church organs), coupled with a more precise control for fine-tuning.

On the Technosaurus Selector VCO module, the rotary range switch is marked 64, 32, 16, eight, four and two (feet). This control sets the rough range of pitches, while other control voltages (usually from a keyboard that may be three, four or five octaves long and so extend the range accordingly) will determine the actual note played by the oscillator. The 64 (feet) setting represents an extremely deep bass sound, modifiable from the keyboard until it is so low as to apparently change from a musical note into an individual series of clicks. This low setting would generally be used to create very deep bass drones and organ bass-pedal effects.

The 32 (feet) setting is more appropriate for bass guitar, synthesizer bass, and similar applications, while the 16 (feet) setting is appropriate for repeated sequences and accompaniments. The 8 (feet) setting is used for melody parts (perhaps for synthesizing flutes, oboes, strings, and similar instruments), while the 4 (feet) setting would be appropriate for high strings, whistles, and prominent lead parts. The two (feet) setting, particularly when extended higher up the keyboard, is extremely high

pitched and only occasionally used in musical applications, or to create higher overtones for lower-pitched sounds.

As described, the overall range control is usually teamed with a fine-tuning control. This is the case for several reasons. Firstly, the very wide range that has to be covered by an audio oscillator usually puts quite a demand on the basic circuit design, and at very high or very low ranges, slight retuning may be necessary. Secondly, random drifting in pitch, sometimes caused by the temperature (or lack of temperature) in the room, has been a long-standing problem for analog synthesizer designs despite many attempts to cancel out its effects by making parts of the circuit run hot compared to average room temperature, and so retuning may be periodically necessary due to temperature-induced drift. Thirdly, when several oscillators are playing together, it is not always optimal to have them perfectly in tune with one another, so the facility to slightly detune an oscillator is usually required. This creates slight pulsing or "beating" in the sound, the result of the oscillator wave shapes adding to or cancelling out each other's output volume as they slip in and out of phase with one another, and this is an important reason why 2-oscillator synthesizers sound much more powerful than single oscillator designs. Listen to track 4 on the website for a demonstration.

On the Technosaurus Selector VCO, the tune control covers a narrow range, just sufficient to correct tuning or to introduce intentional detuning. On some other synthesizer designs, the range and fine-tune controls are combined (on the EMS VCS3 and Synthi A an expensive multi-turn or "vernier" potentiometer is used, which gives better resolution), while on others, such as the Sequential Pro One, the range control covers a somewhat reduced range of octaves, while the fine-tune control covers a much wider range.

1948 Transistor invented by Bardeen, Brattain and Shockley at Bell Labs

The next control generally found on a VCO governs the selection of output wave shape, which determines the basic tone of the sound created. The wave shapes generally available from simple analog circuits are sine, sawtooth, triangle, square and its variation the pulse, and as discussed in Chapter 1, these have distinctly different tones: the sine is extremely smooth and bland, like a flute; the sawtooth and triangle are slightly more cutting; the square wave is distinctly harsher; and the pulse wave sounds progressively thinner and weaker, as its "on" times become shorter.

1950 Robert Moog launches his own business building Theremins

On many synthesizers the wave shape has to be selected for each oscillator individually, so there will usually be a rotary switch giving a choice of one or another – or, in some cases, a variable control that changes the wave shape from one to another continuously. The rather featureless sine wave is sometimes omitted (although the Minimoog does offer it) and the square wave is sometimes fully variable in width using a separate control, or sometimes (again as on the Minimoog) offered in a variety of fixed widths. If there is more than one oscillator on a synthesizer they can almost always be set to different wave shapes, and this is ideal so that a sound could have a smooth element and a more cutting element. But the most luxurious option would be to have all wave shapes available simultaneously, in different amounts, from all oscillators, and this was the case on the Technosaurus Selector. Sawtooth, triangle, sine, and square waves each had their own level controls (so there's no overall oscillator level

Aspects of analog sound

control as you'd find on most synthesizers), while a separate pulse width control alters the tone of the square wave. Listen to track 2 on the website for a demonstration of the different oscillator waveforms typical of any modular synthesizer.

As mentioned in Chapter 1, a special technique is applicable only to the square wave, referred to as pulse width modulation. Since the tone of a regular square wave is fuller than that of a thin pulse wave, varying this pulse width creates a change in tone, and changing it continuously creates a continuous shifting in tone that can sound similar to chorus or phasing. On most square wave oscillators the pulse width is made voltage controllable, so it can be modulated with an LFO for chorus-like effects, or by an envelope shaper so that the tone of the oscillator varies through the course of a note. It would even be possible to vary pulse width from the keyboard, so the oscillator tone becomes thinner on higher notes, or from a sequencer playing a series of notes, so that it appears to vary in regular patterns; on the Technosaurus Selector VCO there are two inputs for pulse width modulation (PWM), each with its own level control, so two or more of these techniques could easily be applied simultaneously.

Some synthesizers that only have a single oscillator (such as the old Roland SH101) try to strengthen their sound by adding a sub-oscillator. This comprises the sound of the main oscillator divided by a frequency-dividing circuit (more on this later), usually to a pitch either one or two octaves below the main oscillator. This can add to the strength of the sound, particularly in terms of creating strong bass, but it does not have the same effect as a second oscillator since the sub-octave cannot generally be slightly detuned from the main oscillator sound and there is usually no choice of wave shape being limited, due to the nature of the frequency-dividing circuit, to a square wave. On some synth designs, such as the Roland SH3 or ARP Explorer (check track 46 on the website), several sub-octave sounds of different octave values can be faded up simultaneously; this is rather similar to playing several pipes on a church organ together.

We've looked at how the rough and fine pitch of the oscillator and its volume and tone can be selected. The next task is to actually play it in a musical fashion and perhaps to add some expression. Generally, the oscillators on any synthesizer design are controlled from a keyboard and the connection is internal and scaled, so that a one-octave range on the keyboard creates a one-octave change in pitch from the oscillator (on the ARP 2600, for instance, this scaling has to be set by the player). Sometimes (as on the Minimoog and Sequential Pro One), it is possible to disconnect one or more oscillators from the keyboard (on the EMS VCS3 and Synthi A, the oscillators are not controlled from the keyboard at all unless specifically connected, so they drone on one note or are controlled from other sources), but on many integrated synthesizers the connection between keyboard and oscillators is permanent. This led Bob Moog, from his very earliest designs, to establish as standard that a change in voltage of one volt from the keyboard circuit would cause a change in oscillator pitch of one octave, so it can be assumed that a five-octave keyboard creates a voltage five volts higher from its highest key than from its lowest key (but see the note later regarding other arrangements). On very few synthesizers it is possible to readjust this scale so that an octave on the keyboard does not correspond to an octave change from the oscillators, or even (as on the ARP 2600) to reverse the scale, so that playing higher up the keyboard produces a lower note. On the Technosaurus Selector the keyboard voltage, scaled to one volt per octave, is connected to the oscillator through a socket specifically marked CV.

But there is another obvious purpose in controlling the pitch of an oscillator, which is to create a small, regular change in pitch known as vibrato. This is a common technique that can be produced on stringed instruments, such as the guitar or violin, by slightly varying the length or tension of the string in a regular manner, which makes a sound richer and apparently more expressive. A musical-sounding vibrato will vary the pitch by perhaps one-tenth of a semitone (the interval between two notes on the piano keyboard) or less, and at a rate of perhaps 7 Hz (seven times per second), in a regular manner represented by a sine wave or perhaps by a triangle wave. Any great variation from these figures – in terms of depth, speed, or waveform – will sound distinctly unnatural, although this may be exactly the effect required. The regular variation of pitch is also known as frequency modulation. Listen to track 3 on the website for a demonstration.

On some synthesizers the oscillators can only be modulated, apart from by the keyboard, using the LFO. But there are other ways of controlling pitch too, so having more than one input for oscillator pitch modulation (or frequency modulation, FM) is preferable. On the Technosaurus Selector VCO there are three inputs for oscillator pitch modulation: one fixed and two with level controls marked FM1 and FM2 to adjust their range. If one input is used for vibrato modulation from an LFO, the others can be used (remembering that the additional control input from the keyboard is internal and permanently connected) for alternative control, perhaps from an envelope generator. Chapter 1 explained how envelope generators give a control voltage output that varies over time during the course of a note; applying the output of an envelope generator that starts at a high level and falls away quickly to an oscillator will make it shoot up in pitch then fall rapidly, an effect often compared to an electronic tomtom or "syndrum".

On the Technosaurus Selector VCO there are two further facilities: a CV Off switch, which as we discussed can disconnect the normal control signal from the keyboard; and a Sync switch, which puts the oscillator into synchronisation with another. This means that the frequency of the oscillator is locked to the frequency of another oscillator. This is not a very interesting facility in itself – in fact, it removes some of the interesting effects that can be obtained by playing two slightly detuned oscillators together – but as soon as any attempt is made to change the pitch of the synchronised oscillator, it begins to generate interesting and sometimes quite striking overtones. If the change in pitch is an octave or more, these overtones can become extremely harsh and metallic. Controlling the pitch of the synchronised oscillator with an LFO creates regularly varying overtones; controlling it from an envelope generator creates a harsh, clanging sound on each note; and controlling it from a pitch bender (without bending the pitch of the oscillator to which it is synchronised) creates an expressive, screaming effect that can help generate very distinctive lead or bass lines. Oscillator synchronisation was available on the Moog

1963 Don Buchla develops modular synthesizer for Morton Subotnick

1963 Walter Carlos composes early pieces for tape and electronics

Prodigy and is a commonly found custom modification on the Minimoog. Examples of oscillator synchronisation on the website include some of the sounds in tracks 51, 52, 68, and 79.

On the SE Boomstar the VCOs have footages marked from 32' up to 2' as well as a Lo position. Since OSC2 can modulate OSC1, both can also be modulated by the LFO, and OSC Sync is available, a wide range of tuned or more abstract sounds are possible. In this way, many of the possibilities of a fully modular system can be offered by a smaller instrument without the need for patching cables; it is also possible to route MIDI control information to many parameters to vary the sound with performance controllers. Examples are on the website, tracks 83–85.

1964 Paulo Ketoff designs Synket in Rome, later used in performances in New York

VOLTAGE-CONTROLLED FILTER

In the standard analog synthesizer configuration, the audio signal from the oscillator or oscillators is passed straight to a voltage controlled filter. As explained in Chapter 1, this circuit acts to cut out certain frequencies from the incoming signal, doing very little to a plain sine wave, but smoothing the tone of a sawtooth, triangle, or square wave, or any combination of these, which contains some higher harmonics. If a powerful low pass filter is set to its lowest cut-off level it can completely silence the signal fed to it; often a source of confusion when first learning the control settings on a new instrument, as it is assumed that either the oscillators are not turned up or that the keyboard is not working.

1964 Robert Moog gives a talk at the Audio Engineering Society on the Voltage Controlled Amplifier

Many integrated synthesizers have a single low pass filter, while modular systems sometimes have separate modules for low pass, high pass, band pass, and sometimes band reject (or "notch") filters. Some filters, known as multi-mode filters, can be switched from one mode to another and on the Technosaurus Selector the VCF2 module could be switched to low pass, high pass, band pass, or notch (demonstrated on track 5 of the website). On the Boomstar SEM version the filter is specifically designed to emulate that of the classic Oberheim SEM module and only has low pass mode, though there is a separate band pass switch, as well as a notch setting for a much wider variation in sound.

The voltage-controlled filter is usually modulated by an envelope and by other voltage inputs. On the Technosaurus Selector the envelope level is controlled from one control and there are three scalable voltage control inputs marked FM1, FM2, and FM3. Exactly as for vibrato modulation of an oscillator, these can be fed by an LFO, a sine, or triangle wave, giving a regular variation in tone, referred to sometimes as "wow" or, if very gentle, as a tremolo. Examples are on tracks 6 and 7 of the website.

The envelope modulation input is used to create one of the most distinctive sounds in the whole field of analog synthesis – a sound that changes very markedly in tone as the note progresses, something highly unusual in the world of acoustic instruments. This is particularly striking if the filter's resonance is turned to a high setting. Check out track 26 on the website. Track 27 has a similar effect "upside down", with the envelope sweeping the filter upwards instead of downwards.

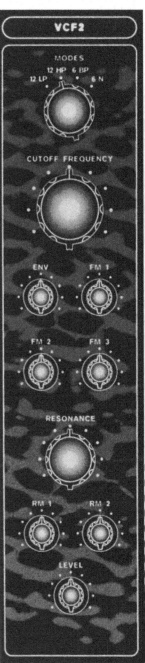

On some filter designs, resonance can also be voltage controlled, and on the Technosaurus Selector VCF there are two scalable inputs for resonance modulation marked RM1 and RM2. Finally, there is an output level control that controls the audio output level from the filter, which you can use to "make up" some output volume if you strongly filter the incoming sound. On the Boomstar, switches route the output of VCO2 or the LFO to control the filter and determine the degree of keyboard tracking, MIDI controllers can also be routed to the filter.

ENVELOPE GENERATOR

Envelope generators differ from VCO and VCF designs in that they control other functions but generally do not have voltage or audio inputs other than for a voltage trigger or "gate", a brief, relatively high voltage "spike", which starts the envelope generator (EG) going through its cycle (though more advanced EGs with voltage controllable parameters are also available). The most obvious source of this spike is a keyboard. Every time a note is played the keyboard generates a trigger or gate that sets off the envelopes, which ensures that, usually alongside adopting a particular pitch, the new note also has its own loudness and filter envelope. But it is also possible to trigger envelope generators from other sources, sometimes from an LFO on every cycle, sometimes from an external source such as a sequencer, as described later.

The Technosaurus Selector VCF module alongside the Boomstar SEM's filter section.

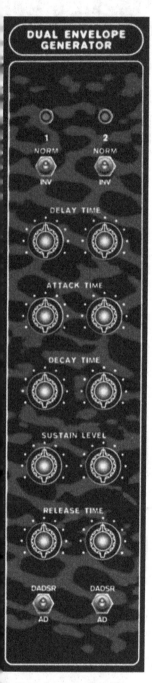

As discussed more fully in Chapter 1, envelope generators typically have four parameters for attack, decay, sustain, and release (ADSR). These are all periods of time, other than sustain, which is a level expressed as a percentage of the full envelope level reached at the end of the decay phase.

As mentioned, envelope generators can sometimes be more complex and on the Technosaurus Selector system there is a dual envelope generator in which each envelope also has a delay phase, so it can be described as DADSR (there is something similar on the Korg MS20). This simply means there is a variable time period after the note is played, before the envelope triggers, so, for example, you could create a rather muted sound which abruptly brightens up after a second or so. On the Selector each envelope can also be switched to a simpler AR format, and there is a gate input, and normal and inverted voltage outputs for each envelope. For convenience, there is an LED light to indicate each time the envelope triggers. On the Minimoog, the envelopes only have ADS stages, if you switch in the release phase it has the same length as the decay stage. Bob Moog corrected this limitation and others on both the Crumar Spirit and the Moog Voyager.

On the SE Boomstar the filter envelope, as mentioned before, can be looped, while the VCA envelope can be repeatedly triggered from the LFO, so smaller integrated instruments aren't denied this wide variety of options. On some synthesizer designs there is a choice of overall rates for the envelope, speeding up its period to cycle, while some keyboards will offer a choice of ways in which the envelopes can be re-triggered: on the high, low, or any new note, repeating with a key down only, or repeating at the speed of an LFO, as found on the ARP Odyssey. It is annoying to sometimes find modular designs in which the LFO output is not enabled to repeatedly trigger the envelopes, though it's sometime possible to shape and amplify the LFO waveform into a trigger pulse that the envelope generators will accept. See the section on keyboards for more details.

Envelopes are generally applied to the VCA to alter volume during the course of a note and to the VCF to alter the tone during the course of a note. Examples of varying volume are on tracks 8 and 12 of the website, examples of varying tone are throughout – on tracks 22, 29, 34, 35, 63, for instance – as this is a very common technique of analog synthesis.

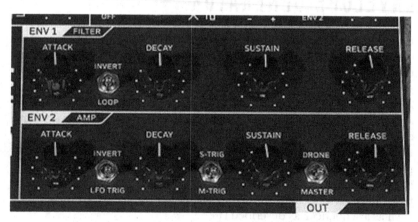

The Technosaurus Selector dual EG module and the Boomstar's envelope section.

VOLTAGE-CONTROLLED AMPLIFIER

The VCA generally has few functions as it is normally connected directly to one of the envelope generators, so it often appears just in the form of a final volume control. On the other hand, if it is required to create a genuine tremolo, in which the output volume varies rapidly just as the oscillator pitch would vary rapidly in a vibrato, then it is useful to be able to access the VCA independently. On the Technosaurus Selector the VCA was a module separate from the envelope generators and comprised an input level control with a dual-colour LED that indicates if it is overloading (which would cause audible distortion), four audio inputs, three voltage inputs with scaling controls for amplitude modulation (AM), exponential/linear selection, and normal/invert switching. In this way the VCA can be used to combine the output of two, three, or four oscillators, control their levels together using an envelope generator (the relative levels of the oscillators are mixed at the oscillator output, since there is no audio mixer as such in the system illustrated) and add tremolo using an LFO, at the same time as using another level control such as a pattern of differing levels sent from a sequencer. The invert setting can be used to set up two VCAs to opposite phases to create stereo panning.

On the SE Boomstar, typically of smaller integrated instruments, the VCA isn't separately marked, though there is a master Volume control and on the top panel a minijack socket for external voltage control marked AMP AM – amplifier amplitude modulation – offering either control of volume from a voltage pedal or regular tremolo using an external LFO. There is also a Drive setting, which slightly overdrives the output VCA for musical distortion, an incidental though popular facility of the original Minimoog, here made much more explicitly available.

LOW-FREQUENCY OSCILLATOR

As previously discussed, the LFO is simply a voltage-controlled oscillator designed to operate at lower speeds and intended for use as a control source rather than an audio source. On the Technosaurus Selector LFO, the wave shape was selectable using a rotary control for sawtooth, triangle, square, or sine waves, but these are also available simultaneously at independent voltage output sockets. The speed is variable and the pulse width of the square wave can be varied manually or automatically using

The Technosaurus
Selector VCA module.

a voltage at the PWM input socket. Unusually, the LFO speed itself is also voltage controllable, so this is more properly a VCLFO or voltage-controlled low frequency oscillator. This can be an interesting facility because the speed of the modulation created by the LFO (whether on a filter or an oscillator) can itself vary regularly and all sorts of complex patterns can be created.

On the Boomstar, although the LFOs shape control is a continuously variable knob, it offers nine individual shapes, including Random, with flexible routing to the pitch and pulse width of both oscillators, the cutoff of the filter, and the repeat of Envelope 2. The LFO rate can also be synchronised to an incoming MIDI clock. So again, an example where most of the facilities of a full modular system are readily accessible, though some more advanced features such as voltage controlled speed and waveshape are not.

The differing waveforms output by an LFO are useful for different purposes. The sine or triangle is most often used to create vibrato on any oscillator, tremolo/wah-wah effects on a filter, or chorus-like pulse width modulation effects when applied to the pulse width control input of a square wave oscillator. A square wave LFO output, however, produces trills when sent to an audio oscillator (tuned ones, if they are scaled correctly), while if there is a thin pulse or other LFO wave shape available, this has fewer obviously musical uses. On the EMS synths the envelopes could re-trigger automatically (referred to as a trapezoid generator) and in this case themselves become LFOs capable of outputting extremely unusual shapes, this facility re-appears on the Boomstar, given its ability to play the filter envelope in repeated Loop mode.

Some LFOs, for example, on the Korg MS20, also output a random waveform, useful for creating bubbling, windchime, and similar sounds. The availability of this output from the LFO waveform selector often obscures the point that it is created by a combination of the LFO, a source of white noise, and a sample-and-hold circuit all working in concert (more on this later), though in modern designs while the audio oscillators may be truly analog, the LFOs are often digital and the random waveform is generated using any one of a number of digital techniques (and may not be all that random, sometimes it's possible to hear an apparently random pattern actually repeating over a fairly short period).

Listen to tracks 3, 6, and 7 on the website, which illustrate simple modulation of an oscillator's pitch from an LFO, simple modulation of a filter's cut-off frequency from an LFO, then more complex filter modulation from several LFOs simultaneously. Very fast modulation of either pitch or filtering can sometimes create human voice-like effects, as heard on tracks 9 and 15; the second sound of track 31 features square wave pitch modulation, while most of the sounds elsewhere that exhibit regular variations of pitch or tone will also depend on the use of an LFO.

On the Technosaurus Selector LFO module there was also a noise source, switchable from white to pink, which appears at an audio output socket with output level governed by its own control. While several synths group the LFOs and white noise source together, there is no particular necessity for doing so.

WHITE NOISE SOURCE

Apart from oscillators (and possibly a filter set to self-resonate) the white noise generator is the most common sound source found on analog synthesizers. Just as white light comprises all colours of light mixed together, white noise consists of sounds at all frequencies randomly mixed together. This random sound, which resembles a sea or wind noise, can also be heard from a radio set tuned between stations and is also generated as an unwanted artifact of many electronic designs. The simplest white noise generator circuits just use a "noisy" diode, amplifying its imperfections as much as possible to create a strong noise signal. But there are several other ways to create white noise, playing random numbers at suitable speeds from a digital sound generation circuit can work equally well.

There are also two less commonly found variations: pink noise, which offers equal energy level at all frequencies (found on the Delta Music Research modular system, for example, and used for testing acoustic systems); and red noise, which boosts the lower frequencies, and so sounds damped compared to white and pink noise (found on some of the EMS synths).

White noise is endlessly useful above and beyond the obvious creation of sea, wind, and explosion effects. All sorts of percussive and acoustic sounds contain an element of white noise and it often needs to be incorporated into more complex imitative analog sounds. For instance, the "chiff" on the start of a flute sound is best created with white noise passed through a short envelope and an imitation of human whistling benefits from the addition of white noise. All sorts of drum and percussive sounds need a white noise element, indeed, satisfactory short high-hat cymbal sounds can be created using nothing more. A synthesized snare drum sound (as heard on the Simmons and other electronic kits) typically comprises a tuned oscillator sound, to replicate the sound of the skin being struck, balanced against a burst of white noise to replicate the sound of the snare, which is a rattle made of metal wires. Check out the website tracks of the JHS Drum Synth. On the SE Boomstar the white noise section isn't obvious, but it's there, as one channel on the small built-in mixer. Small integrated monophonic systems like this will often find use for percussive effects, so those which leave out the white noise option are missing a trick. Vocoders also need a white noise input to help generate sibilant ("s" and "t") vocal sounds.

Since white noise in an electrical circuit comprises random voltages it can also be used as a source of random effects. But the random elements in white noise arrive much too rapidly to be of much use in musical terms, so it's necessary to pick and choose from them, as we see next.

SAMPLE-AND-HOLD

The term "sample-and-hold" is usually associated with the creation of random sound effects, though this module can be used for much more. It is

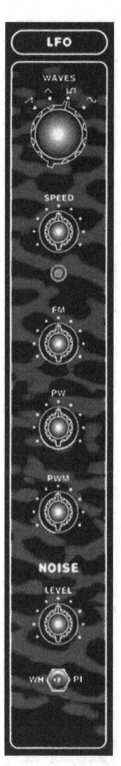

The Technosaurus Selector LFO module included a White Noise output.

The Technosaurus Selector
sample-and-hold module.

not generally appreciated that most analog synthesizers already have one sample-and-hold circuit associated with the keyboard; when a new key is played, the keyboard sample-and-hold circuit detects and resets itself to the voltage produced, then holds that voltage as an output to the oscillators, setting them to the pitch of the new note until a different one is played. If the note of the oscillators drifts or "droops" it is often because the keyboard sample-and-hold circuit is faulty, sometimes because a capacitor (an electronic component that stores electrical charge) is slowly leaking away its charge.

But it is true that an independent sample-and-hold module is most often found in very close conjunction with a source of white noise, which is effectively a producer of rapidly varying random voltages. Some synthesizers do offer a white noise source but no sample-and-hold facility – the Sequential Pro One is an example – and on the Moog Prodigy there are, unsurprisingly, no random modulation options because there is no white noise source at all. On the Minimoog the white noise source can be patched directly to control the oscillator frequencies, but with no independent sample-and-hold circuit this just produces a rather jittery and unmusical variation in pitch (Moog later produced an optional sample-and-hold unit, the Model 1125 used by George Duke and others, which is actually quite versatile).

An independent sample-and-hold circuit, as its name suggests, typically looks at the random voltage coming from a white noise source, which is varying very rapidly, matches its output to that voltage and holds that output voltage steady for a variable amount of time. The speed at which the sample-and-hold circuit updates the output voltage is controlled either from an internal or an external LFO, or other source. When the output of the sample-and-hold is applied to an oscillator the result is a series of random pitches, when applied to a filter, the result is a series of random tones. This latter technique is very useful, particularly when a repeated note is playing, as the impression can be given of a much more varied series of sounds. There's a famous example in the middle of Emerson, Lake & Palmer's track *Karn Evil 9*. Examples on the website include tracks 27, 39, and 57.

On most integrated synthesizers that are equipped with a sample-and-hold circuit, a white noise source is permanently connected as its only input signal. But to do this is to lose a lot of the potential of the circuit. On the Roland SH3, ARP Odyssey, and just a few other synths, the sample-and-hold can also be fed from an LFO or another source. The resulting output depends very much on the wave shape of the LFO input; inputting a sine wave results in a series of output voltages that rise and fall repeatedly, generating regular patterns when applied to a filter, or a series of notes from an oscillator, though not necessarily in any regular musical scale. A sawtooth wave would create a repeated rising series of notes; an inverted sawtooth a repeated falling series of notes. This is sometimes referred to as a glissando and so the sample-and-hold circuit can become a rather less musically controllable substitute for some kinds of sequencer or arpeggiator.

On the ARP Odyssey (and the more recent Korg clone) the sample-and-hold circuit is particularly flexible and can become quite baffling. It can be fed through its own dedicated source mixer by the output of either of the two oscillators, or by the white noise source, and its output sent back to either of the two oscillators or to the filter, stepped along either by the LFO or by each new note played on the keyboard, a particularly versatile arrangement which can unfortunately also be rather confusing. Also on the Odyssey, there is a lag control in the sample-and-hold output. This makes the sampled output level glide from one value to another with variable time rather than stepping sharply (this is effectively a portamento or "slew" circuit inserted in the sample-and-hold output). It can create some very unusual effects, as used, for example, by Klaus Schulze on the randomly percussive bass backing pattern of the track *Totem* from the album *Picture Music*. Roland's SH3 and SH3a also had a terrific sample-and-hold circuit which could be driven by the LFO for interesting patterned effects, some modules optionally offer Track & Hold, which retains a positive voltage until the module receives a negative voltage.

On the Technosaurus Selector, the sample-and-hold module has facilities for an external or internal clock with an LED speed indicator, use of an external voltage or an internal random voltage source, and a control for the speed at which random voltages are internally generated, so there is no need for a separate LFO to determine sample-and-hold speed. On the SE Boomstar, while there's a white noise source and the LFO offers a random output, there is no access to a sample-and-hold circuit as such; a typical compromise found on small integrated instruments as compared to full modular systems.

WAVE SHAPER

This is an infrequently seen module that alters the wave shape of incoming oscillator signals and so creates new tones and combinations of tones. By clipping off the highest or lowest parts of the wave shape, it creates frequency doubling, chorusing or phasing effects. On the Technosaurus Selector there was a dual wave shaper module, on each wave shaper the amount of clipping can be set manually or voltage controlled. The waves can be made symmetrical again after clipping and the two modules can be used in series or in parallel.

RING MODULATOR

The ring modulator is one of the earliest electronic music circuits and although it has been refined many times it still takes its name from its simplest incarnation, which comprises four diodes (devices that prevent the passage of electrical current in one direction only) arranged in a ring, though some will tell you that it takes its name from the fact that it can help make the sound of a bell.

The ring alignment of diodes has an odd effect on signals presented to it: it creates new signals, one at the combined frequency of the incoming

The Technosaurus Selector wave shaper module.

signals and one at a frequency found by subtracting one frequency from the other. This means that in order to work, the ring modulator needs two input signals, usually referred to as A and B. Although the frequencies in the resulting output signals are mathematically related to those in the input signals, this does not mean that they are musically related, in most cases they will be significantly "out of tune" with the originals. When the original signals are also mixed back in the result will usually be a dissonant, metallic, or bell-like "clanging" sound. Examples on the website include track 58, realised using the particularly versatile ring modulator of the Yamaha CS80, which has continuously variable voltage-controllable parameters.

The output of the ring modulator also depends very much on what signal is fed into it, but usually it is used to create abstract and metallic rather than musical sounds, particularly cymbals and gongs. It can also be useful when applied to human voices, creating what is usually described as a robot-like voice (this is how the voice of the Daleks in the sci-fi series *Dr Who* was traditionally created) and can also create some exciting effects from sound sources such as the guitar, particularly if the "B" input is changing while the guitar plays (the Big Briar Moogerfooger ring modulator pedal is intended for just such applications).

On the Technosaurus Selector there was a dual ring modulator module; the A input of each section can be doubled in frequency to instantly create more obviously dissonant sounds if the A and B sources are initially rather close together in frequency. The two units can be used in series or in parallel. On the SE Boomstar (and on earlier synths going as far back as the Moog Sonic 6) there is a ring modulator output in the mixer section, not separately accessible, but still useful for creating such metallic sounds.

Can's Irmin Schmidt used the *Alpha77* including ring modulator, filters, and tape delay.

The ring modulator has been used by quite a few avant-garde jazz musicians, such as Chick Corea and Herbie Hancock, particularly to process the Fender Rhodes electric piano, but the masters of the device were the German rock band Can, influenced through its use by contemporary classical composer Karlheinz Stockhausen. Can's key-boardist, Irmin Schmidt, had ring modulators as well as filters and tape echoes built into a large modular processing unit (not really a synthe-sizer) called the Alpha 77. His main instruments with the band were the Farfisa Pro Duo organ and Pro Piano, which often become unrecognis-able through the use of filtering and ring modulation, producing deep droning sounds, and high pitched metallic bell-like solos from the piano resembling Balinese gongs. Drummer Jaki Liebezeit, guitarist Michael Karoli and bassist Holger Czukay (all now sadly deceased) also had their instruments processed through ring modulators at various times, nota-bly on the track *Chain Reaction/Quantum Physics* from the *Soon Over Babaluma* album. The band also favoured Farfisa's Sphaerasound, which was a stereo panning device with distinctive overdrive characteristics. There are some amazing YouTube videos of the system in action, notably a manic performance of *Vernal Equinox* from the BBC's *The Old Grey Whistle Test* show in 1975.

SUBHARMONIC OSCILLATOR

A subharmonic oscillator or sub-oscillator is often found built into smaller analog synths with only a single oscillator or bank of oscillators, with the intention of "beefing up" the sound with added bass without the expense of building a second independent oscillator. Usually set one or two octaves below the main oscillator's pitch, or sometimes switchable between the two, or even mixable at both sub-frequencies, the sub-oscillator generally offers only a square wave due to the nature of the pitch-dividing circuit used and cannot be finely detuned from the main oscillator's pitch. So while it adds some "beef", it does not add to the warmth or fuzziness of the sound.

But there are some more complex subharmonic oscillator designs too, for example, the octal subharmonic oscillator of the Technosaurus Selector system. Most analog synthesizers work by subtractive synthesis: by taking a sound source and filtering it, removing overtones until the sound is as required. In the alternative method of additive synthesis, more commonly found on digital instruments, very simple waves (usually sine waves) at different frequencies are added together until the resulting wave shape offers the required tone. Often, additive synthesis instruments, such as the Kawai K5, have no filter at all because the overall tone of their final sound is defined in this way alone.

The octal subharmonic oscillator offers additive synthesis but based only on tones lower than the basic (or "fundamental") tone. These under-tones can be faded in independently of one another; it's also possible to modulate the oscillator with an LFO and to modulate the level of all the odd-numbered tones or all the even-numbered tones. Finally, there is a high pass filter that can be set manually or also modulated from an exter-nal source, this acts to cut out all or some of the undertones as required. This is a very unusual type of module, obviously capable of creating some extremely deep bass tones but also on higher octave settings capable of

1965 Robert Moog completes Ph.D.; R.A. Moog Inc. company launched

offering some very unusual textures that change in an unexpected manner depending on how the subharmonic levels are modulated.

1966 Robert Moog completes voltage-controlled filter design

RESONATOR

The resonator is a variation on the filter circuit and has been found on various synthesizer designs, including the Korg PS series and the Polymoog, which had a useful three-band resonator. It is effectively a band pass filter with its centre frequency and the amount of boost at that frequency ("resonance") adjustable. It can be used to pick out and emphasise required elements of a sound, whether high, low, or medium in pitch, or to create very strong whistling effects around given frequencies. On the website, compare some high-resonance sounds, such as those on track 16, with smoother low-resonance sounds elsewhere.

On the Technosaurus Selector there was a triple resonator module with an input level control, three resonator circuits with frequency adjustable from 32 Hz to 8 kHz manually or under voltage control from an LFO, envelope or other source, each with a manual emphasis level control, an overall amplitude (loudness) modulation input with a range control, and a useful overall bypass switch.

The Technosaurus Selector frequency shifter module.

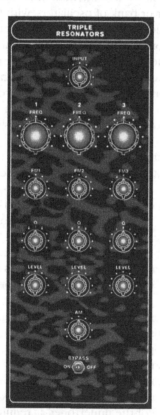

The Technosaurus Selector triple resonator module.

The Technosaurus Selector subharmonic oscillator module.

FREQUENCY SHIFTER

Not always found even on the larger modular systems, the frequency shifter (pitch shifter or harmoniser) simply increases or decreases the frequency of the incoming signal, sometimes mixing back in some or all of the original signal to create interval or detuning effects.

This has many applications for external sound sources as well as for the basic sound of analog oscillators. Depending on how the pitch-shifting circuitry is designed (there are many digital methods of implementing this, as well as analog methods), the pitch-shifted signal will be more or less faithful to the original, and more or less distorted as the amount of shifting increases.

1967 Sales of the first Moog modular synthesizers, Systems I, II, and III

MORPHING FILTER

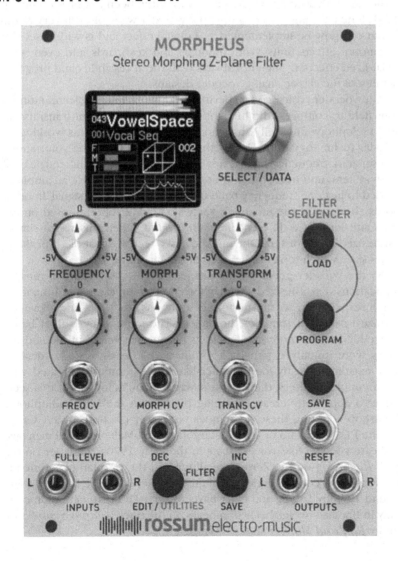

The Morpheus morphing filter for Eurorack from Rossum Electro-Music.

The morphing filter was again an unusual feature of the Technosaurus Selector. There is one cut-off frequency control that can be controlled manually or using any one of three scalable voltage control inputs for control from LFOs, envelopes, or other sources, and one resonance control with its own scalable voltage control input. There are several different filter modes available, including low pass, high pass, band pass, and notch, and the filter can be voltage controlled to change its mode from one to another. As this can be controlled either from an LFO or from an envelope generator or other source, the effect can take place gradually and repeatedly, or once during the course of every note. Morphing filters now appear as part of quite a few Eurorack systems, Dave Rossum's Morpheus is a more modern example.

VOCODER

Vocoders have not commonly been found as modules in modular systems as they are rather complex and have more often been found as complete stand-alone instruments. For a time the vocoder was not even very popular but can now be implemented as a digital effect and is widely seen in rack-mount effects units, synth modules, and keyboards, and even small pedal-sized effects, though not always with the accessibility and programmability of the classic analog vocoder designs.

The vocoder (voice coder) derives from work in the telecommunications field to compress the human voice for more efficient transmission over a telephone line. Bell Labs, among other companies, was working on vocoder technology in the 1950s, this involved splitting the voice sound into separate frequency bands (just as found on the graphic equaliser of a hi-fi system) and using the output from each band to drive an amplifier applied to just one frequency band of an alternative input signal. In other words, the tone of the incoming voice was being superimposed on the pitch and level of a different incoming sound. When this was a simple synthesized drone or buzz, the unexpected effect was that of a "speaking" synthesizer, which took on the pitch of the incoming drone.

So the vocoder needs two inputs, usually the voice and, in musical applications, a synthesized sound, referred to as the carrier and the modulator respectively, so the performer only has to speak, and playing a keyboard usually determines the melody. The more analysis filter bands a vocoder has, the more intelligible its output becomes, but high-quality filters were initially expensive to build and so early vocoder designs were extremely costly. Simple voice-like effects could be created with just a couple of filters, the "Sparky's Magic Piano" songs used this technique to apparently create a singing piano with the BBC Radiophonic Workshop adopting these techniques very early on. Both Walter Carlos and the T.O.N.T.O. duo of Robert Margouleff and Malcolm Cecil managed to create vocoder-like voice effects in the early 1970s, the latter on the track *Riversong* on the *Zero Time* album, but the German experimental duo Kraftwerk commissioned a full custom vocoder to be built for them around 1973 by Leunig and Obermayer at the PTB instrumentation company in Braunschweig, this first appears on the track *Ananas Symphonie* on the *Ralf and Florian* album.

Kraftwerk's early custom-built vocoder as heard on *Ralf and Florian* and *Autobahn*.

When the unit came up for sale on eBay in 2006 it was bought for around £10,000 by the British producer, Daniel Miller, of Mute Records. Sennheiser in Germany and Harold Bode in the USA, who worked with Robert Moog, later created expensive vocoder designs and after Kraftwerk's worldwide success with the *Autobahn* album in 1974, they were able to buy these more advanced factory built units.

EMS also went into the vocoder business, producing 1000, 2000 and 3000 models with varying numbers of bands and controls, these were improved later by EMS Rehberg in Germany. Moog built Bode-designed vocoders and Roland later introduced a less controllable rack-mount, the SVC350, as well as a string synth keyboard with choir sound and vocoder circuits, the VP330 Vocoder Plus widely used by Vangelis and the Spanish band Neuronium, and more recently paid tribute to in the small VP03 module from the Roland Boutique line from late 2016. After finding high-profile use with the Electric Light Orchestra (on *Mr Blue Sky*), Herbie Hancock (on *Rockit*) and of course with Kraftwerk, the vocoder

1967 Release of Cosmic Sounds' *Zodiac* album produced by Paul Beaver

was in demand and more affordable models were introduced by Korg (the VC10 keyboard was part of the MS semi-modular synth system and used by Klaus Schulze, Neuronium in Spain and many others), by PAiA (in the form of the 6710 rack-mount vocoder kit) and by Electro-Harmonix (the 2U rack-mount E-H Vocoder was a stylish though not very intelligible design). Mike Oldfield used the Roland VP330 Plus and other vocoders to create wordless vocals – as he had done with unprocessed singers on *Hergest Ridge* and *Ommadawn* – but also used the vocoder to process his drum machine, for example on *QE2*, creating odd comb-filtered percussive patterns.

The Korg VC10 was a rare early example of a compact keyboard vocoder.

Kraftwerk's use of the vocoder, though, most clearly demonstrated the instrument's strengths and weaknesses. A simple "robot" voice is easy to produce by using a simple drone as source, as long as this has plenty of mid and high overtones to process. A rich, buzzy, string-like sound played in chords produces a convincing "singing" string section, while a full bass sound, such as those produced by a Minimoog, with plenty of bending and modulation techniques, was used by bands such as Imagination in the UK and The Jonzun Crew in the USA to create expressive, funky voice-like bass lines. For the result to be truly intelligible, a lot of extra work had to be done. Convincingly recreating sibilants ("s" and "t" sounds) required some use of white noise, either as an external source or as sometimes found built into the vocoder. For absolute intelligibility there was often no substitute for mixing in a small amount of the original unprocessed voice, and the vocoder, though capable of some fascinating voice-like effects, never gave the convincing ability to sing to someone who couldn't already do so.

After finding a great deal of use with techno and dance bands, the vocoder rather fell out of fashion, when digital synths such as the Yamaha DX7 and Roland D50 largely replaced analog designs the appearance of new analog vocoder models dried up. However, it was quickly realised that a digital signal processing chip intended to create reverb, delay,

1968 Release of Walter Carlos's Moog album *Switched On Bach*

and other effects could just as effectively realise a vocoder effect, so the vocoder started to reappear as a gimmicky effect on inexpensive studio rack-mounts, then later on virtual analog synths from Quasimidi (the Sirius), Novation (the Nova and A-Station), Korg (the MS2000 and MicroKorg), Access, and others (the Access Virus kb becoming the vocoder of choice for Karl Bartos after leaving Kraftwerk), though all of these lacked the individual control of filter bands available on earlier analog designs.

After the analog revival some new analog vocoder designs did appear, including simple rack-mounts from MAM, FAT, and Next, which often included the simple monophonic sound source that was enough to produce straightforward "robot" and similar voices, and a modular vocoder also appeared as part of the Doepfer A100 modular system. This took the vocoder right back to its first principles, with separate analysis and synthesis sections, white noise generation, and a module to control the hold, slew rate, and other parameters within the vocoder circuit design, which affected its intelligibility. This sort of design emphasises the way in which the vocoder can readily be used to process sounds other than voices; this is always a rewarding area for experimentation, though basic vocoder techniques can be tried out with a very inexpensive unit such as a Zoom RFX1000 rack effect or an Alesis Akira or Metavox pedal, or indeed in software, since a very comprehensive programmable vocoder is now included, for instance, in Steinberg's Cubase SX sequencer software and in the EVOC synth within Apple Logic.

1968 Elektra Nonesuch releases *Guide To Electronic Music* LP

SEQUENCER

The sequencer is one of the most underused and least understood elements of the analog synthesizer. Non-synthesizer players often talk about an instrument being "played through" a sequencer, or think that a sequencer in some way picks what notes to play. In fact, this unit, which can have any one of a number of designs, is simply an alternative to a keyboard in presenting different voltages to the oscillators, filters, or other modules. Due to the limitations of early analog technology, the number of different voltages available usually numbered eight, sometimes 12 or often 16. Each voltage would be represented by a rotary potentiometer and the sequencer would step from one potentiometer to the next, send the appropriate voltage to the oscillators or elsewhere (usually accompanied by a trigger to start off an envelope), and usually reset itself to the first voltage as soon as it had reached the last.

The speed at which the sequencer stepped from one voltage to the next would be determined by an internal or external LFO or other clock. Because of the extreme accuracy of the sequencer's clock compared to human playing, the effect was usually one of an extremely precise and repetitive pattern playing. Since the sequencer was usually limited to 16 notes at most this tended to lend itself to creating a repeated bass line; whole styles of music appeared based on just this technique. Examples on the website are in tracks 9–18.

1969 White Noise album *An Electric Storm* using mostly tape manipulation released in UK

Sequencers appeared as part of the Moog modular systems, as part of the ARP 2500 (an unusual ten-step design), as part of the Roland System 700 modular system, and later as part of the Roland 100 and Roland 100M systems, the Korg MS system (in the form of the triple 12-step MS10 sequencer), the Oberheim range (in the form of the dual eight-step mini-sequencer found in some Oberheim 2-Voice keyboards or as a stand-alone unit), the stand-alone dual 16-step ARP Sequencer and elsewhere. It must be said that the limitations of eight-, ten-, 12- or 16-step sequencing were appreciated fairly early on and great efforts were made to provide sources of analog voltages controlled by more extensive digital memories. The KS keyboard from EMS, with its 256 event sequencer, was one early example, while Tom Oberheim came up with the DS2 digital sequencer very early on and Roland created the CSQ range, which took analog voltages in and gave analog voltages out, but relied on digital memories to offer a capacity in the region of several hundred notes.

Some musicians revelled in the apparent limitations of the 16-step analog sequencer, finding its spontaneity and ease of manual control outweighed its apparent limitations. These included: Tangerine Dream, who used Moog sequencers from 1974 onwards; Klaus Schulze, who also used the sequencers of a Moog modular system to great effect, controlling them from a keyboard custom built for him by PPG; and Richard Pinhas, who with Heldon brought the sound of the Moog sequencer to France. Michael Hoenig worked wonders with the ARP sequencer on the album *Departure from the Northern Wasteland*, while Tim Blake made the Roland 100 system sequencer sound much more flexible than it actually was on *New Jerusalem*, the same model also finding use with Vangelis on the *Spiral* album.

All of these musicians had taken the apparent simplicity of the sequencer to another level by using it in an imaginative manner. The first step is to be able to vary the length of the sequencer pattern so it does not keep to an unwavering eight or 16 notes; a rotary switch is usually provided to do this, although on the Korg SQ10 a patch cable has to be used, a strange design decision corrected with a simple internal modification and not found on the more modern SQ-1.

The next step is to gain some control over the sequencer pitch by running it in parallel with a keyboard, either by summing the keyboard and sequencer voltages at the oscillator or, if the facility is available, by feeding the keyboard output voltage into an input on the sequencer itself. This has the effect that whenever a new note is played on the keyboard, the sequencer transposes, continuing the same pattern but in a new key at a higher or lower pitch. This makes it possible to create sequencer-based music that does not have to stick monotonously to the same key.

The third step is to apply the sequencer output to destinations other than oscillator pitch. One obvious destination (for which the third row of potentiometers on the Korg SQ10 was really provided) is the voltage-controlled filter's cut-off, which can be "played" just like an oscillator. This has the effect of creating a repeated pattern of tones rather than a repeated pattern of notes, which can give a lot of variation to the music.

1969 Sale of Moog modular systems to The Beatles and The Rolling Stones

A standard technique used by Klaus Schulze was to apply a 16-note sequence to the oscillators and a parallel 15-note sequence to the filter. This made the pattern of tones different on every cycle, to the extent that it often took some time to realise just how simple Schulze's backing patterns really were.

Another favorite technique applied to sequencer patterns is the ping-pong echo, in which an echoed signal with a delay of around half the time between notes is added to the sequenced pattern. This emphasises any slight difference between one note and the next and gives the impression of an extremely complex pattern emerging – always great fun to play with.

All of these have become standard analog sequencer techniques and are frequently denied to digital sequencers, which often cannot have their pattern length or transpositions adjusted while playing. But there are many other sequencer techniques that are very much underused: how about applying a sequencer output to the pulse width modulation voltage control input of an oscillator, to the resonance level voltage control input of a filter if it has one, or to the decay time voltage control input of an envelope generator if it has one? Alternatively, a sequencer can often be used in "one-shot" mode so that it plays through its pattern once when triggered, then stops rather than looping back to play the pattern again. It's easy enough to apply the output of a sequencer in one-shot mode to a filter's cut-off voltage control input, so every note played contains a short pattern of tones that can appear at a variable speed.

The more voltage control inputs are available on a synthesizer, the more interesting places can be found to apply a sequencer's output, so a modular system obviously has more possibilities in this area than an integrated synth. It is often possible to add inputs, though, for instance, an external filter cut-off voltage control input to a synthesizer, which is not so equipped, and an external sequencer can often be run in parallel with other sequencers, arpeggiators, or LFOs if it has an input for an external clock control.

The Technosaurus Selector system never included a sequencer module, though a small stand-alone unit, the Cyclodon, was available and offered up to 16 steps with CV and gate output and internal or external control of clock speed.

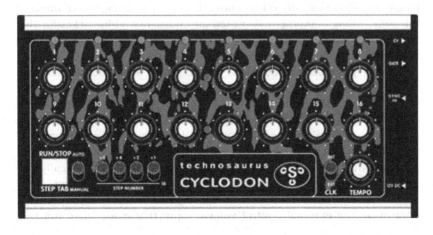

A typical small analog sequencer, the Technosaurus Cyclodon.

In more recent years many Eurorack and other modular system have offered true analog sequencers. From 2015 Korg offered the SQ1, a tiny stand-alone battery-powered dual 8-step design, which is both affordable and flexible, pretty good for improvisation, and a clear reference back to the old SQ10 design.

As with the vocoder, some musicians commissioned custom sequencer designs early on and one notable example was Jean-Michel Jarre, whose use of the Matrix sequencer built by Michel Geiss on the *Oxygene* and *Equinoxe* albums is very distinctive.

KEYBOARD

Almost all analog synthesizers, including modular systems, will have a keyboard of some kind, initially designed to interface using control voltages, later through MIDI, and most recently using USB control (or even wireless Bluetooth). But the keyboard doesn't have to be used in a conventionally musical way, this was one of the fundamentals of Don Buchla's designs. See the interview with Suzanne Ciani in Chapter 8.

The most basic analog keyboard is a series of switches in the form of a piano keyboard that provides two signals when a note is played: a control voltage and a "gate" or "trigger" pulse. The control voltage is easy enough to understand, it has to represent the current note played, which as previously discussed, is usually scaled at one volt per octave. In order to give the note a new volume or filter envelope, it is also necessary to send a second voltage known as a gate pulse or trigger. This is a short, relatively high voltage of perhaps 5V that sets off the function of an envelope generator or other module, may stay high for a time in the case of a gate signal, then quickly dies away.

It ought to be mentioned that different synthesizer designs have used different standards both for voltage control and for the gate or trigger pulse. Bob Moog's design was perhaps the most straightforward, one volt per octave for control voltages and a short 5V positive pulse for gates. But some Moog equipment used an alternative type of trigger, the S-trigger or switch to ground trigger, which actually comprised a brief short circuit to ground. It is not difficult to convert one type of trigger to another, but some Moog equipment also used an unusual two-pin socket for the S-trigger input, making it both physically and electronically incompatible with other designs.

Other synthesizer designs held to the one volt per octave standard with a 5V positive trigger, though EMS used a one-third of a volt per octave scale, which made their designs very difficult to interface with those from other manufacturers. But many Korg and Yamaha designs ditched the volt-per-octave design altogether in favour of a Hz/volt scale, which means that the oscillator pitch increases mathematically rather than musically with increased input voltage. This can be easier to implement electronically but makes these systems incompatible with volt per octave equipment.

Some Korg synthesizers, such as the original MS20, included a scaling control on the external control voltage input that made them not too

1969 EMS model VCS3 synthesizer launched in the UK

difficult to control from external one volt per octave sources. Korg also made a stand-alone interface unit to convert Hz/volt to volt/octave control signals but these are now rare. Running instruments from an inverted or unsuitable trigger usually results in their envelopes starting off on the wrong beat of a pattern so it can be very difficult to play them from an incompatible system.

Once a keyboard with suitable control voltage and trigger or gate output is being used, the next question is how many notes it can generate and when a new trigger will occur. The first generation of keyboards for the ARP 2600 were monophonic (playing only one note at a time) but the second generation was duophonic, playing a high note and a low note simultaneously. The keyboard on the ARP Odyssey was also duophonic, while that of the Minimoog was not. The keyboards of most modular systems are monophonic, although Moog did offer a duophonic model keyboard and Roland offered a four-note polyphonic model, the 4CV, with the 100M system. Duophonic keyboards can be interesting to use but when the thickness of a synthesizer's sound depends on its comprising two oscillators slightly detuned, playing two different notes on them from a duophonic keyboard often simply makes them sound thin. Duophonic or "paraphonic" playing has recently re-appeared on models such as the Moog Sub37 and Subsequent37, and the Erebus from Dreadbox.

A large modular like this Roland System 100M isn't the only way to create complex sounds.

The keyboard may also have differing responses in the way it produces new notes and gate signals. On the original Minimoog, playing a higher key while holding a low one does not produce a new note, but playing a lower one does, though the gate will not re-trigger on the new note; the keyboard is said to have "low note priority" and "single triggering". On other designs, only playing a higher key will trigger a new note, but the envelope will always re-trigger; the keyboard has "high note priority" and "multiple triggering". On other designs still, either a higher or a lower note will sound, as will the new envelope; the keyboard has "new note priority". In some cases the keyboard's behavior is switchable to high, low, or new note priority and from single to multiple triggering. Having good control over this facility can be an important performance technique, particularly on synthesizers that offer no response to keyboard velocity or pressure (aftertouch). On single-trigger keyboards, notes played legato (in other words, with the previous key still down while a new key is pressed) will not trigger an envelope, so will come out softly if the filter or envelope has closed down; notes played staccato, so that each one is released before the next one is played, will re-trigger the loudness and filter envelope, so may be louder or brighter in tone. Selecting multiple triggering defeats this option and makes each note more nearly equal to the next regardless of the exact playing technique.

On many synthesizers the keyboard voltage can be routed other than to the oscillator pitch, most commonly to the filter cut-off frequency. This is referred to as "keyboard follow" and means that the filter opens as notes higher up the keyboard are played so that higher notes are brighter in tone, which is sometimes thought desirable. Often, this is scaled at one volt per octave, so that if the filter is resonating and creating its own musical pitch, it will actually play a correct scale. Sometimes, as on the Minimoog, the keyboard follow facility can be switched in with a greater or lesser effect than one volt per octave. On the SE Boomstar keyboard filter track is switchable from Half to Full.

Occasionally, analog controller keyboards have additional facilities built in; some EMS keyboards featured an additional LFO for modulation (the ks touch keyboard had a 256-note sequencer too), while the Roland 4CV keyboard included an arpeggiator (more on this later). Arturia's modern Keystep, a compact two-and-a-half-octave design, has a built-in polyphonic sequencer and produces MIDI and USB controllers as well as voltages.

Arturia's Keystep is a compact source of MIDI and USB as well as control voltages.

Mention has been made of velocity-sensitive keyboards and these have been rare for modular systems, though the facility did turn up on integrated keyboards such as the Yamaha CS80, Sequential Prophet T8, and so on. Usually, velocity can be routed only to the final VCA output to control the loudness of the note, or to the filter cut-off level to control the tone of the note, depending on how hard the player strikes the key. Oddly enough, aftertouch or pressure sensitivity was a more common facility early on, which for some time was neglected; Roland's second synthesizer product, the SH2000, featured aftertouch, as did the

Multimoog, ARP Pro Soloist, Kawai S100P, and Teisco SX400, and in the analog/digital era the Ensoniq SQ80. Often, aftertouch can be routed to several destinations: to increase the VCA output volume or filter cut-off level, to introduce LFO modulation to the oscillator or filter cut-off level, or to introduce special performance effects such as "growl" (fast filter modulation to the filter).

The Multimoog was an early example of an instrument with keyboard aftertouch.

On the subject of introducing modulation, purely for convenience, the controls for pitch bending and modulation tend to be placed next to the keyboard. Generally, these are made larger and easier to handle than other control knobs so that very expressive use can be made of them. Much has been written about the difference between various types of performance control, although it must be said that the early Moog arrangement – one vertically placed wheel for pitch bend with a centre stopped position and another vertically placed wheel for modulation with a bottom stopped position – has remained as popular as any.

There have been many variations on this theme. Moog tried a left-to-right pitch bender (on the Sonic 6 and Constellation prototype), or a pitch bend strip plus a modulation wheel. ARP offered a simple knob with a dead centre position on the early Odysseys and the pressure-sensitive PPC pads both for pitch bend and for modulation on later models. Roland offered a left-to-right stick for pitch bend and later added a sprung forward movement to introduce modulation – as on the JD800 – but the question remains of how the player is intended to introduce modulation and then leave it on. Korg experimented with joysticks, offering two on the Sigma, while on other instruments such as the Poly 800 taking the opportunity of offering left-to-right movement for pitch bend, upward movement for vibrato modulation and downward movement for filter modulation. The Oberheim OB1 used flipper-like controls, while the OSC OSCar stuck to pitch bend and modulation wheels, but sprung both of them, so again vibrato could be easily introduced but not so easily left in. On the Korg MS10 a centre detent wheel was provided but had to be patched to either pitch bend or to another parameter.

On most synthesizers, whatever the provision of pitch bend and modulation wheels, they are generally not patchable to as many destinations as may be desired. On the Moog Prodigy, Oberheim OB1 and a few other early designs, they could at least be used to control oscillator sync to create screaming pitch bend effects, but on the whole, performance controllers have been a neglected area of synthesizer design, addressed early on only by Bob Moog in his design of the Moog Pitch Ribbon controller and smaller units for his Big Briar company, and by a few other independent companies that have offered add-on pitch bend strips, X/Y control pads or other alternative controllers. Pad controllers were common for a time – on the Korg Z1 and Novation XioSynth, for example – and aftertouch has made a comeback on the Studiologic Sledge and DSI Prophet 12 for example (both instruments being digital but capable of great expressive analog-style sounds).

Portamento or "glide" controls are also generally placed near the keyboard. As explained in the section on sample-and-hold, the portamento control introduces a lag into the charging time of the keyboard sample-and-hold circuit, so a new key on the keyboard results in a sound that slides up to the new note rather than playing it immediately. This gives a Theremin-like effect and is often used for soft flute and other sounds, but if set too high, simply means that the new note does not reach the correct pitch before the next one is played. In general terms, the circuit producing the portamento effect is known as a lag controller and can introduce a delay in the changing of any voltage, so if it can be accessed independently is useful for making any kind of effect "glide" from one level to another.

On the Minimoog the portamento or glide can be switched in and out with a switch next to the keyboard, although its time is varied from a control on the front panel. It can also be switched in and out with a momentary footswitch, which can help create a very expressive performance, and on some synths such as the Sequential Pro One there are two portamento modes, usually referred to as Auto and Manual. In the Auto mode the portamento will always apply; in the Manual mode it only applies if the existing note is held while a new one is played. This, too, can help create some very expressive performances and can be compared to the way in which single or multiple envelope triggering works.

On some early polyphonic analog instruments the glide or portamento facility was omitted because it was thought too difficult to implement (for example, on the Polymoog which as an alternative has a pitch bender strip), or was available only if the instrument was switched to monophonic or unison mode. Several instruments, though, do have polyphonic glide, such as the Sequential Prophet series and the recent Studiologic Sledge. It can be an impressive effect if not overused.

The arpeggiator, mentioned earlier, can be thought of as a simple form of sequencer and was common on early analog synths, neglected for some years, and is now becoming popular again. An arpeggio is a chord of several notes, played with a slight delay between each rather than simultaneously. Repeat an arpeggio and you have a repeated pattern of notes, the length of which is the same as the number of notes you are holding.

This is the simplest implementation of the arpeggiator, a keyboard circuit that examines which notes are currently held then plays them one after another in a repeated cycle, with speed determined by the player. Often, the pattern can be played repeatedly upwards, or downwards, or upwards followed by downwards. The next most commonly found arpeggiator setting is "random", which does not mean that random notes will play, but that the notes that are held will be played in a randomly chosen pattern.

The arpeggiator was a common feature on analog control keyboards, such as the Roland 4CV, then on microprocessor-controlled analog synths, such as the Roland Juno 6 and Juno 60 and the Sequential Prophet 600, and more recently re-appeared on MIDI or USB MIDI designs, such as the Akai LPK25, Yamaha KX25, and the Arturia Keystep. Different arpeggiator designs work in different ways: if you hold a single note, some will play it repeatedly while others will not start playing until two notes are held. Holding two notes an octave apart gives a simple repeated pattern that can be a good basis for working out a composition, while three- or four-note arpeggios can sound complex enough to back whole songs.

Some arpeggiators allow their range to be extended so the pattern plays once in its original octave, then again in a higher octave and sometimes again in a third. A more complex pattern can be achieved if the arpeggiator can play in note order, that is, play the notes in the order in which they are first held down rather than the order that they are found on the keyboard. A "note order" arpeggiator can be just about as versatile as a small sequencer.

Most arpeggiators can "latch" or "hold", often under the control of a footswitch, so they will then continue to play repeatedly even when keys are no longer held down. In this mode it is often useful to synchronise them to an external clock (in the early days, a click or 24-pulse per quarter note lock, nowadays, more usually a MIDI or USB MIDI clock) so they can play along with a drum machine or sequencer.

Once an arpeggio is locked, new notes can be handled in different ways. On the Prophet 600, any new notes simply played the synth in the normal way. On other designs, new notes are added to the locked arpeggio but only as long as they are held. This can be a very expressive performance technique, with plenty of variation available over the top of a basic repeated pattern. On the Roland XP10 there was yet a third mode, a sort of additive arpeggiator, in which new notes are added to the pattern whether they are held or not, which can make the arpeggio become longer and longer as new notes are played.

Apart from arpeggiators built into synths or into control keyboards such as the Oberheim Xk, very few stand-alone models of arpeggiator exist, of which the Oberheim Cyclone was one of the first and most powerful, with many modes, polyphonic note handling, footswitch control and other options, but its compact design and cryptic display make it very difficult to master and use. Eurorack format arpeggiators now exist, though there is no substitute for having an arpeggiator as an integrated part of a controlling keyboard and some MIDI modules or desktop synths, including the Quasimidi Quasar and Technox, and more recently the Access Virus models and Novation Peak released in 2017, offer extensive arpeggiator facilities with a good example of a varying arpeggiator pattern – from the built-in arpeggiator of an OSC OSCar – on track 53 of the website.

The conventional piano design is not the only approach to keyboard controllers though. Don Buchla established this very early on and his controllers, which included several touch keyboards, often didn't have a conventional chromatic scale layout (see the Suzanne Ciani interview in Chapter 8). Touch keys could be used either to play sounds and notes or to trigger, transpose, or otherwise control events. A touch keyboard also appeared on the EMS Synthi ks and on the EDP Wasp, in the latter case certainly for budgetary reasons since it was much cheaper to build than a moving mechanical keyboard design. Then the touch keyboard more or less died out but has made a comeback recently as an easy way to incorporate a small keyboard into the limited space offered by Eurorack designs. Pittsburg Modular, Sputnik Modular, Future Retro, and Verbos among others all make touch keyboard designs, either to stand alone or to fit within the Eurorack format, and Intellijel among many others make touch pads, which can provide great expressive control too. The future of keyboard controllers is perhaps best seen in the Roli Seaboard, a multi-dimensional touch surface through which expression can be created with many different types of movement. More on the Roli in Chapter 8.

MIDI INTERFACE

Most modular analog synthesizer systems are now offered with a MIDI interface but its abilities may vary. Generally, one MIDI input channel is supported and notes will always be output but there may also be voltage outputs for velocity, aftertouch, modulation, breath control, pitch bend, clock, and other controllers. Clearly a MIDI interface is one way to integrate an analog system into the more up-to-date world of MIDI and computer-based composition (USB-MIDI and even Bluetooth and other forms of interfacing now starting to become common also). On a large modular system, a multi-channel MIDI interface may be required (particularly if there is to be an attempt to play the system polyphonically) and there are many four-, eight- or 12-channel MIDI-to-CV models on the market, including Befaco's MIDI Thing four-channel design and others from Intellijel, Endorphin, Ladik, Synthrotek, and many more.

ASSORTED MODULES

Most modular systems will also feature assorted modules intended to solve specific problems, sometimes found only on that particular system. The most obvious of these is the "multiple" module, which simply combines or splits voltages, so that two VCO audio outputs can be sent to one filter, or an LFO's output can be split to modulate two different oscillators. Also, there are often mixer modules, acting either as audio mixers to balance the sound of different oscillators, or as voltage mixers to balance the effects of different LFOs or envelopes (often these can be the same module – one of the wonders of analog synthesis is that audio signals and control voltages can frequently be more or less interchangeable).

Other common modules include the lag processor or slew limiter, mentioned previously, in connection with creating portamento or other effects associated with voltages varying smoothly rather than instantaneously, and the voltage inverter, which simply turns a positive voltage into a negative one or vice versa. This is useful, for example, in helping create stereo panning effects as one part of an LFO's split output can be inverted and sent to a VCA, which then reaches its maximum level at the exact moment when another VCA controlled by the uninverted signal reaches its minimum level, the effect being apparent stereo movement. The voltage inverter can also help adjust trigger signals so they take effect at different times, or correctly trigger modules from different manufacturers.

Attenuators are simple but vital – simply for decreasing voltages coming in to any module in order to correctly scale them to the job in hand – while the voltage-controlled switch (VCS) simply directs an incoming signal to either of two outputs (or either of two incoming signals to a specific destination), switching these every time it receives a gate or trigger pulse. The Scale Programmer (or Quantiser) takes varying input voltages and limits the output signal to certain fixed, quantised voltages (a facility that was built into the stand-alone ARP Sequencer); the output scale can be a well-tempered 12-note scale, a major or minor scale, or a number of others.

The Fixed Filter Bank is simply a variant of the hi-fi graphic equaliser, boosting some frequencies and cutting others, but it is not usually voltage controllable. The Pitch-to-Voltage converter takes an external audio signal and derives a control voltage from it, usually also generating a trigger when it reaches a variable threshold level. The reactions of this module to complex sounds can be unpredictable but simple incoming signals, such as a flute or a guitar playing single notes, can make it possible to play the modular system from an external acoustic instrument. A P-V convertor was built into the original Korg MS20.

In a modular synthesizer, the art of creating sounds lies in the basic understanding of the components of sound, the imagination used to patch together the individual modules available, and the parameter levels and performance techniques chosen. On integrated keyboard synthesizers the possibilities may not be so extensive, though on some designs, such as the ARP Odyssey, Sequential Pro One, or Arturia MiniBrute, the routing possibilities are so extensive that they approach the complexity of a small modular system.

Probably the most basic way of using any modular system is to configure it to reproduce the layout of a standard small keyboard synth. A commonly found configuration for a very small modular system is two oscillators, one filter, one amplifier, two envelope generators, one LFO, a white noise source, and a mixer. Perhaps not coincidentally, this is very similar to the specification of the original Minimoog, which was first mocked up using a handful of existing Moog modules, though patching such a system as per the internal patching of the Minimoog – with oscillators and white noise source passing through the mixer to the filter and amplifier, one envelope controlling the amplifier and the other the filter, and the LFO modulating the oscillators or filter – would generate very few possibilities not readily available on virtually any small keyboard synthesizer.

A typical patch sheet, here aiding sound setup on the non-programmable Sequential Pro One.

PRO-ONE PATCHES

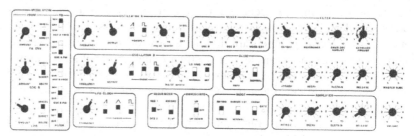

SOLO ORGAN

NOTES: Vary Keyboard Amount in the Filter section for brightness. Vary Envelope Amount for the "key click." Adjust Cutoff for "dark/light" tone. Tune OSC B up one octave + 5th. Adjust Filter Resonance carefully. Add Mod Wheel for vibrato effect.

In order to gain more benefits from such a system a few alternative modules and a little extra patching would be necessary (on the semi-patchable Korg MS20, for example, there are plenty of patching possibilities, though without a second LFO or other facilities, not much you can do with them). Adding a second LFO, a sample-and-hold module, and a ring modulator would probably be the next most obvious steps. Let's look at the possibilities these would offer. Firstly, the additional LFO could sweep the filter slowly while the first was kept solely for the task of creating a conventional vibrato. Alternatively, the output of both LFOs could be routed to the highly resonant filter to create complex abstract effects. Check out track 56 on the website, which is created on an integrated rather than a truly modular synthesizer, though an unusually flexible one, the Moog Sonic 6, which has two LFOs. Next, the output of the sample-and-hold driven from the white noise source could be applied to the filter to create randomised tones, or the ring modulator driven by both oscillators could be used to create dissonant tones for processing through the filter. Listen to tracks 15, 16, and 18 for examples.

To become more experimental, try routing the output of one or both of the LFOs to the sample-and-hold and using the result to modulate the filter. Repeated patterns of tones will be created depending on the LFO wave shape, rising or falling staircase patterns, glissandos or pseudo-sequences. Using the output of an audio oscillator rather than an LFO to modulate the other VCO or the filter will create dissonant, metallic sounds. At some points during the adjustment of the filter these often become very human voice-like. This facility is available on a good few integrated instruments such as the SE Boomstar, though it was first prominent on the Sequential Prophet series, where it is described as Poly Mod and is capable of some very unusual sounds (check track 64 on the website). Also found on the Prophet is audio oscillator modulation of the filter cut-off value, which can create similar ring modulator-like effects, and filter envelope modulation of the audio oscillators, which causes them to rise and fall in pitch with the state of the filter envelope. This can create falling "SynDrum"-style percussion effects and much more but is a facility sadly denied to unmodified examples of integrated keyboard synths such as the original Minimoog.

On a fully modular system, most parameters will also be capable of variation depending on the note played on the keyboard (on integrated keyboard instruments usually only available for the filter cut-off point, when it is known as "key follow"). If, though, there is a voltage-controllable LFO (VCLFO) available, it could be patched to play faster for high notes than for low notes. If the same LFO is also clocking the sample-and-hold, this can mean that unusual, fast patterns can be superimposed over the playing of high notes, while these only occur very slowly over low notes (or vice versa, of course). It's also useful to try triggering the sample-and-hold only from the keyboard triggers, so a new random level of pitch, filtering or another parameter, such as decay length (if voltage-controllable envelope decay is available) is selected for every note played.

As suggested in the section on sequencers, you could also try integrating a sequencer module in unusual ways. Rather than allowing the sequencer to be stepped along by its own internal clock, send a gate from the keyboard to the sequencer's step input. This will generate a new sequencer output level with every note played and can be patched to filter opening, pulse width, or any other parameter. Try sequencing patterns of levels for less than obvious parameters, such as pulse width, resonance, or LFO speed. On the EMS VCS3, even the springline reverb level can be voltage controlled and so could theoretically be sequenced too.

Using a modular system to create several different effects simultaneously can become confusing. If it's possible to remove and rearrange all the modules of the system, this could be done with a view to making one section of the system specialise in conventional tuned sounds, one in abstract sounds, one in percussive sounds, and so on. In other words, you could organise different areas of a modular system to do the typical jobs of a Minimoog, an EMS Synthi, and a SynDrum, for example. Modules from one section may still be temporarily needed by another section and, here, patch cables of different colors could help to avoid confusion (though these are much easier to obtain for quarter-inch jack systems than mini-jack systems); failing that, the jack plugs themselves or the ends of each cable can be labelled with coloured flags.

Inevitably, though, the result of any long experimentation will be complex patches and, equally obviously, these can be very hard to reproduce later on. Patch diagrams can be made, but the precise settings of various parameters can become so critical to the final sound that sometimes it cannot be exactly recreated. At this point, the user either has to settle for recording the sound, sampling it, or facing the fact that it is going to be slightly different each time it is used. This is one of the great drawbacks of modular systems, which led to their originally falling out of favour, but is also one of the major joys of using this kind of equipment.

Early on in the analog revival, some who had no access to a fully modular system considered modifying an existing integrated instrument. Some were easily modified: the original Roland SH101 could have an external filter input added relatively easily, while ten cables taken from readily accessible points on the circuit board of the Sequential Pro One could give independent access to white noise, all three waveforms of both oscillators, an independent control voltage input to the second oscillator, an output voltage from the filter envelope and an output voltage from the LFO. These days, though, the original instruments are considered too valuable to modify and modern reproductions from Arturia, Roland, Behringer, and others have been designed with more flexible access in mind,

Korg's Volca series provides flexible access on small instruments without going modular.

as well as for extra modulation and control sources from MIDI and USB, and those who really demand full flexibility can more affordably go into the Eurorack or other small modular arena, or investigate ranges of small hybrid analog/digital instruments, such as the Korg Volca series.

Rules of electrical safety should always be observed when carrying out any kind of instrument modification work (or in building powered cabinets for Eurorack) but this kind of modification can effectively turn an older keyboard synth into a semi-modular system. For convenience, some can even be separated from their keyboard at the same time, rack mounted, then run from a MIDI-to-control voltage converter; the old Moog Prodigy can be separated in this way without too much difficulty.

Extra facilities can also be added to models such as the original Minimoog – oscillator synchronisation is easily added and the author added powerful pulse width modulation to a Moog Sonic 6 very simply – but sometimes it is thought undesirable to modify the more luxuriously constructed classic instruments. In these cases it's better to own two or more small analog synthesizers and take advantage of whatever interfacing capabilities they may have, treating them as the independent parts of a small modular system.

Certainly, manufacturers have worked towards make more flexible facilities available and even when an instrument is constrained by budget to having only monophonic performance or no built-in effects, there are now improved patching and access facilities, for example, on the Arturia Microbrute and Minibrute, so musicians can now fully integrate either classic instruments, modern reproductions and more affordable recent releases, or fully flexible modular systems.

3

The birth of analog, the manufacturers and the artists

1970 Moog Musonics (later Moog Inc.) established in Buffalo, NY

In the first chapters we saw how an analog sound circuit works, what it can do and how it can be turned to musical use. Attempts to use this kind of circuitry for musical purposes date back almost as far as the use of electricity itself; early experiments in the electrical and audio fields showed that an electronic "musical instrument" was a possibility, simply because an electronic oscillator like the Hartley design could create a range of apparently "musical" pitches. Controlling the pitch to create a meaningful musical performance has been the problem ever since; novelty effects such as the "singing arc", in which an electrical discharge through the air creates a musical tone, having been abandoned as impractical fairly early on.

By the 1870s, though, with the telegraph system using Morse code bleeps spread almost throughout the USA, pioneers such as Elisha Gray were experimenting with adding elementary keyboards (at that time, basically a simple collection of switches) to collections of oscillator circuits. In 1874 Gray showed his simple "musical telegraph". By 1896, Thaddeus Cahill had completed work on the enormous Telharmonium, originally intended to give concert performances along the telegraph wires. The Telharmonium created sound using an electromechanical system of cogged wheels that closed electrical contacts at a variable rate and was capable of combining simple sound waves to create more musical tones. Cahill used the term "sound synthesis" to describe this process.

The Telharmonium, though, was huge, taking up most of a room in its final form. The rotation of the scores of cogged wheels was noisy and created excessive heat. Cahill's system did have a spiritual successor though. In 1929, the Hammond organ company introduced the first of their tone wheel organs, which similarly used cogged wheels and which created complex tones by mixing together, through a series of sliding "tone bar" (later "drawbar") controls, the simpler tones created by the individual tone wheels.

The Hammond organs had adopted electromagnetic design elements rather than noisy contact switches and, although still heavy, were compact enough to be adopted into the home market, as well as for stage, church, and theatre use. The second generation of Hammond organs, from 1939, introduced valves, which are effectively electronically controlled switches operating inside a vacuum tube. This allowed the instruments to become a little lighter again. Early use of the valve had also made practical, following its introduction in the 1920s, instruments such as Russian emigrant Leon Termen's "Theremin".

The Theremin, which has experienced a great revival in recent years with Moog's part-digital Theremini released in 2014, avoided the problems associated with keyboard control by offering a single (monophonic) sound, controlled not by a keyboard but by the player's proximity to two capacitance-sensitive aerials. Varying the distance of the hand from one aerial plays a note and creates vibrato, while varying the distance of the other hand from the other aerial controls the volume, makes breaks in the playing, or creates tremolo.

Although the Theremin was a dead end in terms of mass use for perhaps 50 years, the instrument experienced great popularity on its launch in the USA, with Termen giving grand recitals and multi-instrument performances. The Theremin sound was basically a sine wave, but with enough side bands to give it some additional character. It is difficult to play the instrument well: the pitch is continuously variable rather than being limited to individual notes, so a rather legato, flowing style tends to become imposed on the player and so it never became a domestic product, although some composers, such as Edgard Varèse, found it of interest and wrote extensively for the instrument.

Several other electronic instrument designs introduced in the early part of the twentieth century also had some impact with composers, while hardly entering the domestic market at all. Maurice Martenot's "Ondes Martenot", developed in France, was available in various versions, the basic one offering a monophonic keyboard producing a sound not unlike that of the Theremin, but more easily capable of playing individual notes. The player wore a ring around one finger which completed the electronic circuit and was able to create vibrato and other performance effects, an Ondes Martenot style controller the *French Connection* becoming available more recently from Analogue Systems in the UK.

The instrument was available in a simpler form as the Ondioline but larger and more complex versions of the Ondes Martenot also offered various optional sound generators, such as a tuned gong with rows of electro-mechanically operated moving pins behind it, playing the keyboard moving the individual pins to create a metallic, percussive sound.

Other contemporaries of the Ondes Martenot included the Selmer Clavioline, a miniature keyboard, monophonic valve-based instrument manufactured in various countries under licence from 1934 until as late as 1965, initially designed to be attached underneath a piano keyboard, and the Trautonium, developed in Germany by Friedrich Trautwein, championed by composer Oskar Sala, and later expanded into the more powerful Mixtur-Trautonium and simplified into the Monochord.

Again, these instruments got some use from avant-garde composers and later still in the film industry. In the 1950s, on the soundtracks of films such as *The Day The Earth Stood Still,* the Theremin sound created by virtuoso players, such as Clara Rockmore, started to become synonymous with unearthly and sinister happenings, while still having little or no impact on the domestic market, perhaps because of the difficulty in mastering the instrument and its limitation to monophonic playing.

More ambitious composers, for example Louis and Bebe Barron on their 1956 soundtrack for the science fiction film *Forbidden Planet,*

1970 Keith Emerson plays the Moog modular system on stage with The Nice

Oskar Sala at the keyboard of
the Mixtur-Trautonium.

1970 Launch of Alan R.
Pearlman's ARP with the
Model 2500

found greater rewards in the areas of *musique concrète* (creating music
through cutting, manipulating and re-recording tape), because flexibility
and controllability of other electronic instruments was simply still not
available. The only electronic instrument that had any mass impact on
the domestic scene remained the Hammond organ, renowned for its play-
ability rather than for any advanced sound creation potential. In Europe
composers, such as Karlheinz Stockhausen, shared the frustration of the
Barrons, settling for avant-garde music created with conventional instruments, *musique concrète*, and some use (as on *Studie 1 & 2*, 1953/4) of
simple test tone oscillators and noise generators, mostly equipment readily found at the WDR radio studio in Cologne.

MOOG

On to this scene, in the early 1960s, came the man most often credited
with the invention of modern electronic instrumentation, Dr Robert
(Bob) Arthur Moog. Moog (a name of Hungarian origin) studied electronics at Cornell University, during which time he built himself a Theremin;
a hobby that he turned into a small business. But the limitations of the
Theremin were clear, and Moog, although no musician himself, wanted to

develop something more flexible. The fact that he apparently solved the problems involved one at a time probably led to the "modular" nature of the early synthesizers, because Moog's innovations really came in three parts.

Bob Moog at work with the company's early modular system and keyboard.

The launch of modern electronic music is generally dated from Bob Moog's paper on "Voltage-controlled modules for electronic music", given at the Audio Engineering Society of America in 1964 and quickly followed by his patenting of designs for a voltage-controlled oscillator (VCO), a voltage-controlled amplifier (VCA) and a voltage-controlled filter (VCF). But what was the advantage of voltage control? Earlier electronic designs had relied on various electrical parameters to determine the pitch that the player wanted to create. On the Theremin, pitch was determined by capacitance: a parameter notoriously variable depending on temperature, humidity, and even upon the clothes the performer was wearing. Other instruments had relied on current to determine pitch, but again, current flow can depend on the room temperature, steadiness of the power supply and many other factors.

1970 John Simonton publishes electronic drum machine designs in the USA

The advantage of voltage control is that it can be regulated fairly reliably. Once a voltage in an electronic circuit is set it can be held at a fixed level, varied up or down rapidly or gradually, and generally combined and manipulated in any one of a number of ways. By making every major parameter of his sounds – pitch through the VCO, tone through the VCF and volume through the VCA – subject to voltage control, Moog had made it possible to create musical tones that were steady, predictable and controllable, and all performed from the simple set of switches, which, in practical terms, is all that constitutes the keyboard of an electronic instrument.

1971 Moog Inc. releases Minimoog Model D, leading to an eventual 35000 sales

The impact of Moog's innovation was almost instant, although that isn't to say that others had not been working on the same problems. Donald Buchla is credited with having designed a working modular synthesizer,

the Buchla 100, in 1963, to a commission from the avant-garde composer
Morton Subotnick. However, Buchla was interested in unusual forms of
control, such as triggering the synthesizer directly from tapes or touch-
sensitive pads, and also had limited interest in conventional chromatic
musical scales. His instruments – which persist in relatively limited numbers to the present day – rarely had conventional keyboards, labelled and
allocated most control functions differently from competing designs, and
as a result never became more than (sometimes very expensive) rarities.
More on Suzanne Ciani and the legacy of Buchla in Chapter 8.

While Donald Buchla was reacting to commissions from extremely
avant-garde musicians, Bob Moog, who understood his own limitations
as a musician, was making sure he consulted with as many working performers as possible in order to make his completed designs more widely
accessible. Amongst these were Herb Deutsch, who helped Moog design a
prototype double-keyboard modular synthesizer, and Walter (later Wendy)
Carlos. When Bob Moog set up his first workshop at Trumansburg, New
York, in 1963, making Theremins, he was still working on his Ph.D. at
Cornell University. Moog was introduced, by his New York sales agent,
to Deutsch, a music instructor who had worked on experimental tape
compositions and who showed great enthusiasm for Moog's prototype
voltage-controlled modules. Between them, Moog and Deutsch developed
the specification for a complete modular synthesizer system and showed
the ideas to Myron Schaeffer, head of the University of Toronto electronic
music studio.

A stand cancellation at the New York AES Convention gave Moog
an opportunity to show off his early designs, leading to enough orders
(mostly from avant-garde composers, such as John Cage) to launch the
business seriously. In summer 1965, Moog completed his Ph.D. thesis
and went into business full time, employing as many as ten people, and
later meeting Walter Carlos, a student of Vladimir Ussachevsky at the

Columbia-Princeton Electronic Music Centre. Carlos began to order Moog modules, making suggestions for new modules and for modifications, and recording some Bach pieces using them, leading to a commission for a whole album of synthesized classical music from CBS. Giving a talk to the AES in 1968 on organising electronic music studios, Moog played the last movement of Carlos's recording of Bach's *Brandenburg Concerto No. 3* and got an instantly positive reaction. CBS launched Carlos's *Switched On Bach* album jointly at a press party for Terry Riley's minimalist epic, *In C*, and while Riley played live on a Farfisa organ previewing his forthcoming multitrack keyboard masterpiece, *A Rainbow in Curved Air*, Bob Moog demonstrated the modular synthesizer to an uncertain audience of CBS executives. The Bach album, however, was a huge success, selling over a million copies very rapidly, and Moog's 40 or so regular customers were suddenly joined by all the other major record labels, each wanting to buy the largest modular system possible for similar album projects.

> "I'm not a musician, but I was interested in musical circuits and ended up designing a lot of custom products for musicians. These were people who were earning money from music – either by creating albums, or advertising music, or working in university music departments. Finally, we had enough different products to create a complete musical instrument, and people would start to hear the synthesizer, maybe just for a few seconds on a TV ad or theme tune. But it was *Switched On Bach* that really made us, and after that album became popular there were rock musicians like The Beatles and The Rolling Stones using the Moog modular synthesizers."
>
> (Bob Moog, interview with the author, 2004)

Moog's early modular systems were impressive indeed. From 1967 to 1972 he produced the I and IC, II and IIC, III and IIIC systems all (relaunched in very limited numbers in 2018), each with different complements of VCO, VCA and VCF modules. These were mounted in large wooden cabinets, were often bulky and heavy, and sometimes required long periods of warming up before staying in tune reliably. But the modular systems instantly found favour with musicians, such as Paul Beaver and Bernard Krause in the USA (Beaver producing the first commercial Moog album, *Zodiac* by Cosmic Sounds), George Harrison of The Beatles (his Moog-based album *Electronic Sounds* remains available on CD), and Mick Jagger of The Rolling Stones (who appears with a Moog modular in the movie *Performance*). Some found the instrument fascinating, while others found it baffling; the fact that it produced no sound at all until the various modules were connected together with patch cables was a sticking point for some, while others were surprised to find that such an expensive instrument could only produce one note at a time. It was because of the Moog modular's monophonic playing limitation that many musicians were forced to start developing multitrack tape techniques, as Wendy Carlos had, in order to turn single Moog lines into complete compositions.

1971 EMS launches Synthi A briefcase model used on Pink Floyd's *Dark Side of the Moon*

1971 Yamaha launches SY1 and SY2 monophonic synthesizers

The Moog modular system. Tangerine Dream in 1975 in the short-lived Edgar Froese-Chris Franke-Michael Hoenig line-up with Moog modular and Minimoog synths, plus Farfisa organ and (on top) two examples of the Synthi A from EMS.

1972 Tangerine Dream release the *Zeit* double album using EMS synths and the Moog modular from fellow German band Popol Vuh

By 1970, a lot of early Moog systems were on the second-hand market, less ambitious players than Carlos having become frustrated with their dissimilarities to other electronic keyboards, including monophonic playing, and the company also had competition in the form of ARP. Moog allowed an ambitious businessman to buy out RA Moog Inc., rename it Moog/Musonics (and later just Moog Music), move the company to Buffalo and launch the Moog Sonic Six, a portable synth based on a non-Moog design, the Musonics Sonic V. The Sonic Six was interesting because it was built into a large briefcase with an internal speaker (and so widely used for lectures and educational purposes, by Bob Moog himself amongst others) and had two LFOs that could be controlled by the envelopes, offering many unusual sound possibilities; some examples using a Sonic 6 specially modified by the author are on the website, tracks 56 and 57, and on the author's CD album *Analog Archives*.

ELP's Keith Emerson on stage with the Moog Modular and Hammond organ.

In England, the Moog modular system had also been taken up by Keith Emerson, who had established a reputation for innovative keyboard playing with the "classical rock" group The Nice. On the first Emerson, Lake & Palmer album, Emerson's solo on *Lucky Man* (added at the last minute to the recording) introduced tens of thousands of listeners to the gliding, sweeping sound of the Moog, and his on-stage use of the instrument on albums, such as the live *Pictures at an Exhibition* (available on DVD) gave the instrument very wide exposure.

1972 John Simonton's PAiA publishes the company's first kit-built synth designs in the USA

> "The first person brave enough to take a Moog modular on stage was Keith Emerson. He got one into the studio just in time to use for the closing solo on *Lucky Man* on the first Emerson, Lake & Palmer album, and the way he tells it, he was just fooling around using an empty track towards the end of the song. Then the rest of the band and everyone in the studio said it was fantastic – he wanted to play it again, but it was sonically so striking that they made him keep it. And just that one solo did us a lot of good."
>
> (Bob Moog, interview with the author, 2004)

Keith Emerson's recollection of the events differs only slightly:

"I'd already used a Moog modular on stage with The Nice. Someone played me the *Switched On Bach* album by Walter Carlos and the only Moog synth I could find to look at in London was owned by Mike Vickers, who was playing with Manfred Mann. I tried it out at his place and persuaded him to let us use it for a concert at the Royal Festival Hall. So for that first show using a Moog, Mike was hiding behind it and popping out to change patch cables. But it went pretty well, and I wanted to get hold of one, so Tony Stratton-Smith wrote to Moog asking if they could send us one. They replied it was much too expensive and needed a lot of training, and it wouldn't be fair to bands like The Beatles and The Rolling Stones who had bought them, but they'd be happy to see us out at their studio in New York for a demo. So time went by and after the end of The Nice, I was working on the first ELP album at Advision studio, and they had a Moog there, which we added to the end of *Lucky Man*. And they all tell me what's on the record was the first take – I'm not sure if it was the very first take, or whether I knew they had the tape rolling, but Eddie Offord, who was producing, had helped me set up this sound, and I knew I wanted to use the portamento for a gliding effect, and some other things – so they called me into the control room to listen and they all thought it was great. I wasn't so sure, but of course it went on to become a very influential solo. Then before the first ELP gigs – a warm-up at the Lyceum in London, then a show in Portsmouth Guildhall before the Isle of Wight Festival, which was going to be very big – I had my own Moog modular delivered from the USA. Bob Moog knew I wanted to use it on stage, so he'd built this little preset box which called up most of the levels for eight different sounds, though there were still a lot of adjustments I had to make on stage, as well as tuning it all the time. So I called in Mike Vickers again, because we had so little time to get the sounds ready and we were going to do *Pictures at an Exhibition*, which I'd been arranging from the Mussorgsky composition. So I had these few sounds ready, and I added a tuner with a digital readout so I could try to keep the thing in tune, and it worked pretty well. Then after that it was a case of expanding the system more and more, and using the sequencer and other modules, and to finish it off we had this oscilloscope display built in, which was actually a fake, but it did look great."

(Keith Emerson, unpublished conversations with the author, 1994)

There's a lot more detail on this period in Emerson's autobiography, *Pictures of an Exhibitionist*.

Meanwhile, a Moog IIIC system was also hugely expanded by producers Malcolm Cecil and Robert Margouleff into T.O.N.T.O. (The Original New Timbral Orchestra), with the addition of keyboards and modules from other manufacturers plus some smart, forward-sloping wooden cabinets. The instrument featured on several albums by Stevie Wonder, Gil Scott-Heron and others, later being seen (though not heard) in Brian de Palma's movie *Phantom of the Paradise*.

Despite the relative success of the modular systems and the existence of the Sonic 6, Bob Moog also had his own more portable synthesizer in mind, which after going through various supposedly futuristic prototype cabinet designs as Models A, B and C, settled down to become the Mini-moog Model D. Hiring a marketing manager, Moog started to get the Minimoog out of the specialist electronic music studios and into the general music stores, which mostly were not used to stocking anything as complex as an analog synthesizer. The Minimoog came onto the market in late 1970 and early 1971, selling to high-profile players, such as Chick Corea, Keith Emerson, Rick Wakeman, and Jan Hammer. Wakeman says he got his first one for around a tenth of its new cost, the original purchaser (actor Jack Wild) assuring him that it was not working properly because it would only play one note at a time; eventually Wakeman was using up to five on stage, each set to a different sound. Other expressive Minimoog soloists have included Tim Blake and, on the early Camel albums, the late Peter Bardens, who consistently used every available modulation possibility in his solos.

1972 Hit single worldwide for Hot Butter's version of *Popcorn*, with Stan Free on Moog

Moog's portable Minimoog became the classic analog synthesizer, re-issued in 2016–17.

The Minimoog has become perhaps the all-time classic analog synthesizer, both for its sound (a particularly cutting one, perhaps because the oscillators would overdrive the filter slightly, giving an unobtrusive, warm, smooth distortion – some examples are on the website, tracks 29–31) and for its logical design and layout. A comfortable performance instrument, with its control panel sloping towards the player, the Minimoog was pre-patched, needing no untidy cables, and more or less defined the logical signal path for all the instruments that followed; oscillators and white noise leading to a filter, through envelopes to an output amplifier. The inclusion of an A440 tuning oscillator acknowledged that tuning could drift on early models, a situation improved by an oscillator revision in 1978. Options included a pitch bend ribbon rather than a wheel and the splitting of the keyboard from the control panel, to which it could be attached with a long multi-core cable.

"Keith Emerson really defined the sound of the Minimoog on *Tarkus*, but it was also taken up by Herbie Hancock, Chick Corea, Rick Wakeman and by Jan Hammer with the Mahavishnu Orchestra. He developed this technique of pitch bending like a guitar, and no-one really knew it could do that."
(Bob Moog, interview with the author, 2004)

1972 Hit single for Edgar Winter
with *Frankenstein,* largely using
ARP synths

The modular systems were developing too, a IIIP (a portable IIIC in a flight case) selling to Florian Fricke of Germany's Popol Vuh, who added two optional sequencers. Apparently, though, he tired of the instrument fairly quickly, selling it to Klaus Schulze, who in his many live improvised concerts, including one at the London Planetarium in 1977, became one of the most prominent champions of the instrument (a YouTube clip of Schulze performing the piece *For Barry Graves* on German TV from around this time shows very clearly how Schulze used the system, transposing it from a custom-built keyboard while sequences played, with sounds added from an ARP 2600 and solos on a Minimoog and Micromoog). From 1972 to 1981 the modular systems were revised into the somewhat smaller System 15, 35, and 55, selling to Larry Fast, Hans Zimmer, Jan Hammer, and many others. Bob Moog had also been commissioned to work on a small preset monophonic synthesizer, the Satellite (played by Vangelis among others – see website tracks 43–45), and the company made a highly lucrative licensing deal to have it built into certain models of Thomas and Cordovox organs, as well as having large numbers of stand-alone units built, each bringing a relatively generous royalty payment to Moog Music. A slightly extended Satellite, the Minitmoog (now very rare), added touch sensitivity to compete with models such as ARP's Pro Soloist, as well as a second oscillator for a much stronger sound.

The modular synthesizers and Minimoog were still not creating huge profits but the Thomas deal briefly made Moog Music appear extremely lucrative and it was sold to the large music conglomerate Norlin. Bob Moog, more interested in working on research and development, was apparently happy with this:

> "The agreement was that if I stayed four years with the company they would pay me something for my share, which was going to be a nest-egg for my family. So after four years I had about as much as I could take of the corporate world; I didn't really get on with it, and in 1977 I left. Not much later the whole of Norlin went under, but by that stage the Moog division had 300 employees, and was turning out branded products like Maestro effects and electronics for Gibson guitars as well as synthesizers."

Bob Moog was trying to develop the huge Constellation, which would play polyphonically as well as offering bass pedals and other facilities, but this proved impractical using the technology of the time, and the project was split into several components: the Lyra monophonic lead synthesizer was abandoned, the bass pedal section became the Taurus, which sold to Genesis, Yes, Steve Hackett, Gandalf, Rush, The Police, and many others, and the Apollo became the Polymoog.

Herb Deutsch remained marketing manager for Moog but the company's most recent launch, the Polymoog, had been much delayed, having needed hundreds of engineering changes before working acceptably well. The Polymoog was the company's effort to overcome the monophonic playing limitation that had affected the modular systems, the Satellite, the Minimoog, and its smaller spin-offs the Micromoog (used by Kraftwerk and PFM among others) and the Multimoog (which added pressure sensitivity

Moog went portable with the Liberation, played by Tom Coster and Jean-Michel Jarre.

to the keyboard and was played by Steve Winwood and Saga, some example sounds featuring on the website, tracks 54 and 55). The oscillator circuits in these instruments were relatively expensive and unstable, and it was difficult to see how to control several of them simultaneously to accomplish polyphonic playing. Both Moog and ARP had launched duophonic control keyboards, playing one high note and one low note simultaneously, but assigning more than two voltages simultaneously appeared a difficult task.

1972 Davoli organ co. in Italy launches the Davolisint used by Vangelis and PFM

Attempts to solve this problem on the prototype Moog Constellation, which was used only by Keith Emerson, resolved into the eventual approach of making the voltage-controlled oscillator and voltage-controlled amplifier circuits simple enough to be combined onto a single integrated circuit chip, with one whole chip assigned to every key on the instrument, so it would actually be fully polyphonic like a grand piano capable of playing as many notes simultaneously as the performer could hold down. With integrated circuit design in its infancy this was a major and expensive undertaking. Excessive heat problems amongst others delayed the launch of the Polymoog greatly. By the time it was launched several competitors were on the market, notably the Yamaha CS80, and the Polymoog began to seem extremely expensive. Admittedly ambitious, it had taken a draconian approach to solving the problems of polyphony and programmability, also using a separate board full of components to define the tone and envelope of every preset sound (such as piano, clavinet, harpsichord, and strings), and throwing in advanced facilities such as a 71-note weighted keyboard with velocity response, a three-band resonant filter bank, sample-and-hold, and much more.

Moog introduced touch
controls on the Source
alongside programmable
memories.

1972 E-Mu launches with their
modular synthesizer designs in
the USA

The result still sounded disappointingly thin but was an improvement over what was available at the time and the Polymoog was quickly taken up by Abba, Genesis, Kraftwerk, Klaus Schulze (notably for strings on *Body Love* and many other albums), Chick Corea, Geoff Downes with The Buggles (on *Video Killed The Radio Star*, which later became the first track played on MTV), Blondie, Keith Emerson, Larry Fast, Jools Holland with Squeeze, Patrick Moraz, Yellow Magic Orchestra in Japan, and many more. But a simplified less variable version, the Polymoog Keyboard, did not prove popular (though Gary Numan extensively used its only new and most distinctive sound, Vox Humana) and the company looked to other areas for profits, in 1980 launching the Liberation, a monophonic synthesizer with a simple organ-like polyphonic section, which strapped around the player's neck like a guitar. Despite use by Tom Coster, Herbie Hancock, and others (and later by Jean-Michel Jarre during his *Concerts in China*), the Liberation was not a great success either, partly because it was excessively heavy.

From 1979 to 1981 the Micromoog and Multimoog were replaced by the Prodigy, a budget monophonic two-oscillator synth for the entry-level market, notable for offering the cutting, distorted sound of oscillator synchronisation, and from 1981 to 1983 by the Rogue, an even more cut-price model. The Radio Shack chain of stores in the USA also commissioned a Rogue-like model, the Realistic (Tandy) MG1, which added back the simple polyphonic section of the Liberation, and during the same period Moog introduced the innovative Source, designed as a microprocessor-controlled programmable Minimoog using touch membrane switches and offering an arpeggiator and simple sequencer. Costly (due to the extensive research and development involved), never quite matching the Minimoog's sound quality and consequently unpopular at launch, it later came into favour with bands such as Portishead and The Shamen.

From 1980 to 1985, Moog introduced the synth that the Polymoog should have been: the Memorymoog had solved the voice assignment problem, finding a way to assign just six voices to the keyboard but making them much more powerful than the Polymoog voices. Even so, they were not exactly the equivalent of a polyphonic Minimoog; by this time, oscillators, filters, and amplifiers designed from individual components had been replaced by facilities on a single chip, the Memorymoog using the very popular line of chips offered by the US company Curtis. The problem of memory storage had also been solved, as well as programmable memories the Memorymoog had an arpeggiator and various other useful facilities. But the Memorymoog was again late, large, expensive, and not very reliable, and sadly was introduced just before the launch of the MIDI connection standard. These factors, combined with the departure of Bob Moog some years earlier, started a decline for the company that ended with its closure around 1987.

Meanwhile, Bob Moog himself had been commissioned by the Italian keyboard manufacturers Crumar to design a new monophonic synthesizer, the Spirit, on which he tried to incorporate many of the features that, for the sake of simplicity, he had omitted from the Minimoog. Bob Moog also worked for Musonics on an ultimately unsuccessful guitar add-on, the Gizmotron, and then joined the keyboard company Kurzweil as a consultant:

> "After leaving Moog I was made an offer I couldn't refuse, which was to become Vice President in charge of New Product Development for Kurzweil Inc. Then after a few years they went bust – I hope you're not starting to spot a pattern here. But by that time I had the taste for building instruments again. I started my own company, Big Briar, and went back to building Theremins. Then I started to build effects pedals which were like individual synth modules – a filter, then a phaser, a ring modulator and others. I couldn't get the Moog name back even though it had lapsed from use – even in the USA it was very difficult to show that people associated me with the Moog name. So we called this line Moogerfoogers."

The Moogerfooger line proved very popular for processing digital and computer-based sounds with portable analog technology. With the launch of a small control and interfacing unit finally built into a complete modular synth (minus oscillators) that could even be rack-mounted. Obviously a complete new synth design was now on the cards, and this was prototyped as the Big Briar Performance Synth while Bob continued to fight to regain the rights to use the original Moog and Minimoog names. Before Bob Moog's passing in 2005 he re-entered keyboard manufacture with the Voyager and Little Phatty, of which more later.

ARP

One of Moog's earliest competitors was ARP, founded by designer Alan R. Pearlman. Pearlman had already run and sold one successful electronics company in 1967 and the following year, influenced by the release of Wendy Carlos's *Switched On Bach*, looked at the possibilities of launching a new electronic instrument line.

1972 Rick Wakeman releases *The Six Wives of Henry VIII*, strongly featuring the Minimoog

1972 Jazz keyboardist Paul Bley launches the Bley-Peacock Synthesizer Show

Pearlman had cash available from his earlier company and the advantage of being able to examine and improve upon Bob Moog's existing modular system design, and so the launch of ARP was a relatively strong one. The first product, shown in 1970, was the huge ARP 2500, which if anything had even better quality oscillators than the Moog. They certainly stayed in tune more reliably and the 2500 quickly found favour with university music departments.

Rather than using the patch cables necessary on Moog systems, the ARP 2500 grouped oscillators, filters, and amplifiers in the centre of a control panel and provided a matrix of patching switches near each group. This made setting up new sounds faster and avoided obscuring the panel with patch cables but had the disadvantage of creating some crosstalk, with signals from some parts of the instrument leaking through to other sections. But apart from these problems the 2500, available as a central section with two optional "wings" containing sequencers and other accessories, and with a single or dual keyboard, was successful, finding users, such as Roger Powell (originally an ARP demonstrator, who later joined Todd Rundgren's band Utopia), and later with Pete Townshend of The Who, and with Vangelis.

Alan Pearlman's business partners wanted to expand the company rapidly, taking loans in order to do so and launching new products, sometimes without sufficient quality control. A relatively simple polyphonic instrument, the Omni, became a massive seller, however, moving 4000 units between 1975 and 1980. The ARP 2600, a much simplified prepatched 2500, which was still a much more versatile sound creator than the competing Minimoog, was successful and endorsed by players, such as Stevie Wonder and Weather Report's Joe Zawinul, and was later used by Depeche Mode. With a built-in speaker and spring reverb, its sliding controls gave a better visual indication of settings than the Minimoog and later it had a duophonic keyboard designed by Tom Oberheim.

Smaller ARP models were also successful: the Odyssey offered just two oscillators and a three-octave keyboard, competing closely with the Minimoog though not having such a distinctive sound. Favoured by Dave Greenslade, Caravan's Dave Sinclair, and Deep Purple's Jon Lord for its precise, cutting sound quality, it offered sound possibilities not available on the Minimoog, as demonstrated by Ash Ra Tempel and by Klaus Schulze, who created a whole backing rhythm with one on *Totem* from the *Picture Music* album, and went on to use a distinctive Odyssey sample-and-hold chime sound on many albums and concerts (some examples are on the website, tracks 24 and 25).

1973 Roland launched in Japan with SH1000 and SH2000 monophonic synthesizers

Some of ARP's quality control problems were worsened by the encapsulation of parts of the circuitry in resin, allegedly to aid temperature stability but perhaps to cover up copyright-infringing circuitry. Certainly this made servicing difficult on models, such as the Odyssey and Axxe, which simplified even further upon the Odyssey, offering just one oscillator to compete with the Micromoog; a later intermediate model, the Solus, offered two oscillators and was built into a flight case, omitting some of the more expensive facilities of the Odyssey (some abstract Solus sounds are on track 50 of the website). ARP more or less introduced the idea of revising instruments while keeping their names the same; as well as

An early success for ARP – the 2600.

Smaller ARP models included the Explorer 1 and the Pro DGX.

updates to the 2600 and Omni, both the Axxe and Odyssey were available in three or four variations, with changes to cosmetics, the oscillator and filter design, new input and output connections, and on the last models, Proportional Pitch Control (PPC), a set of three rubber pressure pads for modulation, pitch bend up and pitch bend down. While the PPC controls seemed a neat idea they were, by all accounts, not popular, and in the final Axxe and Odyssey models the keys protruded way beyond the front edge of the casing and were constantly being snapped off.

ARP's stand-alone analog sequencer, brilliantly used by Michael Hoenig (on *Departure from the Northern Wasteland*) among many others, also underwent various revisions. ARP instruments were used distinctively by artists, such as The Who's Pete Townshend (on *Baba O'Riley*, inspired by the repetitive minimalist music of Terry Riley) and Edgar Winter on *Frankenstein*, while minimalist Philip Glass used the small Explorer 1, a preset

1973 Release of *Tubular Bells* by
Mike Oldfield, no synthesizers
on this first version

and semi-variable organ-top design monophonic synthesizer good for powerful bass lines (website, tracks 46–48). Although ARP never made huge
profits due to its borrowing from various banks, the financial turnover was
high, and the company's executives mixed with stars of the music business
and film industry. Product specialist Phil Dodds delivered a huge custom
ARP to the set of Steven Spielberg's *Close Encounters of the Third Kind*
and was persuaded to stay on to play its operator in the movie.

However, not satisfied with the success of the Omni, the preset Pro
Soloist, its keyboardless version the Little Brother expander, and the
improved Pro DGX version that competed with the Moog Satellite, ARP
decided to take on the synthesizer polyphony problem with its Quadra
design launched in 1978. Huge, late, and expensive, the Quadra, most
prominently used by Tony Banks with Genesis, combined the Omni, two
Axxes for lead and bass synthesis, and basic programmable polyphonic
synthesis, but was not ARP's ultimate effort in polyphonic synthesis,
which took the form of the Centaur VI. This was intended to be both a
keyboard and a guitar synthesizer (given the argument that there were
four times as many guitarists as keyboard players in the USA) but its
design eventually comprised 115 circuit boards and the project was
abandoned at massive cost.

**The Quadra marked the
beginning of the end for ARP,
though popular with Genesis.**

The Avatar, a stripped-down guitar synthesizer based on the Centaur
but more obviously similar to the existing Odyssey with an added six-
input guitar distortion circuit, seemed like the company's saviour, but its
sales at launch in 1978 were disappointing. By this time ARP desperately
needed a quick success but their 16-preset electronic piano rushed out
in 1979 suffered from design flaws and caused huge cash-flow problems,
since almost every unit sold was quickly returned for service. Licensing
and re-badging reliable designs, such as the existing Solina string synth
and compact organ-like Quartet from European companies, did not help
significantly.

Alan Pearlman was, by this time, in favour of selling the company, but
it was unattractive to most prospective purchasers. Eventually, a deal was
made with CBS, which ran the Fender and Rhodes instrument lines to

at least take on ARP's research and development department along with its latest design, the Chroma, an advanced, weighted keyboard computer-controllable digital synthesizer. ARP finally closed in 1981, with CBS going on to market the Chroma relatively successfully to players including Peter Vetesse with Jethro Tull, though its stripped-down, five-octave derivative was saddled with one of the longest hierarchies of names in the history of synthesis: the ARP CBS Fender Rhodes Chroma Polaris. Much more on Korg's ARP re-issues in Chapter 8.

1973 UK release of *In a Covent Garden* by Brian Hodgson and Dudley Simpson as Electrophon

EMS

Synthesizer development was by no means confined to the USA, with designers in the UK, Germany, Italy, France, and Japan quickly catching on to the possibilities in the field after the success of the early Moog instruments.

The first easily portable synthesizer was launched in the UK by Electronic Music Studios (EMS) in 1969, even before Moog had introduced the Sonic 6 or Minimoog. The EMS VCS3 – it stood for Voltage-Controlled Studio model 3, although models 1 and 2 had never appeared – was designed by David Cockerell under the influence of Peter Zinovieff, an avant-garde composer who had become interested in computer music.

Zinovieff was running a computer music studio based on a DEC PDP8 computer and was interested in computer-controlled analysis of sounds, for which Cockerell built him a 64-band filter bank. But wishing to provide some easily interfaced analog sound sources for the computer to control (and perhaps to raise money for the studio's development), the pair, along with composer Tristram Cary, decided to create and market a small analog synthesizer, which at first lacked a keyboard because neither Zinovieff nor Cary was particularly interested in composing using traditional scales.

The VCS3 has always looked traditionally British, perhaps more like a product of the 1950s than of the late 1960s. In a squarish wooden casing, it stood on a smaller horizontal panel that supported a pin matrix to connect the various sound-generating and processing elements, together with a joystick that allowed several parameters to be varied at once (a useful device that did not otherwise appear on synthesizers for many years). Each matrix pin contained a resistor, so voltages could be mixed freely, though with no pins inserted the synth made no sound at all but with an imaginative patch, using its three oscillators, resonant filter, built-in spring reverb and speaker, the VCS3 could create anything from simple musical tones to indescribable sound effects and slowly developing, self-generating sonic landscapes. A moving-needle meter that could indicate the state of the envelope voltage completed the rather archaic appearance of the machine.

1973 Keio Organ becomes Korg, launching the MiniKorg 700 synthesizer as used by Kitaro

But with the addition of an optional keyboard, the DK1 "Cricklewood" or later the duophonic DK2, the VCS3 (also known as the Synthi or, after the area of London where Zinovieff's studio was located, the Putney) started to sell to rock bands as frequently as to avant-garde

The EMS VCS3 marked the beginning of the UK's synthesizer manufacturing history.

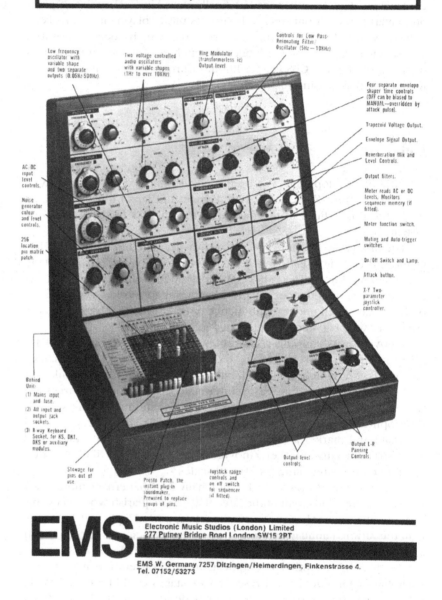

SYNTHI-VCS3, Mark II

Low frequency oscillator with variable shape and two separate outputs (0.05Hz-500Hz)

Two voltage controlled audio oscillators with variable shapes (1Hz to over 10KHz)

Ring Modulator (transformerless ic) Output level

Controls for Low Pass-Resonating Filter: Oscillator (5Hz – 10KHz)

AC-DC input level controls.

Noise generator colour and level controls.

256 location pin matrix patch.

Four separate envelope shaper time controls (OFF can be biased to MANUAL—overridden by attack pulse).

Trapezoid Voltage Output.

Envelope Signal Output.

Reverberation Mix and Level Controls.

Output filters.

Meter reads AC or DC levels. Monitors sequencer memory (if fitted).

Meter function switch.

Muting and Auto-trigger switches.

On/Off Switch and Lamp.

Attack button.

X-Y Two-parameter joystick controller.

Behind Unit:
(1) Mains input and fuse.
(2) All input and output jack sockets.
(3) 8-way Keyboard Socket, for KS, DK1, DKS or auxiliary modules.

Stowage for pins out of use

Presto Patch, the instant plug-in soundmaker. Prewired to replace groups of pins.

Joystick range controls and on off switch for sequencer (if fitted)

Output level controls

Output L-R Panning Controls.

EMS

Electronic Music Studios (London) Limited
277 Putney Bridge Road London SW15 2PT

EMS W. Germany 7257 Ditzingen/Heimerdingen, Finkenstrasse 4.
Tel. 07152/53273

composers. Pink Floyd used it extensively on *Dark Side of the Moon*, the early Tangerine Dream featured it on *Alpha Centauri*, The Who processed an organ sound through its filter on *Won't Get Fooled Again*, Peter Bardens of Camel soloed on it on *Never Let Go* from the band's first album, and instruments also went to Klaus Schulze, Brian Eno with Roxy Music (notably heard on *Virginia Plain*), The Moody Blues, Curved Air, and Todd Rundgren.

The tuning of the VCS3 always tended to drift, so many players went on to use more performance-oriented instruments, such as the

SYNTHI-AKS-80

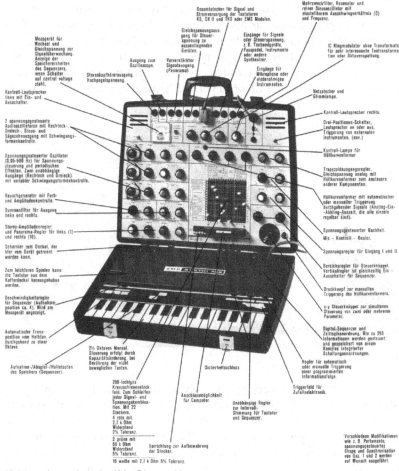

With the addition of the touch Keyboard Sequencer, the EMS Synthi A helped define the sound of Pink Floyd's *Dark Side of the Moon.*

Minimoog, as soon as they were able to, most VCS3s being relegated to duties involving the creation of abstract sound. But they excelled at this. Tim Blake, who was responsible for demonstrating and selling the instrument in France, used a double VCS3 with Gong and on his Crystal Machine solo albums, and Jean-Michel Jarre featured the VCS3 extensively on his debut album, *Oxygene*, eventually collecting and rack-mounting six of the machines.

1974 EMS launch the "Delaware" model used by the BBC Radiophonic Workshop

The progressive rock band Yes later used a prototype VCS3 with a mini-key, built-in keyboard, but although extensively pictured in advertisements, this was not a very successful design. What was successful was the Synthi A or "Portabella", introduced in 1971, with essentially the same electronics as the VCS3 squeezed into a highly portable plastic briefcase. Musicians exported them as hand luggage without anyone noticing, EMS advertised them being played by nuns, brass bands, symphony orchestras and many other unlikely users. Later the space in the lid of the briefcase was filled with a capacitative touch-keyboard with built-in 256-note digital sequence recorder, the KS or Keyboard Sequencer, giving the complete package the title Synthi Aks.

The EMS Synthi A was a VCS3 in a briefcase but the pictured keyboard version never went into production.

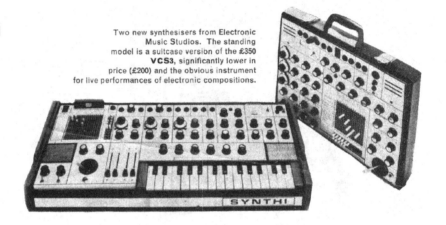

Two new synthesisers from Electronic Music Studios. The standing model is a suitcase version of the £350 **VCS3**, significantly lower in price (£200) and the obvious instrument for live performances of electronic compositions.

The VCS3 and Aks were also widely used for education – EMS published tutorial handbooks explaining their facilities in a highly systematic manner – and later designer Tim Orr was commissioned to create the Synthi E, a simplified version specifically for the educational market, using battery power, patch cables instead of the costly pin bay, and a pitch ribbon instead of a keyboard. Orr went on to develop the early filter bank designs by creating three vocoder models for EMS, some of which were later refined and produced in Germany by EMS Rehberg, which also modified the Synthi E into the slightly improved Synthi Logik.

1974 Tangerine Dream sign to Virgin label for *Phaedra* album using Moog modular systems

In early 1972, the VCS3's power supply and output amplifier were redesigned and the external audio signal could be made to trigger the envelopes on reaching a certain level. Some integrated circuits were introduced as the early models had been made entirely with discrete components; David Vorhaus of White Noise still has the prototype VCS3, which does not even feature any printed circuit boards but is assembled entirely on Veroboard. Various attempts were also made to develop an expanded model VCS3, each called the VCS4 (Brian Eno appears to have used one, another completely different one was used in Zinovieff's own studio), but the next major development introduced in 1974 came to be called the Synthi 100.

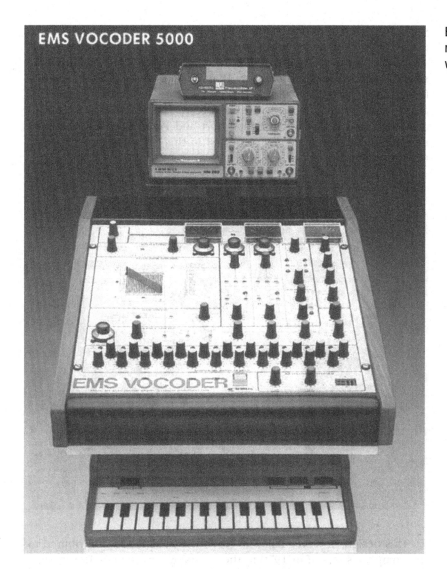

EMS VOCODER 5000

EMS developed one of the most powerful lines of vocoders.

Also known as the "Delaware" and essentially comprising three VCS3s with a total of 12 oscillators, eight filters, a three-track monophonic digital sequencer and two 64 × 64 pin patching bays, the Synthi 100 was so large as to frequently require composers to demolish walls in order to fit one into their studio. Regardless of this, units were sold to the British Broadcasting Corporation's Radiophonic Workshop, which was responsible for creating a large amount of widely heard and highly influential music for television, as well as to Swiss composer Bruno Spoerri, Wolfgang Dauner, and Karlheinz Stockhausen (with an added EMS vocoder) for use on his compositions, *Zodiac* and *Sirius*. Mute Records founder, Daniel Miller, later became an owner for his personal studio.

As well as very large systems, EMS built some small devices, such as an early pitch-to-voltage converter in 1971, which would allow any instrument to trigger a voltage-controlled synthesizer, and later standalone slew generators, ring modulators, quadrophonic effects mixers, an eight-octave fixed filter bank, and others. The VCS3 and Synthi A were also improved with the addition of a socket for "Presto Patches", small banks of pre-wired resistors that avoided the use of the pin bay for patching, but that still left the user having to set the position of all the rotary controls. Later on, the Presto Patch socket was instead described as a computer interface port and Peter Zinovieff's interest in computer music was seen to persist in the 1979 design of the EMS Computer Synthi.

Karlheinz Stockhausen with the EMS Synthi 100.

1974 Kraftwerk release the *Autobahn* album, featuring Moog and ARP synthesizers, a custom vocoder and electronic percussion

This design, however, failed to reach the market and contributed to the company's sell-off in 1979 to the company Datanomics. In the same year, perhaps in an attempt to keep up with developments from Moog and Yamaha, EMS launched the Polysynthi, co-designed by Graham Hinton. Superficially impressive – with a huge and colorful front panel, fully polyphonic playing, built-in voltage controllable analog echo unit, four-octave pressure-sensitive keyboard, and a rear panel port promising a polyphonic sequencer to come – the Polysynthi was extremely disappointing, only between 30 and 50 were made. Its polyphony derived from a simple organ-like circuit; it had only a single filter to process however many notes were playing, the pressure sensitivity actually worked by simply hinging the whole keyboard downwards, the internal wiring was extensive and difficult to service, and the matching polyphonic sequencer never appeared, though EMS had developed a "Universal Sequencer" used with very limited success by keyboardist Tim Cross on the ambitious "Exposed" tour of Mike Oldfield.

Datanomics gave up control of EMS in 1982 after the failure of the development of the Datasynth, a digital programmable synthesizer design, and control passed to Robin Wood, who had been with the company from its very early days. Designer David Cockerell, who had also turned in the EMS Synthi Hi-Fli, a plastic-cased analog guitar effects processor used by Pink Floyd, Ash Ra Tempel, and Groundhogs, among others, had incorporated an Electro-Harmonix delay line into the EMS Polysynthi, and at that time went to work for that company in New York. Robin Wood concentrated on repairing VCS3s, later releasing the EMS Soundbeam, a motion detector that has had some use in creating music for handicapped people.

While EMS has never really gone out of existence, the company never surpassed the glory days of the VCS3, although one of its prominent designers, David Cockerell, certainly flourished, later working on the Akai range of digital samplers.

OBERHEIM

Back in the USA, Moog and ARP were experiencing further competition. Tom Oberheim had designed the duophonic keyboard for the ARP 2600, having earlier produced a design for the Maestro brand, the RM1A ring modulator, and later a popular phase shifter. He also worked as a sales representative for ARP but soon, in what seemed something of a leap in technology, decided to develop through his own company a small digital sequencer that could control the Minimoog or ARP 2600. This, the first real Oberheim product, was the DS2. Having had some success with the DS2 sequencer, Oberheim wanted to develop his own instrument for it to play, but didn't want to take on the design of anything as large and complex as the contemporary Moogs and ARPs. His solution was the Oberheim Synthesizer Expander Module, or SEM, the first complete synthesizer voice without a keyboard, a design which proved popular enough to re-issue in three version with varying degrees of MIDI around 201.

The SEM comprised a small white desktop box with two oscillators, a filter and two envelopes, and could be played from the DS2 sequencer or patched using mini-jack sockets to a Minimoog or ARP 2600 to offer them more oscillators and more powerful sounds. Players, such as Jan Hammer, rapidly took up the SEM, developing an overdriven sound imitating a lead guitar using an SEM and a Minimoog that featured on many of his albums.

The SEM had a rich sound that was subtly different from the competition, Oberheim's filter having a 12 dB rather than a 24 dB cut-off characteristic. In order to exploit this further Oberheim simply designed a multi-voice keyboard controller and started piling up SEMs to create the Oberheim 2-Voice (recently re-issued in a new version) and then 4-Voice synthesizers. These synthesizers originally had no overall modulation and pitch bend controls, and a choice of modes for voice assignment to the keyboard (including "rotating" voice assignment), so despite the introduction of a programmer that recorded some of the settings on each SEM, they could easily become difficult to control. Although the 2-Voice sometimes offered a small, but useful, dual eight-step analog sequencer as well, the user was invariably left with several settings to

1974 Isao Tomita releases Snowflakes Are Dancing (The Newest Sound of Debussy)

make by hand. Since each SEM could sound subtly different, it was hard to create a polyphonic synthesizer sound in which every voice was actually the same.

However, in some ways this was useful, creating subtly shifting sound textures, and the Oberheim 4-Voice machine produced between 1975 and 1979 was used by Stevie Wonder, Jan Hammer, and Pink Floyd, among others. From 1977 to 1980 Oberheim built up huge eight-voice machines with one or even two keyboards, initially just by adding four more SEMs to the rear of a 4-Voice, and later by restyling the cabinet with a new keyboard designed with the staff of E-Mu; these instruments were played by Chick Corea, Chris Franke, Herbie Hancock, Rush, and Yellow Magic Orchestra among others, and even 12-voice instruments were introduced for ambitious users such as Patrick Moraz.

From 1977 Oberheim also marketed the OB1 but this apparently caught on very slowly and in later years its significance was often overlooked. The OB1 was the first programmable monophonic synthesizer with a full complement of memories accessed using touch-sensitive metal buttons (memory expansion boards were also available) plus a novel set of performance controls comprising two "flippers" that could be used for modulation, pitch bend, or sync bend (some examples are on the website, tracks 38 and 39). Again, not all the front-panel parameters could be memorised – LFO and tuning settings still had to be handled by the player – but the main reason for the OB1's lack of impact was probably the introduction of other polyphonic synthesizers at around the same time. Oberheim, in fact, did get onto the market with a much more straightforward polyphonic design in 1979: the OBX, produced until 1983, was played by Prince among many others, and offered four, six or eight typical Oberheim two-oscillator voices in a single integrated unit, rather than using separate SEMs.

The OBX was quickly followed in 1980 by the OBsx, a preset polyphonic instrument with 24 and later 56 memories, then the OBXa, which restyled the casing, offered split or layered sounds, and eventually offered up to 120 memories. It was this generation of Oberheim synths that really defined the Oberheim "sound", seeing use by Eddie Van Halen on the hit single *Jump* for a much-imitated rich synthesized brass effect. Oberheim also launched a polyphonic digital sequencer (DSX) and a sampled-sound drum machine (DSX), both equipped with a custom "parallel buss" interface for interconnection that predated MIDI, so the total "Oberheim System" of keyboard, drum machine, and sequencer, could become an attractive and very powerful proposition. Unfortunately, this was rapidly outdated by the launch of MIDI; the OB8, introduced in 1983, was used by Tangerine Dream, Sting, Steve Roach, and Hector Zazou, and again offered 120 patch memories and keyboard split, but by the end of its production run was offering only MIDI as the standard interface.

The following year Oberheim embraced MIDI fully with the Xpander, temporarily abandoning keyboard designs, since this large desktop module had none. The Xpander remains an amazing instrument, offering six dual-oscillator multi-timbral voices that can be controlled either from MIDI or from "old-fashioned" control voltage and gate inputs, plus filters with many modes, including 12 and 24 dB low pass, high pass, band pass,

1974 Wolfgang Palm launches PPG in Germany, modifying synthesizers for Tangerine Dream

The polyphonic OB8 was a major early success for Oberheim.

band reject, and others, and an incredible flexibility in assigning modulation sources to destinations, including offering five LFOs per voice and a total of 30 envelope generators and 90 VCAs, many existing only in software. As a result, the Xpander was capable of incredibly complex sounds comparable to those from a large analog modular synthesizer, as well as standard analog requirements such as rich strings, brass and lead patches. An even more powerful instrument, the Matrix 12, followed the next year, doubling the number of voices and reintroducing the keyboard.

Unfortunately, Oberheim's finances were not going well. After a takeover by the ECC company they launched the simpler Matrix 6 in 1986, offering many of the same modulation possibilities in a much more performance-oriented synthesizer with simplified controls. The oscillators this time were digital rather than analog, so a fraction of the richness of prior instruments was gone, and unfortunately the advanced programming possibilities of the Matrix 6 were not so easy to access – this applied even more so on its rack-mount version, the Matrix 6R – and as a performance instrument it offered nothing special.

Probably Oberheim's most impressive early product – the huge Matrix 12.

1974 Korg launches the 800DV
(Dual Voice) or MaxiKorg
synthesizer

The Matrix 6 voicing circuits, though, were used to develop the Matrix 1000, a keyboardless 19-inch rack-mounting module that overcame the problem of accessing complex programming by offering no hardware programming facilities at all. The Matrix 1000 offered a huge 1000 memories, programmed by the most imaginative Oberheim owners and specialists around the world, 800 of these in non-editable ROM memory. The remaining 200 could only be edited using a Matrix 6 or 6R or a computer software package but most users found, with such a wide choice of analog sounds in the Matrix 1000 (despite it again using digital oscillators), that they were not concerned about editing it at all. As a result, the Matrix 1000 (although not offering stereo or split, let alone multitimbral playing) survived several more business changes at Oberheim, as well as a few design revisions, and appears to have been in almost constant production from 1988 to 1999, an astonishingly long time for any synthesizer design.

The German company Access eventually offered a stand-alone programming box for the Matrix 1000 (as they did for the Waldorf MicroWave module), meanwhile, Oberheim/ECC were also attempting to enter the stand-alone MIDI accessories market with a range of devices called Perf/X, which included an arpeggiator, drum pattern controller, and MIDI keyboard splitter, plus a neat five-octave MIDI master keyboard with arpeggiator, the Xk.

These devices were not popular, though, and Oberheim became more or less dormant, showing little new except a looping digital delay, the Digital Echoplex. In 1991 a prototype OberM006 rack-mount synth was shown, the name intended to give the impression that it offered both Oberheim and Moog characteristics, which it did, featuring both 12 and 24 dB filters. But the unit, developed over a period of several years with design input from Donald Buchla and eventually appearing as the OBMx, was extremely expensive, despite existing in versions of between two and 12 voices and used by very few artists with the notable exception of The Human League.

Later Oberheim releases, including Hammond-style organs, MIDI master keyboards, and several digital effects units, were actually re-badged designs from the Italian company, Viscount. There was one synthesizer product, the four-octave OB12, prototyped in Oberheim white but eventually marketed with a fetching blue-panelled finish. This was a virtual analog synth with digital effects, including split keyboard sounds, a modulation ribbon, programmable morphing from one sound to another, multiband equaliser, and many other useful facilities. This was an excellent and versatile synth, available at the time of writing for a few hundred pounds second-hand, though it seems not to have been widely liked and did suffer from some software bugs and an unreliable LCD display.

Tom Oberheim had, by this time, long before left the company to which he had given his name, returning briefly to synth design in 1993 with the Marion Systems MS2 and Pro Synth, of which more in Chapter 6, but has experienced a much more substantial resurgence another 20 years further down the line with re-issues and new designs partly in conjunction with Sequential/DSI's Dave Smith, for more on which, see Chapter 8.

SEQUENTIAL CIRCUITS

Founded by Dave Smith in 1978, Sequential Circuits Inc. started life in Smith's garage in San Jose, California, making, as the company's name may suggest, small stand-alone sequencers. Their first product was the Model 800, a digital design not dissimilar in concept to Tom Oberheim's DS2; their second was the Model 700 Programmer, a stand-alone memory unit intended to store envelope settings and LFO speeds to help make large modular synthesizers at least partly programmable. Dave Smith's dream, however, was to create a powerful programmable polyphonic synthesizer of his own, perhaps based on the Minimoog. The Model 700 Programmer design held hints of this, with wooden end cheeks, modern-looking rotary controls and bright LED digits, giving the design an up-to-the-minute look.

Smith's first attempt was highly problematic, though. Titled the Prophet 10, the design generated too much heat to be stable. Just after launch the drastic decision was made to recall the handful of existing instruments and cut them down to just five voices. Some believed that the renamed Prophet 5 would just not offer sufficient polyphony for most players, with eight-note Yamaha CS80s and fully polyphonic Polymoogs already on the market by the time of its launch in 1978. But the Prophet 5 had considerably improved on either of these designs sonically and so the limited polyphony tended to be ignored. It really did sound like a polyphonic Minimoog, with two oscillators based on Curtis chips plus a strong filter design creating every variety of sound from cutting lead lines to smooth and rich strings.

Sequential's groundbreaking Prophet 5 polyphonic synth.

The Prophet 5 also featured 40 reliable programmable memories plus PolyModulation, which routed the second bank of oscillators or the filter envelope to the first bank, creating clanging, metallic ring modulator-like sounds. As a result, the Prophet was used by everybody, from pop outfits, such as Hall and Oates, Abba, and Phil Collins, to progressive rockers Rod Argent, Genesis, and Yes, to experimentalists David Vorhaus (one of the first in the UK, used on White Noise III *Re-Entry*), Tangerine Dream, Larry Fast, and Terry Riley (on *Songs for the Ten Voices of the Two Prophets*), as well as techno-pop bands, such as OMD, Japan (notably for Richard Barbieri's ring modulator-like sounds on the single *Ghosts*), Yellow Magic Orchestra, and Soft Cell.

Unfortunately, the early Prophet 5s were not very stable. Many revisions were made to correct their tuning, memory stability, and other factors. Revision 1 and Revision 2 Prophet 5s used SSM chips, which became in short supply, and after drastic changes on Revision 3 these were replaced with the more dependable Curtis chips. This led to more reliable (and today much more easily serviced) instruments, although many people claim that the less dependable Revision 1 and 2 models sound warmer. But sufficient units were sold to get the company off to a strong start and Sequential (as they were later renamed) offered first a polyphonic sequencer with built-in microcassette dump for the Prophet 5, then a Remote keyboard (1982) interfaced to the Prophet 5 using a connection system, which was to become one of the forerunners of MIDI.

In 1980 Dave Smith was even able to return to the original Prophet 10 concept, this time turning it out as a huge, double-keyboard instrument that sold only to a few very high-end players, such as Tony Banks (Genesis) and Vangelis. Almost everybody continued using Prophet 5s, though, and in 1982 Smith was highly influential in the specification of the MIDI interconnection standard and introduced the first commercial instrument design to use it, the Prophet 600. The 600 was not such a great success, despite offering six voices against the original five and adding polyphonic portamento. Although the oscillators were still analog, some of the control functions were digital and the filter could clearly be heard stepping down in small increments as it closed. The built-in real-time polyphonic sequencer did not synchronise to MIDI and so was little more than a notepad, while the membrane switches that had replaced the Prophet 5's keypad were thought of as potentially unreliable. The addition of an arpeggiator was not seen as very significant, competition from Japan was seen as more affordable and analog was in any case starting to take a back seat as Yamaha launched their all-digital DX7 in 1983. But over the years the 600's sonic versatility has proven to outlast that of almost all its contemporaries. The author regularly recorded with Prophet 600 serial number 693, which included the only firmware update ever issued by Sequential, to improve its originally very basic MIDI options. Other users include Keiko Kumagai with Japanese progressive rock group Ars Nova; some Prophet 600 sound examples are on the website, tracks 63–65.

Sequential decided to continue to look mass-market, having had some success in 1981 with the Pro One, which offered a single Prophet voice with extensive modulation capabilities, an arpeggiator and sequencer, but which was being marketed relatively late for a monophonic instrument with no MIDI retrofit available. They started with a series of more affordable MIDI polyphonics, beginning with the Six-Trak in 1984. This was one of the first analog instruments to offer multi-timbral playing – in other words, each available voice could create a different sound simultaneously – and it was equipped with a limited built-in sequencer to help create complete (if very simple)

The author's studio around 1985, featuring the earliest MIDI instruments – Sequential Prophet 600, Yamaha DX7, and Roland Juno 106, along with a Micromoog, Logan Vocalist, Moog Satellite, Minimoog, Moog Sonic 6, Apple IIe running Greengate DS:3 sampler, Syco Analog-to-MIDI interface, dual Roland TB303s and TR606, Hammond DPM48 drum machine, and Oberheim MiniSequencer.

compositions. After the advent of MIDI, Dave Smith also foresaw the coming application of home computers to music, the Six-Trak had software available to edit its sounds using a Commodore 64 computer. Sequential also launched a Commodore 64 MIDI interface with sequencing software built in, then simplified the Six-Trak into the Max, which was very much stripped down, lacking performance controls, and aimed mainly at computer users.

They also expanded the Six-Trak into the five-octave Multi-Trak, which added chorus and separate audio outputs for each voice, and the eight-voice multi-timbral Split 8 and Pro 8 (the former largely available only in the USA and the latter built in and available only in Japan under a licensing deal), and with the Drumtrax launched a powerful sampled sound drum machine that could form the heart of an excellent composition system with the Max or Six-Trak. The company also badged the simple Prelude and Fugue ensemble synths and the Piano Forte from the Italian manufacturer SIEL.

One of Sequential's early attempts at capturing the mass market, the Multitrack.

For various reasons these models did not help the company's commercial position, and Sequential made one further attempt to go upmarket with the Prophet T8, first seen in 1983. As was so often the case with launches from US manufacturers, this was a case of too late, too expensive and too unreliable. Adding eight-voice polyphony, a weighted velocity and individual aftertouch-sensitive keyboard and MIDI to the Prophet 5 design, it was expensive on launch compared to Japanese competitors, and despite being used by players such as Howard Jones and The Thompson Twins, effectively exhausted the company's finances.

Sequential's last synthesizer product was the Prophet VS and VS Rack, an advanced digital oscillator synthesizer with analog (IC) filters, which again seemed expensive and complex at launch. Sequential had also gone into other areas with the Studio 440 drum machine/sequencer, plus the Prophet 2000 keyboard sampler and Prophet 2002 and 3000 rack-mount samplers. In 1988 the company was taken over by Yamaha, which cleared out the Prophet 3000 and Studio 440 cheaply; many of Sequential's designers also went to Korg and some of the Prophet VS concepts were seen again in the Yamaha SY and Korg Wavestation ranges of keyboards.

The closure of Sequential, manufacturers of perhaps the world's most successful analog polyphonic synthesizer in the Prophet 5, effectively saw the end of large-scale analog design in the USA but after an interval of several years in the software field, Dave Smith himself re-entered synthesizer design with the highly successful Dave Smith Instruments (DSI) and later new Prophet models, of which more later in this chapter and in Chapter 8.

YAMAHA

So far we have not considered what was happening in the field of synthesizer design in Japan, where one of the biggest players was a company which predated all the American and British synth designers by more than half a century. Yamaha was founded in 1887 and built up a strong tradition of building pianos and reed organs, as well as diversifying into fields such as audio equipment, motor cycles, sports equipment, and leisure facilities. This long-established corporate financial stability and product diversification, plus advance planning which can look as far as 15 or 20 years into the future, led to our very rarely hearing from the direction of Japan about anything like the business problems and financial failures that plagued the companies we've already discussed.

Given this corporate strength, Yamaha was in a good position to capitalise on the success of Moog, starting with instruments to accompany its line of domestic electronic organs that had launched with the D1 in 1959. Yamaha's earliest synthesizers dating from around 1971 were the SY1 and SY2, which offered preset sounds selected with organ-like tabs, some control over filtering and vibrato, a pressure-sensitive keyboard, and on the SY2 a high pass filter and resonance control. Now rare, these were picked up by a few institutions, such as the BBC's Radiophonic Workshop, though Yamaha had more widespread success with a striking line of portable organs, including the YC45D used by Pink Floyd, The Osmonds, Eric Clapton's band (an example now owned by the author), Tangerine Dream, and most notably on albums, such as *Persian Surgery Dervishes* and *Shri Camel*, by Terry Riley, who had his YC45D retuned to Just Intonation. The YC45D featured a long pitch bend strip, which worked only on some monophonic special effects sounds, but which was to emerge again to great effect on later Yamaha synthesizer designs.

The relative lack of interest in the SY1 and SY2 seemed to discourage Yamaha from going too deeply into the field of synthesizer design, so it was something of a surprise when in 1975 the company launched the massive GX1. This headed the company's Electone organ series but was in fact an enormous polyphonic synthesizer. Weighing 387 kg and finished in striking white fibreglass, the GX1 had two 61-note polyphonic keyboards plus a 37-note mini-key monophonic top keyboard, a 25-note pedalboard and two large speaker columns, and despite a launch price of perhaps US $60,000, quickly sold to top artists, such as Stevie Wonder, Abba, John Paul Jones with Led Zeppelin, Richard Wright with Pink Floyd, and less well-known figures, such as Juergen Fritz with Triumvirat, and the late Rick Van Der Linden with Ekseption. The author later owned the near identical and equally massive "domestic" version the EX1, which was capable of the same polyphonic synthesizer sounds, and released a CD *"EX1GENE"* using only this very versatile instrument.

But the GX1's most prominent champion by far, in a move away from Moog instruments, was Keith Emerson, who introduced the instrument with a literal fanfare on Emerson, Lake & Palmer's massively anticipated *Works* album. Despite some criticism of the album as a whole, Emerson's GX1 playing on *Fanfare for the Common Man* and *Pirates* was ground-breaking; the instrument had keyboard touch sensitivity that could control filter response and vibrato speed as well as depth, while the mono keyboard turned out cutting, expressive lead lines. Emerson was filmed on tour with ELP in Canada playing complex lines that ran from one GX1

Yamaha's mighty GX1 with matching powered speakers, as played by Keith Emerson.

keyboard to another, with the modular Moog system now very much side-lined behind him, and a new keyboard legend was born.

The GX1 could be reprogrammed only using a small programming box and included a simple drum machine; only about 50 instruments were ever made, some proved to be unreliable due to the huge amount of internal wiring. Keith Emerson eventually bought a second instrument from John Paul Jones when his first one became unreliable; since eight men were needed to lift it the GX1 did not tour well. The benefits of the GX1's huge development costs did, however, filter down very quickly to (slightly) more average users. Yamaha introduced the CS50, CS60, and CS80 in 1976 and 1977, these offered many of the advantages of the GX1. All were polyphonic and pressure sensitive, each had highly controllable LFOs and ring modulators capable of helping create a wide range of unusual metallic sounds, as well as smoothly modulating strings and brass, and each had useful performance options including, on the CS60 and CS80, the pitch ribbon that had first appeared on the YC organ range.

Very much the star of this range, though, was the CS80, the instrument that gave so much effective opposition to the Polymoog from 1977. It offered 22 preset sounds, including strings, brass, and "funk" (a sort of synthesized clavinet), plus four programmable memories varied using sets of tiny sliders under a hinged panel. The CS80 played with eight-voice, two-layer polyphony, had highly expressive individual note aftertouch, a luxurious weighted keyboard, and rather thin filters and oscillators but rich chorus, tremolo and ring modulator effects. Weighing 100 kg, so rather susceptible to damage during transport despite being built into a flight case, the CS80 quickly found favour with bands, such as 10CC and Genesis, with jazz players, Herbie Hancock, David Sancious, and Robin Lumley, and with progressive rockers, Patrick Moraz, Eddie Jobson (very notably on albums from the progressive rock/jazz fusion band UK, such as *Night After Night*) and Peter Vetesse. But perhaps the greatest exponents of the CS80 have been Klaus Schulze, who frequently opened concerts with great clanging ring-modulated sounds bent by the pitch ribbon on the CS80 (something similar is on the website, track 58), and Vangelis, who has often declared it his favourite instrument of all time, and who has purchased at least six machines in case of reliability problems. His typical CS80 sounds, heard for instance on the *Blade Runner* movie soundtrack,

The Yamaha CS60 and larger CS80 were developed from GX1 technology.

Staff at one of London's original keyboard stores Rod Argent's show off an early Oberheim 8-Voice with Oberheim MiniSequencer, ARP 2500 and Yamaha CS80.

Patrick Moraz was an early champion of Oberheim synths, including the rare twin manual 12-voice in the centre, with Yes and The Moody Blues. Also seen are grand piano, Yamaha CS80, Fender Rhodes piano, Micromoog, Minimoog, EMS Synthi As, Hohner Clavinet, organ, ARP 2600 and Pro Soloist, Polymoog, twin manual Mellotron.

are smooth synthetic brass textures, long bending swoops and, again, metallic ring-modulated effects.

Yamaha followed up the success of the CS80 with a range of much more affordable synthesizers: the CS10 and CS30/30L (1977/8) offered monophonic playing, one or two oscillators respectively, and an analog sequencer on the CS30; the CS5 of 1979 simplified these designs, offering a single oscillator; and the CS15/15D of the same year offered two oscillators again, the D model emphasising preset sounds. While these instruments found plenty of users, such as Soft Cell, Depeche Mode,

Vangelis favoured Yamaha's CS80 but also played many other instruments, including the Prophet 10, Emulator, Roland Pro Mars, Jupiter 4 and Vocoder Plus, and Minimoog.

China Crisis, Human League, and 808 State, Yamaha had decided not to follow the Moog design of making their keyboard range correspond to one volt per octave, preferring a logarithmic Hz/volt range, which made their instruments much more difficult to interface to other products; in addition, they usually had pitch bend sliders, but rarely overall modulation

Yamaha's compact CS5 offered analog synthesis on a budget.

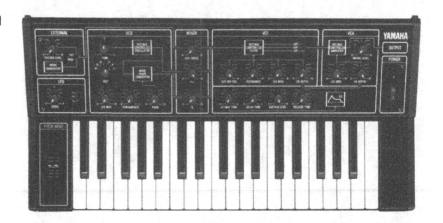

controls, so they were not as performance oriented as competing instruments from Moog or even ARP.

Yamaha were, however, more interested in offering faster programmability and more memories, as on the CS20M and CS40M (1979), the latter a rather large synthesizer that did at last introduce pitch and modulation wheels, but which was still monophonic. Apparently unable to easily surpass the polyphonic synth design of the CS80, Yamaha then decided to go in for a range of ensemble keyboards offering string, brass, and organ sounds, plus some polyphonic synthesizer abilities, which were very limited. The SK10, 15, 20, and 30, released between 1979 and 1981, have never become collectable; the SK50D ensemble of 1980 is of some interest only because it has a huge and impressive-looking double keyboard design. Yamaha seemed rather to have lost their way when, in 1981, they released the CS70M, an attempt to recapture the success of the CS80 with dual six-voice polyphonic ability, added memories, polyphonic sequencer and an envelope to control LFO speed, but the styling came straight from the rather domestic-looking SK range and the programming method involved using fiddly magnetic strips; the CS70M was not a great success.

Surprisingly, that's the end of Yamaha's early analog history, apart from the release in 1982 and revision in 1984 of the tiny CS01, a minikey portable monophonic synth for entry-level users featuring a digital oscillator and breath controller input, with a surprisingly pleasing resonant filter and white noise. That is because the company's next launches in 1982 were the GS1 and GS2, huge and expensive baby grand piano-like instruments with a completely opaque sound generation system that turned out to be the first appearance of digital FM synthesis. The release of FM, as many keyboardists will already appreciate, marked the end of the first phase of interest in analog synthesis. Yamaha's small CE20 and CE25 "Combo Ensemble" keyboards released the same year also concealed the details of this patented system fairly well, but in 1983 Yamaha showed the groundbreaking DX7 programmable digital FM synthesizer and the smaller DX9, and the full implications of the system became clear. FM sounds were more clinical, more precise and in some ways more realistic

The huge Yamaha SK50D combined synthesizer and organ technology.

than analog. After perhaps 15 years of increasing familiarity with analog sounds, a change seemed at the time to be a very good idea.

Yamaha showed the huge DX1 digital FM synthesizer in 1984, then a rack-mount set of modules, the TX816; a desktop module, the TX7; a powerful professional synthesizer the DX5, effectively two DX7s layered together; and then further small entry-level instruments, such as the DX21, DX27, and miniature key DX100. Attempts were made to program some of these instruments to imitate analog effects but these were never very successful and it was to be several years more before Yamaha had any further involvement with analog synthesis of any kind.

KORG

Yamaha's early contemporaries in Japan were Korg, originally the Keio Organ company, which made rhythm units for Yamaha organs. They produced a drum machine called the Doncamatic in 1963 and an experimental synthesizer in 1968. Their first commercial synth releases, though, were the Korg 700 in 1973 and the 700S in 1974; these were flat, boxlike, black-finished monophonic synthesizers (also known as MiniKorgs, perhaps to associate them with the Minimoog, and under the brand name Univox in the USA) with selector buttons along the front panel for organ-top mounting, no performance controls as such, and a combination of low

The compact Korg 700 – a favourite of Kitaro.

pass and high pass filters oddly referred to as a "traveller" and operated from a linked pair of sliding controllers.

The 700S had two oscillators against the 700's one, but the Korg sound was still rather thin, since their filter and oscillator designs did not match those of Moog or ARP. But the instruments were relatively cheap, particularly in their home country, where they helped launch the career of solo synthesists, such as Kitaro, who had been a member of the Far East Family Band. Stevie Wonder also used a 700, as did Vangelis on *Heaven and Hell*. In 1974 and 1975 these designs were followed up by the 800DV (the so-called MaxiKorg), a dual voice instrument more or less comprising two 700s mounted one above the other and capable of playing in unison or duophonically. You can hear a good example of this on the website, track 76, on which the upper and lower sounds of the synth are allocated to the left and right channels. Korg also launched the 770 and 770S, which offered two oscillators and angled the control panel away from the player, as on the Minimoog; the SB100 or Synthe-Bass, a cut-down 800DV with a shorter keyboard intended only for bass tones; and in 1975, the 900PS, a three-octave organ-top instrument with preset tabs featuring an unusual metal rod running under the keyboard, which the player could touch to introduce modulation effects.

The next year, Korg launched two string ensemble instruments, the PE1000 and 2000, notably used by Tim Blake on the *New Jerusalem* album. In 1977, some of the best elements of the 700 and 900PS were combined into the M500 Micro Preset, a very compact monophonic instrument with preset sounds such as flute, brass, and synth, as well as white noise and a few variable parameters (some examples are on the website, tracks 40–42). With a speaker added on the M500SP version, it proved very successful on early singles from Orchestral Manouevres in the Dark. Later in 1977, Korg revealed a much more ambitious project: the PS3300 was a fully polyphonic synthesizer featuring a four-octave keyboard with three oscillators assigned to every note. The result was a huge instrument, the size of a modest Moog modular system, but one which was highly flexible; among other facilities, it was possible to retune each note of the chromatic scale in order to play alternative scales and tunings. Each oscillator bank was independent so the instrument was effectively a triple modular system with many patching points as well as a pre-patched signal flow. The 3300, which featured a

A compact and very simple early Korg synth – the 770 lacked even a pitch bender.

keyboard that was separate from the synthesizer unit and which could be placed at a distance from it, was played by Keith Emerson, while the simpler 3100 model offered the same polyphony with just one oscillator per voice and an integral keyboard. In 1978, an intermediate model, the PS3200, was introduced, which left off some facilities such as voltage-controlled resonance but which did offer programmable memories.

Keith Emerson endorsed Korg's PS3100 and PS3300 in the early 1980s.

Korg's powerful PS3100 was fully polyphonic.

These were upmarket instruments, though, and in 1978 Korg brought in a range that introduced thousands to the delights of analog synthesis. The MS10 and MS20 were well designed in that they offered a simple ready-patched signal path that could also be overridden and added to using patch cables. This design gave the MS series many of the advantages of modular synthesizers without their attendant problems. Since the instruments' control panels angled away from the user, as on the Minimoog, they were immediately attention grabbing too. The single oscillator sound of the MS10 was rather thin, and it had just one performance wheel, normally assigned to pitch bend, however, this could be re-patched to other functions, such as modulation depth, and the instrument featured white noise and all the analog basics (some examples are on the website, tracks 19 and 20). The MS20 added a second oscillator, an external input that converted pitches to voltages, and many more patching options, which made its claim to offer the best of pre-patched and modular synthesis more convincing. However, both instruments used a Hz/volt system like Yamaha's, so were not easy to interface to non-Korg products (though Korg did make some interface accessories to make this easier). Korg added to the MS range with a matching dual 12-step analog sequencer, the SQ10, and also with a matching keyboard vocoder, the VC10. The following year, 1979, they released a now rare and sought-after expander module, the MS50, which was like a keyboardless MS10, adding facilities including a ring modulator and voltage-controllable sample-and-hold. The MS20 design was revived more recently, first with the MS20IC, which was a USB-equipped keyboard for the iMS20 software, and later the self-contained MS20Mini in black or white.

Also in 1979, Korg released their last monophonic analog design for some time, the preset touch-sensitive Sigma, which featured two joystick performance controls. The Lambda was a 48-note (F-E) layerable polyphonic strings and piano synthesizer with three VCOs per note, a phaser, 96 independent envelopes but just a single filter, while in 1980 the Delta was a four-octave string synth with simple polyphonic facilities, white noise and Korg's first 24 dB filter. The company was now much more interested in polyphonic designs, in 1979 launching the Trident, which layered a programmable polyphonic synthesizer with strings and brass. The eight-note polyphonic instrument was large, with multiple audio outputs, a flanger, and keyboard split, but it wasn't possible to edit a memorised sound – the player had to go back to manual mode and recreate it from scratch.

Korg's MS10 was a small monophonic synth with plenty of patching possibilities.

The X911 released in 1980 was ostensibly a guitar synthesizer, in other words, a monophonic synth played from a guitar (which needed to be picked cleanly one note at a time), but comprised the pitch-to-voltage circuit from the MS20 driving a small preset monophonic synth plus a gentle

Korg's Delta was a typical
small string sound synthesizer.

guitar distortion circuit. Also equipped with control voltage and gate inputs,
the X911 was a useful little monophonic analog synth expander. You could
do tricks, such as playing it from an analog sequencer then transposing the
sequence by playing a single guitar note, as guitarist Mark Francombe did
with the UK experimental band Uncle Ian and the Tooth Decay.

But the X911 was simply a diversion before the release in 1981 of the
Polysix. This was a clear attempt to compete with contemporary American
manufacturers at a more affordable price; it even used the SSM chips seen
in early Prophet 5s, offering six-note polyphony with just one oscillator
per voice but good sub-octave generators and chorus, plus truly independent filters and amplifiers for each voice, 32 memories, an arpeggiator, pitch
and modulation wheels. The Polysix was successful, and so the 1982 Korg
MonoPoly seemed a throwback; it was essentially a large monophonic
synthesizer with an arpeggiator on which all four oscillators could play in
unison, with a thin-sounding polyphonic mode thrown in almost as an incidental. In the same year, the Korg Trident was updated into a Mark 2 version
with more memories and better parameter editing, but this was expensive
and was overshadowed in 1983 by the launch of the Poly 61.

Korg's preset Sigma was
unusual in having two joystick
controllers.

The Poly 61 tried to economise by using digital rather than true analog oscillators and by leaving off most of the parameter editing controls in favour of a single data editor. It retained the memories and arpeggiator of the Polysix but had to be rapidly updated when MIDI was introduced and reappeared in 1984 with MIDI as the Poly 61M. The Polysix was also kept in the catalog with a MIDI retrofit but Korg's next success was a simplified and relatively very affordable synth. The Korg Poly 800 had only a four-octave keyboard, a battery power option and guitar strap fixings so it could be worn around the player's neck, but it had MIDI, eight digital oscillators with sub-octaves, memories, chorus, an option of layering the oscillators together for stronger four-note polyphonic sounds, white noise, and a polyphonic step-time sequencer. It was a surprise to find that the Poly 800 had only one filter, so on some patches, all sustaining notes would be filtered again whenever a new note was played. Most users were happy to live with this, though, and a keyboard-less version, the EX800, became one of the first stand-alone MIDI modules, though in a rather unusual format that would take up a massive six units of rack space, plus additional space for the input and output sockets, which were all on the top panel.

A few Poly 800s were released with reverse-colour keyboards (off-white and black keys), and a Poly 800 Mk II model was released in 1986, replacing the chorus with a more versatile programmable digital delay. This marked the end of Korg's initial analog phase, since their next releases, the DW6000 and DW8000/EX8000, had digital sound generation. The five-octave DW6000 was capable of many analog-like sounds, though; it had simply added some more complex digitally stored waveforms resembling those of acoustic instruments to the usual analog-style complement of sine, square, and sawtooth. The DW8000 went even further, adding further alternative waveforms as well as velocity and pressure sensitivity, and a programmable digital delay similar to that on the Poly 800 Mk II, and along with a resonant filter this turned out to be a remarkably flexible instrument. Third-party companies offered keyboard split and memory expansion facilities for the DW8000. The EX8000 rack-mount version, a two-unit-high MIDI module, remained for many years a dependable and widely used instrument.

Korg went for affordability with the battery-powered Poly 800.

Following the introduction of the Korg DSS1, a large and powerful digital sampler/synthesizer keyboard with resonant filters and effects, Korg's next moves really did go far beyond the bounds of analog technology. The Korg M1 introduced the workstation concept, offering powerful sequencing and digital effects as well as keyboard and drum sounds, but more to the point, the sounds were based on very short digital samples, the synthesis part of each program becoming almost incidental. Certainly Korg appeared to find it very difficult to realise decent resonant filters digitally. The whole of their next generations of synthesizers – the M1R and M3R rack-mounts, the T1, T2, and T3 keyboards, and so on – had great difficulty in offering many "traditional" twangy analog patches, though their versatility in other types of sound creation made them a huge success, the distinctive sampled "house music piano" and other M1 sounds becoming very popular.

ROLAND

A history of Roland is practically a history of the analog synthesizer itself, as the company has been by far the most prolific and perhaps the most successful in the field, with over 30 relevant products to consider in the period 1973–1986 alone.

At different times Roland was also making effects units, recording equipment, drum machines, and electronic percussion, many different types of sequencer, guitars, and guitar synthesizers, electronic pianos, organs, and much more, and many of these products interfaced in interesting ways with their analog synthesizer products.

Roland products probably have the best reputation in the world for reliability and for accessibility. Their inputs were generally standard or easy to modify, spare parts support has been as good as could possibly be expected given the rapid advance of technology, and the legacy of one Roland product in particular was almost single-handedly responsible for the analog revival of the late 1980s.

Blessed with Japanese corporate finance and long-term vision, as was Yamaha and to some extent Korg, Roland's history contains little or none of the corporate struggle experienced by most US, British, French, and German companies. Their user list is so huge as to make a mention of individual players almost pointless, although to be fair, the company's entries in any list of all-time classic instruments would be in the areas of drum machines rather than keyboards, with the possible exception of the TB303 Bassline (for special reasons, which we'll see later) and, in the opinion of many, the Jupiter 8 synthesizer.

Roland began as Ace Electronics, making organs, drum machines and amplifiers from the mid-1960s, and introduced its first synthesizer models in 1973. Like Korgs and Yamahas, they were organ-top designs, preset models with some variability and unusual facilities, such as "random note", named SH1000 and SH2000, the latter offering pressure sensitivity that could make it very expressive.

A whole series of variable monophonic synthesizers followed: the SH3 in 1974, the SH3a in 1975, the SH5 in 1976 (some sounds are on the

Roland's SH5 was a typical
early fully variable analog
synthesizer.

website, track 69), the SH7 and SH1 in 1978, and the SH2 in 1979. These offered one or two oscillators, various lengths of keyboard, a filter sound that was thinner than US contemporaries but nevertheless quite versatile, often more experimental facilities, such as sample-and-hold and white noise, and on the SH5 a ring modulator. Plus, usually, but by no means always, inputs for external control (which had not been found on the early SH1000 and 2000), and a tendency to lack performance controls, which made them less expressive than, say, the Minimoog or ARP Odyssey.

• Control Panel Standard Setting (By this setting, performance is ready.)

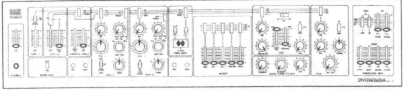

Interfacing the SH5, or how to
use a synthesizer with other
instruments.

• Terminals at Rear Panel

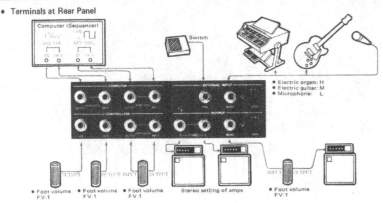

So few musicians chose these as soloing instruments but hundreds of them came into use in studios around the world. Simultaneously, Roland developed a line of simple polyphonic keyboards: the RS101 and RS202 string synthesizers in 1975 and 1976, the RS09 in 1978, and later the same year the more complex RS505, which paralleled an analog guitar synthesizer design, the Roland GR505, in offering monophonic plus simple polyphonic synthesizer facilities, as well as a string synthesizer section.

Roland was also interested in modular synthesizers, testing the water in 1975 with the System 100. This featured: a small, fully variable, single oscillator monophonic synthesizer, the 101; a dual 12-step analog sequencer, the 104; a keyboardless expander synthesizer with sample-and-hold, the 102; plus optional powered speakers and mixer. The whole system could be built up one unit at a time and found favour with players such as Duran Duran's Nick Rhodes (who re-purchased a whole system from the author in the late 1990s), Vangelis, who created many of the smoothly transposing sequences on *Spiral* and *Albedo 0.39* using it, and Steve Jolliffe, who played one with Tangerine Dream on their 1978 *Cyclone* world tour.

The Roland System 100 combined relative portability with modular possibilities.

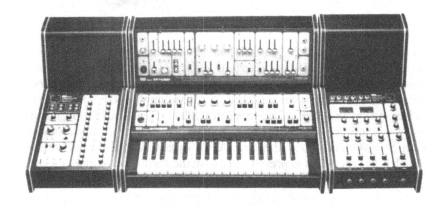

The System 700 was Roland's answer to the large Moog modular systems.

In 1976, Roland also launched the huge System 700, which directly competed with the Moog modular systems of the time. There were 47 different modules available but these were almost always delivered in the same combinations, distributed between six stacking cabinets. The central cabinet featured three VCOs and two VCFs and was pre-patched to act as a powerful stand-alone synthesizer; a five-octave duophonic keyboard was available. The left cabinet added six more VCOs; the right added more VCFs, VCAs, envelopes, and multiplier sockets. The upper cabinets added a nine-channel mixer (most obviously to mix the nine oscillators), a pitch-to-voltage input, phaser, analog switches, more multiple sockets, and so on, and the central upper cabinet comprised just a large 3×12-step analog sequencer. The System 700 was more stable than some of the earlier Moog models and its sound was certainly extremely powerful, although some would say still not as strong as the Moog's. The 700 was used by Isao Tomita, Matthias Becker, Vince Clarke and Depeche Mode, Klaus Netzle, The Human League, Visage, and Hans Zimmer, among many others, and remains a rare and expensive system. A smaller "Lab Series" version comprising just one cabinet with three VCOs was also available from 1977.

At the more affordable end of the market, Roland continued to experiment with programmability and polyphony, showing the four-note polyphonic Jupiter 4 in 1978 and the simpler monophonic MRS2 Pro Mars in 1979. The Jupiter 4 again had a rather thin oscillator sound but with a chorus and use of an interesting arpeggiator it became a popular model with Kitaro, Tangerine Dream, Trans X, Rainer Bloss on tour with Klaus Schulze, Stevie Wonder, and many others. The same year, Roland also showed the VP330, a keyboard vocoder with analog synthesized string and choir sounds (typical Jupiter 4 and VP330 sounds are both on the website, tracks 28, 61, and 62).

The company's next move was to expand the modular synthesizer range with the System 100M. This used tiny modules in custom wooden cabinets that powered up the modules and provided control voltage and trigger interconnections to save some basic patching. Three- and five-module racks were available. Modules included dual oscillators, dual filters, a VCO-VCF-VCA combination that comprised almost an entire synthesizer in itself, mixers, envelope controllers, a phaser, and many more – around 16 different modules altogether – plus three-and-a-half- and five-octave monophonic keyboards and a five-octave, four-note polyphonic keyboard with arpeggiator known as the Model 184 or 4CV.

The compact size of the System 100M modules, which used mini-jack patch cables, allowed some huge collections to be built up – Hans Zimmer apparently had around 200 modules at one stage – and the system was also used by Richard Burgess of Landscape, the BBC Radiophonic Workshop, Chris and Cosey, and again by many of the electro-pop crowd, Vince Clarke, Depeche Mode, The Human League, and so on. With a powerful sound and great flexibility, the system became the basis for several tutorial books and remains a popular way into modular synthesis if you can find one. Modern clone versions are now starting to appear.

Roland concentrated more on portability in 1980 with the SA09 Saturn, a simple polyphonic keyboard, and the SH09, a revision of their earlier monophonic synthesizers, but in 1981 looked upmarket again with the Jupiter 8, which was to remain their flagship synthesizer for some time. A large instrument with eight dual-VCO voices, the Jupiter 8 after a short while featured a proprietary Roland interface called DCB (Digital Connection Bus, these models usually being referred to as JP8a) plus 4/4 keyboard split, hold, an arpeggiator with random setting, solo mode, cross-modulation, LFO delay, high pass filters, and 12 or 24 dB settings on the low pass filters, and saw massive use with Alphaville, Barclay James Harvest, John Foxx, Cocteau Twins, Talk Talk, Jan Hammer (on *Miami Vice*), Simple Minds, Marillion, Tangerine Dream (Johannes Schmoelling replacing a Minimoog with a Jupiter 8 set to solo mode for his lead lines), Michael Jackson, and many others.

A classic polyphonic synth – the Roland Jupiter 8.

The Jupiter 8 was expensive, though, but real strides in affordability were made in 1982 with the introduction of the SH101 monophonic synthesizer, the Juno 6 and 60 polyphonic synthesizers, and the TB303 Bassline. The SH101 was part of a plan to put more synthesizer functions under microprocessor control; in a slim plastic casing with battery power option, the SH101 featured an analog oscillator with mixable waveforms, sub-octave, white noise and sample-and-hold, but scanned its three-octave keyboard digitally to offer a latching, transposable arpeggiator and programmable step-time sequencer.

The SH101 was light enough to strap around the player's neck and an optional "modulation grip" duplicated the pitch bend and modulation facilities for performance with the middle finger and thumb of the left hand. To add to its stage appeal, the instrument was also available in metallic red and blue finishes, as well as the standard gray. This was a relatively affordable yet flexible synthesizer which found use with many entry-level players as well as with professionals such as 808 State, Human League, OMD, and UB40. A miniature version, the MC202, featured a very similar single-oscillator voice, but two more flexible channels of sequencing programmed and played from tiny push-buttons (some MC202 sounds are on the website, tracks 34 and 35). The SH101 was reborn more recently as part of Roland's *Boutique* module line.

For those looking for an affordable polyphonic design, the Juno 6 seemed ideal. It offered simple six-voice polyphony with digital oscillators, sub-octaves, an arpeggiator, chorus, highly resonant low pass filters and a high pass filter, pitch and filter bender, and a modulation button. It was a totally variable synthesizer with no memories, lacking a mono or unison mode and so not very useful for solos, and its basic sound was extremely thin, but with a sub-octave mixed in, slow pulse width modulation and chorus selected, it could sound fairly rich, and the arpeggiator gave some rhythmic possibilities to those who couldn't afford a sequencer. After a while a DCB socket appeared on the rear panel, allowing the synth to be interfaced to various sequencers or to other Roland synthesizers. More annoyingly though for early purchasers, the Juno 60 was fairly quickly introduced. This featured 56 memories as well as having DCB as standard, seeing use with Enya, Cabaret Voltaire, John Foxx, Level 42, Howard Jones, Mainframe (very prominently on the brilliant but neglected album *Tenants of the Latticework*), and The Cure.

In all this excitement the TB303 Bassline, introduced at the same time as the Juno synths, was somewhat sidelined. It was an odd design, a shoebox sized monophonic bass sequencer/synthesizer intended simply to accompany one of Roland's small drum machines, the TR606 Drumatix. Later Roland drum machines never had a corresponding bass sequencer released, so the TB303 had probably been counted something of a failure. The Bassline could record short patterns of notes from a micro-keyboard layout, which could then be chained into complete songs. The result was then synchronised to the drum machine to create a complete rhythm section; there was even an audio input into which to connect the drum machine, which sadly didn't pass through the Bassline's filter.

The Bassline did, however, have control voltage and gate outputs, so its parallel intention was obviously to sequence a higher quality external analog synthesizer or modular system. Its simple built-in monophonic voice could hardly be described as a full synthesizer, in that it only had a choice of two waveforms, controls for filter cut-off and resonance, envelope depth to the filter, decay time, accent level, and a way of assigning accent and glide to selected notes in the programmed sequence. The "flat" bassline sound was a rather uninspiring "plonk" (some sounds are on the website, tracks 36 and 37), not particularly reminiscent of a real bass guitar, while the sequence programming method was extremely difficult to master. With no method of saving sequences externally, and synchronisation only to clock pulses received at a DIN synchronisation input socket, the Bassline was judged uninspiring and was produced for less than two years. So Roland could not possibly have guessed that this machine would be almost solely responsible for the "analog revival" some years later and, before this happened, prices of Basslines had fallen almost to nothing, certainly to £50 or less in the UK.

The introduction of MIDI required much more substantial innovation in the form, in 1983, of the JX3P synthesizer, which cosmetically clearly came from a different design team. This offered six-voice playing from 12 digital oscillators, editing its parameters with a single slider control; there was a built-in polyphonic sequencer, a couple of unusual sound elements, such as metallic and ring modulator settings, and an optional, magnetically attached PG200 programmer, which added a bank of editing controls for those interested in faster access to sound editing. Later, the potential of MIDI was explored again on the Jupiter 6, a six-note splittable (but not layerable) polyphonic design, again with an arpeggiator, which went back to using VCOs rather than DCOs and so had a very high price on launch. This perhaps contributed to its lack of popularity, though it was used by Neuronium in Spain, Men At Work in Australia, The Human League, Pop Will Eat Itself and The Shamen in the UK, and Trans X in Canada, among others. The Jupiter 6 (particularly after its MIDI specification was updated) was very powerful, offering a huge unison mode sound and high pass and band pass facilities on the filters, plus some very attractive red illuminated push-buttons, but it had no chorus and a single audio output, and so is now much less sought after than the Jupiter 8 (and these days Beep Street's Horizon more or less replicates it in iPad app form).

By 1984, the full implications of MIDI had been realised, synthesizers no longer necessarily appeared to need keyboards and the JX3P was "modularised" into the rack-mount, velocity-sensitive MKS30, again using the optional PG200 programmer (a floor-standing guitar synthesizer, the GR700, could also be edited with the same unit). Later in 1984, the Juno range was updated again with the Juno 106; this offered MIDI as standard, just six DCOs with sub-oscillators, 128 memories, Roland's combined pitch bend/modulation controller and the addition of polyphonic portamento. The envelopes on the Juno 6 and 60 had been fast acting and "snappy", but the 106's envelopes were even more so, resulting in a cutting sound that is still sought after today. The arpeggiator was lost though, and the significance of the fact that all the parameter control sliders sent MIDI controller information was not really realised until the use of computer sequencing became much more common.

The HS60 was a home-oriented version of the 106 equipped with built-in speakers, at the same time Roland released the powerful rack-mount MKS80 Super Jupiter. This was the only module Roland had ever released with no keyboard version, featuring eight-voice polyphony from 16 VCOs, originally from Curtis chips and later from chips of Roland's own design. It offered cross-modulation, detune in unison mode for very big lead and bass sounds, and two LFOs per voice; an MPG80, which is a large rack-mountable programmer featuring 29 sliders that again send changes out as recordable MIDI data, is a highly desirable add-on.

In 1985 came the JX8P, a velocity-sensitive hybrid of the JX3P and Juno 106. Because Yamaha was dominating the market by this time with

FM synthesizers, the JX8P was full of programmed attempts to imitate metallic FM sounds, despite using more or less the same voice architecture as the JX3P. Despite its name, it was not eight-note polyphonic either, offering just six-note polyphony, but it did have cartridge memory back-up and a display large enough to allow the user to give names to each program.

The EM101, released the same year, was a desktop half-rack module intended to add preset Juno-like string, brass, bass, and a few other sounds to MIDI pianos, while the MKS7 "Super Quartet" rack-mount offered preset monophonic bass, duophonic lead, and polyphonic Juno sounds, plus sampled percussion sounds from Roland's drum machine range.

Roland's next keyboard release was the Alpha Juno 1, a small four-octave DCO-based synth offering familiar Juno sounds but using membrane switches for patch selection and a single rotating "alpha dial" for editing. The following year, 1986, the five-octave velocity and aftertouch-sensitive Alpha Juno 2 followed, along with a domestic version with speakers, the HS80, and a rack-mount version, the MKS50, plus an optional programmer, the PG300.

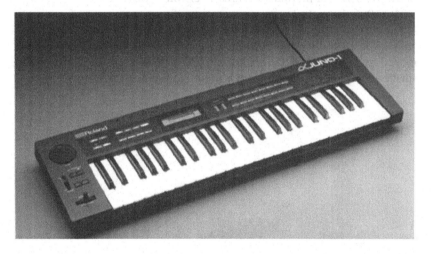

Programmable polyphony on a budget arrived with Roland's Alpha Juno 1.

Also in 1986, the JX8P design was developed further into the Super JX10 and its MKS70 rack version, which while using similar voice architecture had greatly enhanced features. For a start, the Super JX10 was 12-voice polyphonic, splittable, and had a six-octave semi-weighted keyboard. Sixty-four memories could be dumped to cartridge and there was a basic polyphonic sequencer built in. There were various mono unison modes, sync, and chorus, all more easily accessed from an optional PG800 programmer.

This was Roland's last big effort in true analog synthesis, though, as their next releases used a new technique, LA (Linear Arithmetic) Synthesis, based on relatively long digitally sampled waveforms processed by all-digital filters and effects, the first releases being the MT32 desktop module, then the D50 synth and D550 rack-mount, and the budget multi-timbral, multi-output D110 module. As these had resonant filters and waveshapes

Roland's Super JX10 was the company's last early analog product.

Roland's Super JX10 was the company's last early analog product.

including sine, square, and sawtooth, these models were capable of producing satisfying analog-style sounds, but fashion was moving towards greater realism, so they arrived programmed mostly with imitative instrumental sounds, which in the case of the MT32 and D110 later formed the basis of the very conservative General MIDI sound set. Roland's current *Boutique* series of modules features a version of the D50.

By 1986 the eight largest and most significant analog manufacturers – Moog, ARP, Oberheim and Sequential in the USA, EMS in the UK, and Yamaha, Korg and Roland in Japan – were either virtually inactive or had moved away from strictly analog designs. But as we'll see next, many smaller companies had also been working in the field.

4

The growth of analog

As we saw in the previous chapter, the birth of analog was dominated by just eight companies in the USA, UK, and Japan, with histories in the field all spanning 20 years or less. But there were many other participants in the growth of analog, some offering innovative design features in very short production runs, others bringing well-established technology to a wider market or a new level of affordability, and these companies spanned the whole world, from the UK, USA and Japan to Germany, France, the Netherlands and Italy, Russia and even China.

ITALY

One very short production run design that predated the Moog synthesizers, and even the widespread use of the transistor, had started life in Italy. The Synket was designed by Paulo Ketoff in Rome in 1964, using valve technology; Ketoff had worked on an instrument called the Fonosynth and developed a new instrument with three oscillators, white noise, three modulators, a fixed filter bank, and an external audio input. The Synket was later used by John Eaton at the Sonora House arts centre in New York, as much as an audio processor for other more well-known analog synths as in its own right, but only about six were ever made and Ketoff did not attempt to go into mass production, faced with competition from the early voltage-controlled portable analog synthesizers such as the Moog Sonic Six.

1975 Vangelis releases his *Heaven and Hell* album, featuring Roland and Yamaha synthesizers

Also in Italy around 1970, GRS built the VCO3, inspired by the EMS VCS3 and Minimoog, and it saw use by Franco Battiato (when he was still an experimental musician influenced by Stockhausen, rather than a ballad singer). But fewer than 30 were built. Also based in Italy from the early 1970s, Welson manufactured organs and string synthesizers, and had one attempt at a monophonic synthesizer, the Syntex of the late 1970s. This was a preset/variable organ-top style design with two VCOs, a resonant VCF, random sample-and-hold notes and more. Attempts to develop a polyphonic version apparently failed, though.

Mention should also be made of Davoli, which made several organ and electronic piano models, and in 1972 and 1975 turned out the Davolisint and Davolisint B (or Model 75). This looked like a small three-octave organ (the bottom gray octave only selecting oscillator footages), with a bank of selector tabs along a sloping panel, and was more like a monophonic organ with portamento, pitch bend and vibrato, and no envelope release rather than a true synthesizer. It was used to good effect, though, on the German band La Dusseldorf's debut album of the same name; attempts to develop a huge polyphonic synthesizer, the Tauonos, apparently helped close the company.

However, there were a few much more substantial Italian companies that moved successfully from the production of organs and electronic pianos into the synthesizer field. SIEL (Societa Industrie ELetroniche) made the simple polyphonic Quartet for ARP in 1979 (their own version was called the Orchestra), and the Prelude and Fugue for Sequential in 1979. Their own Cruise model of 1981 had simple polyphonic and preset mono-phonic sections, while the Mono had just the monophonic section with more preset sounds added.

But SIEL was also an early signatory to the MIDI agreement, in 1983 it showed the MIDI-equipped Opera 6. Revised several times, this had five octaves of velocity-sensitive keys, two VCOs for each of the six voices, a full set of editing controls, programmable memories and cassette data dump. Later versions used DCOs, in Germany the synth appeared as the Kiwi, elsewhere, following even further revisions as the DK600. There was also a keyboardless Expander 6 version of the DCO models, designed in a desktop style slightly too large for 19-inch rack mounting.

1975 Elka launches the Rhapsody 490 and 610 string synth keyboards

No doubt looking at the relative sales figures for Korg's semi-pro Polysix and budget Poly 800, SIEL then decided to economise with the DK70, a smaller, four-octave, plastic-cased synth light enough to strap around the user's neck and featuring just one filter for all eight DCOs. There was a simple polyphonic sequencer and an optional controller "neck" featuring a pitch bend ribbon. But the DK70 was not successful, the full-size five-octave DK80 (and its EX80 expander version) did little better. Soon afterwards the company was sold and some of its factories were used by Roland for production primarily of electronic pianos.

Elka had a comparatively illustrious history, based in Italy and making organs (also under the Orla brand name), such as the Capri and Panther, and from 1975 the Rhapsody 490 (with four octaves) and 610 (with six octaves). This was a string synthesizer with additional electric piano and harpsichord sounds; the strings and harpsichord played together made for a strong, precise layered texture, which can be heard all over the albums of Tangerine Dream, on Michael Hoenig's *Departure from the Northern Wasteland*, and on Tim Blake's albums, as well as seeing use by Supertramp, Vangelis, Jethro Tull, and Ultravox.

1975 Roland launches the System 100 semi-modular synthesizer system

From 1977, the X605, X705, and X707 were Elka's top-of-the-line organs, featuring dual manuals, a drum machine (except on the 605), the string and harpsichord section from the Rhapsody, plus monophonic and (single-filter) polyphonic synthesizer sections. The author still owns an X705 and performed with it alongside Can's singer Damo Suzuki at London's Queen Elizabeth Hall in 2003. Jean-Michel Jarre used an eccentrically mirror-finished X705 for his Concerts in China, and from 1978 also played Elka's organ-top preset synthesizer, the Solist 505, effectively the monophonic synth section from the organ designs.

Later on, and probably inspired by the success of the Prophet 5, Elka started to design a powerful polyphonic programmable synthesizer. This took quite a while to get onto the market, so the Elka Synthex appeared only in 1982, just in time to get caught out for having no MIDI interface. The problem was solved fairly quickly with the issue of a retrofit kit and later Synthex units featured MIDI as standard.

Elka's mighty Synthex – the first Italian MIDI synthesizer.

Looking like a slightly deeper Prophet 5, the Synthex offered eight-note polyphony from 16 DCOs, band pass and high pass options on the filters, ring modulator, chorus, variable split and layer, ROM and program-mable RAM memories, and a looping four-track monophonic sequencer. The Synthex featured some great sounds (some of the best are on the website, tracks 21–23), partly because of the abandonment of the assump-tion that a two-oscillator synth couldn't benefit even further from having a good chorus added, and found use with Jean-Michel Jarre (very promi-nently on *Revolutions*), Stevie Wonder, Ultravox, Geoff Downes and Keith Emerson. Expensive, though, and sidelined by Yamaha's digital and more acoustic-sounding DX7, it was not a mass market product, attempts to re-issue it in the late 2010s apparently foundered. In 1986 Elka attempted to create something affordable in the six-voice, 12-DCO EK22 but then also went over to fully digital FM synths. The Orla division also made some excellent MIDI master keyboards, though, and the author used for some years the four-zone Orla DMK7, a design rarely matched for compactness, feel, and versatility.

1975 Steiner-Parker in Salt Lake City launches the Synthacon keyboard synthesizer

Farfisa became famous for electronic pianos and stage organs, includ-ing the Pro Piano and Pro Duo organ played by Irmin Schmidt with Can, the Compact and the VIP series organs used by Pink Floyd, as well as for non-musical electronic products such as doorbells. They barely ven-tured into synthesizers, though; the organ-top style Syntorchestra avail-able in silver (stage) or simulated wood (domestic) versions had a clear, thin polyphonic string sound and a simple monophonic synth section, but

The growth of analog

1975 Klaus Schulze releases the award-winning *Timewind* LP

was used to great effect by Klaus Schulze on his early albums, including *Moondawn*, and by Ashra on *New Age of Earth*. A larger five-octave polyphonic keyboard, the Farfisa Soundmaker, had a richer string section and more flexible monophonic synth sounds but was rarely seen outside Italy and Germany.

Italy's other synthesizer success story was Crumar, which marketed a line of electronic pianos, organs, and ensemble keyboards, such as the Compac (not Compact) Piano, Organizers, Multiman and Brassman throughout the 1960s and 1970s. Rather missing out on the era of true analog synthesis, in 1978 they launched the DS1, a two-DCO mono synth with an angled control panel, then the DS2, which added a simple polyphonic section that could be processed through its filter and was played by the US-based improvising jazz musician Sun Ra, among many others.

The Crumar DS2 was a DCO monophonic synth with a simple polyphonic section.

1981 saw the launch of the Crumar Trilogy, a large polyphonic keyboard using frequency divider technology for string and organ sounds. The polyphonic synthesizer section had six envelopes and six filters, but rather than assigning these to voices cyclically, the Trilogy assigned each permanently to a given selection of notes. The design did try to avoid assigning notes which any player was likely to hold simultaneously to a shared envelope and filter, but even so this system could make for some strange phrasing while playing.

The Crumar Stratus, launched in 1982, was a cut-down Trilogy featuring an input for the Master Touch breath controller designed by the US company Steiner-Parker, and the same year Crumar also showed the Composer keyboard, which added a monophonic synthesizer section and which also had a breath controller input. But the synthesizer sections of all these instruments were rather weak. Wanting to create a more powerful monophonic design, the company called in Bob Moog to work on the Spirit.

The Crumar Spirit was a flat-panel, three-octave synth that included many of the features Moog had left off the Minimoog for the sake of simplicity. It used two real VCOs with a different choice of pulse width settings on the square wave of each. Oscillator sync was available (some

Crumar's Spirit, a powerful Italian monophonic synth with design input from Bob Moog.

Minimoog players were having this added to their instruments as a modification), as well as a Drone setting, ring modulator, sample-and-hold (only available on the Minimoog by buying the expensive 1125 add-on module), an inverted envelope setting and an arpeggiator.

The Spirit filter was particularly powerful, with high pass and band pass as well as 12 and 24 dB low pass settings, and an overdrive feature that recreated the warmth of the Minimoog but more at the command of the player. There are three performance wheels – one for pitch bend and two for modulation – and an external audio input. Unfortunately, few people realised at the time how interesting the instrument could be, since polyphony and MIDI had become the fashions of the day, and the Spirit has now become extremely rare and sought after. Although some Spirits had a blank socket marked MIDI on the rear panel, none were actually delivered MIDI equipped.

Crumar did have an interest in producing MIDI instruments, though, but decided a change of brand name was in order and so these were all marketed under the name Bit – though still having Crumar printed on many of the circuit boards. The Bit One, launched in 1984, was a design of Mario Maggi, who had earlier designed the Elka Synthex; with 12 DCOs, six-voice splittable playing, unison mode, SSM and later Curtis chips, and velocity sensitivity but a very limited MIDI specification, it was at least affordable and gave considerable competition to similarly specified models from Roland and Korg.

The Bit One was quickly superseded by the Bit 99, which solved the MIDI problems, added program chaining that was useful for live performances, but lost the powerful Unison mode; the keyboard was relatively compact, it put the performance wheels on the main control panel rather than to the left of it, but the placement of one above the other made their use difficult. The Bit 01 rackmount version was uncontroversial; all the Bit models were capable of producing very clinical, digital sounds as well as the analog basics, and in this sense they were successful. The lines were marketed in the UK by Chase Musicians (through the tiny London Synthesizer Centre store off Euston Road) and by Unique in the USA, and also included a versatile six-octave, three-zone MIDI master keyboard with sequencer, the Bit MMK (MIDI Master Keyboard), known as the Unique DBM in the USA (where the Bit 01 expander module was known as the Unique DBE).

Incidentally, Crumar also had a history of digital synthesis quite separate from these analog product lines. The company had a New York office, which obtained designs for digital synthesis routines from Bell Labs, expanding them into a General Development System (GDS) keyboard and computer on which to experiment with the technology. Although not intended to reach the public, the GDS was played by Christoph Franke with Tangerine Dream (notably on *Thief* in 1981), by Klaus Schulze, including at the *Linz Steel Symphony* concert around the same time, and by Wendy Carlos on albums such as *Digital Moonscapes*. The eventual offspring of the GDS was the DKI Synergy keyboard, which could only be reprogrammed using a GDS or later a Kaypro computer, and which was also modified as a rack-mount expander called the Mulogix Slave 32. But by the time these became available in the USA, Crumar, Bit, Chase, and Unique had all disappeared from the scene.

The Italian analog story is more or less completed by Jen, which marketed a string synthesizer, the Superstringer, in the 1970s, followed by a series of affordable guitar pedals including a simple phaser, showed a single-oscillator synthesizer, the Synt-O-Rama, in 1973, and launched a DCO-based synthesizer, the SX1000 Synthetone, in 1978. This was unusually budget oriented in having a single oscillator and no attempt to strengthen the sound – no sub-oscillator, no chorus and a rather weak filter, and no performance controls or inputs for external control either – although it did have white noise. As a result, its sounds and possibilities were not overwhelming but it was inexpensive and looked relatively impressive – about the size of an ARP Axxe but with large, round, colour-coded knobs like the EMS VCS3, and a simple, logical layout – and so introduced many players on a budget to analog synthesis. In more recent years, it was used by The Orb and Zero 7, and the author modified a couple of units to include audio input through the filter, drone, and VCO filter modulation.

Later the same year, Jen introduced the SX2000 (also marketed through some department stores and catalogs under the brand name Marlin), an organ-top design using similar circuitry though with seven preset sounds, but this seemed something of a step backwards. The company also showed an unambitious electronic piano in 1983 and later the same year premiered the Synx 508, a polyphonic DCO-based synth with memories that used Curtis chips. The design also featured a simple sequencer and chorus, and ensemble or phaser effects, but was not widely marketed. In 1984, Jen showed two new prototypes, Idea 1 and Idea 2, at the Frankfurt Music Fair but these never made it into production.

FRANCE

Meanwhile, over in France, another small company had been inspired by the simplicity of the US-built EML Polybox. RSF was set up by Ruben and Serge Fernandez, initially marketing their RSF Black Box, which added little to the Polybox design other than a few more notes on the keyboard. Polyboxes were imported into the UK by Syco Systems, who sadly suffered a fire that resulted in piles of badly smelling, smoke-damaged Polyboxes sitting around endlessly while the insurance assessors did their job; if they're still around, they probably still smell odd to this day.

RSF had moved on to the Kobol, a now rare, three-and-a-half-octave, 16-memory programmable monophonic synth with oscillator sync and portamento. Non-programmable, but now very much more sought after, was the Kobol Expander introduced in 1981; this was a very unusual rack-mounting design, like having a 19-inch format Minimoog in that it featured two powerful oscillators, an LFO and two envelopes, but with additional control inputs for voltage-controlled resonance and other, rarely seen features. The rare Expander 2 offered further features, such as a ring modulator, and there was also an optional Programmer; soundtrack composer Hans Zimmer was an enthusiastic user.

RSF combined the best of the Kobol and Expander on the Polykobol, a four-, six- or eight-voice programmable polyphonic keyboard with programmable arpeggiator and sequencer that slightly resembled a Prophet 5. In 1983, an Mk 2 version was introduced with an expanded sequencer and a microcassette interface. But when RSF tried to develop a programmable sampled drum machine in competition with those from Roland, Sequential, Linn, and others, the financial effort finished off the company. All of the Kobol models, which were used in France by Richard Pinhas of Heldon and by the ambient band Lightwave, remain extremely rare even in their native country.

THE NETHERLANDS

One of the most respected lines of synthesizer products ever, considering their rarity, must be the Synton line from the Netherlands. In the late 1970s Marc Paping, later joined by Bert Vermeulen, created a modular system, the Studio 500, together with some small stand-alone products such as pitch-to-voltage converters and filter banks. This was developed into the Series 3000 modular system in the 1980s, with more than 20 different modules available, but the company's most successful products were the very-high-quality Syntovox vocoders, used by Kraftwerk among others.

Around 1982, Synton introduced the Syrinx, a two-VCO monophonic synthesizer with two LFOs, two band pass filters as well as a low pass filter, each with resonance, and with a choice of routings for the filters, ring modulator and white noise, and a compact three-and-a-half-octave keyboard made possible by the fact that there were no performance controls to the left of the keyboard. Instead, a diagonally placed bender up/

down pad made from bilayer strips of metal, reminiscent of the ARP Odyssey's PPC, is located towards the left of the control panel. There is no single modulation control as such, and of course the instrument isn't programmable, but the sheer flexibility of its filter design, offering very heavy sounds combined with oscillator sync, or more unusual filtered timbres, made it popular with players, such as Vince Clarke, Daniel Miller of Mute Records, and Francis Monkman with the classical rock band Sky.

The much sought-after Syrinx from Dutch manufacturer Synton.

The Syrinx was available in various panel colors, including a striking maroon. Five further keyboardless units were made from remaining parts in 1995. But the Synton brand itself had by then long disappeared, more recently Marc Paping designed and marketed a small pre-patched modular system, the Fenix, in a very limited edition.

Also in the Netherlands, DGS produced the Scorpion, a two-VCO monophonic synth with a digitally scanned three-octave keyboard, a Scorpion Plus version with a built-in sequencer, and a larger three-VCO model, the Spider (not to be confused with the British company EDP's digital sequencer of the same name), but in very small quantities in each case. E-Pro, a connected company, also turned out a Minisynth model with just a single oscillator.

JAPAN

Teisco and Kawai, two closely linked and frequently interchanged brand names, offered many affordable analog products throughout the 1970s and 1980s, no doubt with the financial support of Kawai's massive worldwide acoustic piano sales. Apart from various ensemble instruments and electronic pianos, Teisco's S-series synths were very interesting. The S60F (from 1980), closely comparable to the ARP Axxe, was a small monophonic synth with a short two-and-a-half-octave keyboard but some nice features such as vibrato delay (perhaps necessary because there was no modulation performance control as such); the pitch bend control was a small, rubber, pressure-sensitive pad and the author modified a couple of S60Fs to add VCO modulation of the filter. The instrument was seen in two cosmetically different versions, the earlier with wooden end cheeks, and is often mistakenly referred to as the S607 due to the odd typeface used for its graphics.

The S60P was a preset single VCO synth, while the S100F added a second envelope and high pass filter to the design, plus unusual possibilities, such as modulating the filter from the VCO. The S110F was even more powerful; duophonic, with fixed filters to create some unusual tones, three rubber pads for pitch bend and modulation control (rather similar to ARP's PPC system), and portamento, it was closely comparable to the ARP Odyssey and is much sought after.

Teisco's major achievement, though, was the SX400, which used polyphonic keyboard technology licensed directly from ARP and was comparable to Roland's Jupiter 4. It was a relatively huge machine for a four-voice instrument, with much of the deep control panel completely blank, but had eight programmable and eight preset memories, a pressure-sensitive keyboard, and an ensemble section that thickened up the sound considerably. Despite obviously lacking MIDI and other facilities (though the author owned a successfully MIDI-retrofitted model), the SX400 would be an interesting find for anyone with a lot of space to spare.

Analog power from Kawai's SX400, but huge, and with no MIDI.

The next developments went under the Kawai brand name. The SX210 launched in 1983 was a striking design, an eight-voice, eight-DCO, five-octave synth with large numbers of push-buttons, a single large rotary

Kawai's SX210 looks enticing but without MIDI may prove a poor second-user purchase.

control for parameter editing and a huge LED display for naming the sounds in its 32 memories (with up to six letters only). With these patch names visible at a great distance, the SX210 looked good on stage and its chord memory functions plus an ensemble to thicken sounds gave it some interesting sonic possibilities.

However, it was being launched just before the popularisation of MIDI and was quickly superseded by the SX240. This looked pretty similar – the display moving further towards the top of the control panel – but it had 16 DCOs, 48 memories, a 1500-note, eight-pattern sequencer with MIDI clock in and out, portamento, and glissando.

Kawai's next effort, the K3 keyboard synth, resembled Korg's DW6000, with digital waveforms giving a wide variety of basic tones, and the K3M rack version had a nice way of accessing different parameters for editing, assigning one memory or one editing parameter to each one of its triple bank of selector buttons. After that time, Kawai went fully digital with the K5 keyboard and large K5M module, complex harmonic synthesizers that tried and failed to go one better than the Yamaha DX7. Later, the company more or less imitated Roland's D series on the K1 and K1 Mk II (and their rack and desktop versions the K1R and K1M), all based on short PCM-sampled waveforms rather than on analog oscillators.

The K3M put Kawai's analog designs into module form.

Slightly more interesting was Kawai's K4, which although fully digital had resonant filters as well as multi-timbrality, plus on the K4R rack version multiple audio outputs; both these instruments are still around and functioning in large quantities, and can be acquired at ridiculously low prices. After a failure with the very budget-oriented Spectra keyboard, Kawai went quiet in terms of professional synthesizers for a while, continuing to make pianos, domestic keyboards, effects units, and MIDI controller keyboards, although their later K5000 series digital synths (which once again use harmonic synthesis and offered some very PPG-like digital sounds) created some interest, as they seemed to offer the best of analog and digital possibilities. The K5000 existed in three versions – keyboard with many editing controls, workstation with sequencer, and module – plus an optional small "Macro" editing box. Users included the DJ Sash.

One Japanese company that didn't make it through to the 1990s went under the various names of Firstman, Hillwood, Pulser, and Multivox. With good US connections the company sold widely in the USA and UK, mostly in the field of inexpensive ensemble keyboards, organs and electronic pianos plus a simple drum machine, though some of the designs, particularly under the Firstman brand, were far ahead of their time, perhaps too much so, as it turned out.

The Firstman FS4V, for instance, was marketed as early as 1981 and was a four-voice, four-octave synth with piano, brass, clavinet, wood, and "funny" presets, a string section with four footages, white noise, LFO delay, and what Firstman called a "senser bar" under the keyboard for creating filter wow and other effects. Around the same time, Firstman were showing the PS86, a five-octave, velocity and aftertouch-sensitive eight-voice synth with chorus and Leslie speaker settings. This had some quite advanced features: as many as three LFOs per voice, high pass filters, envelope repeat, ensemble and "full rotor" effects, white noise, glide, chord memory, and the same "senser bar" under the keyboard, which reduced the necessity for conventional performance wheels.

Even more advanced in some ways was the FS10C, a 32-memory programmable monophonic synth with oscillator sync that had a control panel comprised entirely of touch-sensitive pads (very much in design and specification like the Moog Source), a separate three-octave keyboard featuring a Moog-like pitch bend ribbon, and was reprogrammed using magnetic strips like the Yamaha CS70M.

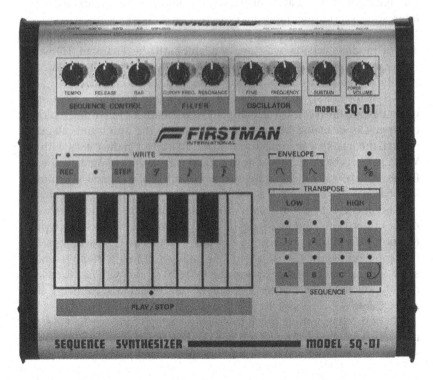

The SQ01 was fellow Japanese company Firstman's intended competition for Roland's TB303 Bassline.

Firstman's SQ01 was a small gray box comprising a simple monophonic synth with a short touch keyboard and a 1024-note digital sequencer, something like a cross between the Roland TB303 Bassline and the EDP Wasp synth with a Spider sequencer built in. This is now very rare and fairly collectable. There was also an SQ10 design, a three-octave keyboard, four-channel, 480-note digital sequencer, and various other products such as bass pedals and an analog drum synthesizer, the battery-powered, flying saucer-shaped, two-VCO Synpuls SD1.

Various organs and bass pedals also appeared under the Hillwood brand, including a two-VCO synthesizer in 1974, but earlier than that the company had marketed the wonderfully named Blue Comets 73. This was a three-octave (F–E) monophonic synthesizer that looked like an organ design, using rocker switches above the keyboard to select some parameters, plus sliders to the left of it to control the envelope, portamento, and other functions.

Rather more designs turned up under the name Multivox. Apart from various string and brass synthesizers, electronic pianos, and bass pedals, the MX880 (also known as the MX2000) was a version of the Roland SH2000 organ-top preset monophonic synth, almost identical except for having two oscillators, while the MX3000 was a polyphonic keyboard competing with the ARP Quadra, featuring string, brass, piano, and monophonic bass and lead synth facilities. The MX8100, a rare keyboard-equipped sequencer, could control Hz/volt as well as one volt/octave synths and could have been useful had it been more widely distributed. As for the Pulser brand, the M75 resembled the Multivox MX2000 in specification but in a compact Micromoog-like rather than organ-top layout, while the Pulser M85 was simply a version of the Multivox MX3000. This brand was also linked with the Italian manufacturer Solton, which marketed some interesting synths, including in 1984 the Project 100, a smart six-voice, MIDI-equipped, 12-DCO programmable design resembling a Roland Juno 106 (a module version was also available). All the Firstman, Hillwood, Pulser, and Multivox designs are now incredibly rare, which would seem to indicate that they did not last very long in a functional state.

No history of synthesizers would be complete without a mention of Casio, despite the fact that the company never marketed anything remotely qualifying as an analog synthesizer. Coming from the watch and calculator market, the Japanese manufacturers made an impact in the 1980s with the VL1, a pocket-sized micro-synthesizer (and calculator!) using digital waveforms, which was followed by a series of affordable digital polyphonic keyboards.

The decision to go into semi-pro and professional synthesizers was a major undertaking and involved the introduction of a whole new type of technology, which Casio chose to call phase distortion (PD) synthesis. This first appeared on the CZ101, a mini-key, eight-DCO, four-part multi-timbral synth with a battery power option; the CZ1000 model with full sized keys quickly followed, then larger instruments with built-in sequencers such as the CZ3000.

All of this has nothing much to do with analog synthesis; the CZ series used a variety of digital waveforms and had no filters. But since PD synthesis could sweep one waveform into another during the course of a note, choosing a bright, resonant tone that cross-faded to a smooth, dull one could create the distinct impression of an analog filter closing down (the proof is on track 77 of the website). Couple this with detuning, a strong unison and portamento mode, ring modulation and white noise effects, and the Casio synths (including the VZ8M and VZ10M modules) could have plenty to offer the analog synthesist; Isao Tomita becoming an endorsee.

Casio's next move was into digital sampling, with the FZ1 and its rack version, both of which had powerful synthesis facilities, but at this point the returns from the relatively tiny professional music market began to seem rather limited to a company used to making millions through sales of watches and calculators. There was a final synthesizer design, the VZ1, which used "interactive phase distortion" (basically several PD circuits that could feed into one another), but despite licensing several of the later synths and samplers to Hohner for distribution under their name (often in a fetching white or silver cosmetic finish and endorsed by Johannes Schmoelling of Tangerine Dream), Casio pulled out of professional instruments – although their domestic keyboard models and digital pianos remained popular – and have shown little sign of attempting to return other than in the low profile XWG1 "groove synth" and XWP1 "performance synth" from 2012, which were generally regarded as being bafflingly complicated.

The hi-fi manufacturer Akai also had a brief early flirtation with analog following their decision to get into the pro musical instrument market around 1984. Several prototype instruments were shown, and of those reaching the market, the AX80 was one of the most interesting. A five-octave, eight-voice, 16-DCO-based velocity-sensitive synth, it featured what seemed like scores of luminescent bar graph displays showing the levels of all the available sound parameters.

Akai's first time round the analog field was brief, the AX80 being one of the highlights.

This was quickly superseded by a much more modest product, the AX60, which had just six voices but could produce layers and splits, as well as having an arpeggiator. This, however, was only marketed in Japan and the USA; Akai were developing their instrument range so quickly that some models became outmoded before they were even promoted. The main reason for this was that Akai had discovered digital sampling, courtesy of the rack-mounting S612 model, and it was decided that the synthesizer products were to be brought into line with the sampler designs and to offer audio input facilities to process their sounds.

The AX73, seen in 1986, was the first to adopt what became Akai's distinctive white styling, and offered a six-octave keyboard that made it attractive as a master controller, plus a multi-pin audio input to handle the individual voices of Akai's samplers. The keyboard was velocity sensitive and the sampler sounds could be assigned to any given keyboard zone, though the synth itself couldn't be split. The control panel was very much simplified and the single-oscillator sounds were rather thin, helped a little by a built-in chorus unit. There was also a module version, the VX90.

Before Akai moved over entirely to sampler production, they showed one more oddity, the VX600 synth. This has a short three-octave keyboard and was really designed to work with the EWI and EVI wind controllers, the latter derived from Nyle Steiner's Electronic Valve Instrument. The VX600 could play six voices using 12 oscillators because it was meant to be able to create chords from the single-note input of the wind controller, but it was multi-timbral as well as polyphonic, and as well as various parameters specific to wind controllers, each voice had three envelopes, two LFOs, and 16 modulation sources, which could be patched to any one of 18 modulation destinations.

This made the VX600 almost as powerful as "matrix modulation" synths, such as the Oberheim Xpander and Matrix 6, which seemed rather over the top for the sound source of a wind controller. Seen in the USA and Germany, but rarely, if at all, in the UK, the VX600 is now much sought after. Although Akai didn't produce any further synthesizers as such, many of their later samplers offered resonant filters and, in conjunction with MESA (Macintosh editing) software, could be as powerful as many analog synthesizers. Later the production of rack-mount samplers ceased, though groovebox-style sampling drum machines in the Akai MPC series, some designed in conjunction with Roger Linn, remained popular, particularly amongst hip-hop musicians. Years later Akai returned to the analog arena with the compact Timbre Wolf and other instruments, about which more in Chapter 8

UNITED KINGDOM

Apart from EMS, the UK had no very major synthesizer manufacturers throughout the 1970s and 1980s, although many attempts were made to launch new companies, often only getting as far as one product release.

There was, though, a great deal of activity in the area of kit-built instruments. At least two companies at the time offered kit-built modular synthesizers: Dewtron (which also made analog bass pedals used by early Genesis and others) until the mid-1980s and Digisound until the early 1990s, their last major undertaking being to provide a huge modular synthesizer to back Tangerine Dream's Christoph Franke during his only solo concert at the London Astoria Theatre in 1991. Kit building was quite a busy scene in the UK in the 1980s (as it was in the USA, mainly through the work of PAiA, and in Europe with *Elektor* magazine's Formant modular system). In the UK, the now defunct electronic components company Maplin provided the designs and parts kits for various instruments published in *Electronics Today International* (ETI) and later in their own magazine *Electronics and Music Maker* (E&MM). Other designs were published in *Practical Electronics* (PE) and some short-lived companies such as Powertran were set up to market them, providing smart, screen-printed, metal control panels that sometimes concealed amateurish wiring horrors behind. The construction quality of these instruments is totally unpredictable, they do not make a reliable second-hand purchase if seen today, but there were some physically very impressive designs available (in different finishes according to the individual builder, often of black tolex, sometimes in wood or even metal) as follows:

- *Maplin/ETI3800, 4600, 5600S*. The 3800 resembled a Minimoog with two oscillators. The four-oscillator 4600 usually appeared in a tall wooden cabinet and had four oscillators, a joystick and a large pin-bay. The 5600S has stereo outputs and a similar design but with the control panel at a flatter angle and usually finished in black tolex.
- *E&MM Spectrum*. Slightly simpler than the 3800 and published in *Electronics and Music Maker* magazine.
- *ETI Transcendent 2000*. Single-oscillator monosynth similar to an ARP Axxe, designed by Tim Orr of EMS, with some excellent functions including random filtering, all built on one huge circuit board with controls mounted directly onto it. Marketed in a reliable kit form by Powertran and with limited point-to-point wiring, a better second-hand bet than other designs, though by no means entirely dependable, and the first analog synth owned by the author, dusted off and revived for the retro album *Sequencer Loops* in 2002.
- *Transcendent Polysynth*. Based on the T2000 voice, a physically very large keyboard that could start as a mono design and have more voices added, and handle a four-voice expander box. Problematic, though, and very few successfully completed.
- *Powertran 1024 Composer*. 1024-note digital sequencer with analog outputs for the T2000 and similar synths.
- *Transcendent Sequencer*. Analog 2×8 step sequencer, possibly the same as that below.
- *Practical Electronics Sequencer*. 2×8 or 1×16 step analog sequencer similar to a Korg SQ10, again problematic and with many corrections published before eventual completion.
- *Practical Electronics MiniSonic*. Square box including analog oscillators played from a stylus and touchpad.
- *PE MiniSonic 2*. Expanded MiniSonic fitted with a conventional keyboard and sounding quite Minimoog-like on a good day.
- *Clef Master Rhythm*. Versatile analog rhythm box published in PE and with a smart, screen-printed, metal casing available.
- *Clef Bandbox*. Chord and harmony accompaniment generator into which a Master Rhythm could be slotted, again with a smart metal casing.
- *Clef MicroSynth*. Neat two-and-a-half-octave highly compact mono synth.

There were many, many other contemporary kit designs, including phaser, distortion and other analog effects, sequencers, and novelties, such as electronic drums and breath controllers. The author was near the centre of these developments, having a Transcendent 2000, PE Sequencer and Clef Master Rhythm as his first analog instruments from 1978 and working on E&MM and the competing *Electronic Soundmaker* magazine during the publication of such novelties as the SynBlo (breath controller) and the first Amdek effects kits (marketed by Roland, which briefly decided to enter the kit-building field). But the kit market did not last very much longer because digital electronics that could not so readily be assembled by the amateur were rapidly being introduced.

A good example and rare exception to the apparent absence of companies from the UK scene was EDP (Electronic Dream Plant), which launched in 1980 with the Wasp. This was designer Chris Huggett's attempt to offer all the basic analog synthesizer facilities at a budget price. The breakthrough was made when it was decided not to include a conventional mechanical keyboard. Like the KS sequencer included with the EMS Synthi Aks, the Wasp offered just a short (two-octave), capacitance-sensitive plastic touch keyboard that could be pasted straight onto part of the main circuit board. This made two analog VCOs an affordable part of the design, but these were digitally controlled to help them stay reliably in tune.

The Wasp also featured white noise, sample-and-hold to the resonant filter, which had high, low and band pass settings, two three-stage envelopes with repeat, and a built-in speaker and headphone output. The whole design was completed with a simple plastic casing, which turned out to be so hard-wearing that a car could be driven over the machine without damaging it; only later did it become clear that the touch keyboard was very prone to wearing thin and that the instrument was oversensitive to temperature and humidity conditions.

The basic sound of the Wasp was surprisingly good though (there are several examples on the website, tracks 26 and 27); the two-oscillator sound was strong and almost Moog-like, and it was highly portable as it could work from battery power as well as an external power supply. There were also two digital interface sockets provided; these looked like the later MIDI sockets but connected only to other Wasps or the cosmetically matching Spider sequencers, which stored 252 step-time or 84 real-time notes and played them back with synchronisation if required to tape, or to almost any incoming click such as a drum machine beat. There are some examples of complex, multi-layered Wasp/Spider sequence programming on the author's album *Analog Archives*. The Caterpillar – a rather basic quality mechanical keyboard capable of playing one, two, three, or four Wasps polyphonically – was also launched later on.

The Wasp/Spider combination was a powerful one that introduced many entry-level synthesists to the field, including Nick Rhodes of Duran Duran. Dave Greenfield of The Stranglers used it extensively on *Meninblack*, despite having ready access to much more upmarket instruments by that time, such as the Minimoog. However, good sales of the Wasp were a mixed blessing for EDP, which had difficulty making them fast enough. The company closed down, relaunching as "EDP (Oxford)" with new products the Wasp Deluxe, a Wasp with three-octave mechanical keyboard, volume balance between oscillators and audio input to the filter, and the Gnat, an even tinier, two-octave, single-oscillator touch keyboard design. Synth composer Adrian Wagner, who released an excellent album, *Distances Between Us*, was involved with the company and made many of their demo pieces.

Other EDP designs, including a proposed guitar synthesizer and a smaller sequencer, the Flea, never reached the market, however, EDP(O) then closed, replaced by Wasp Synthesizers Ltd, which continued to market the Wasp Special and Gnat Special, retaining the wooden casing but going back to touch keyboards, now in black and gold rather than yellow and black. This incarnation of the company also didn't last long; later on, the modifications company Groove Electronics offered to rack-mount pairs of Wasps into a powerful combination called the Stinger, used by The Shamen among others, but very few of these were made. The Wasp's distinctive filters, however, inspired a new rack-mount filter design introduced by the German company MAM in 1999, and in the 2010s Doepfer offered a Wasp Filter for the A100 Eurorack system featuring appropriately black and yellow styling.

EDP's Chris Huggett had even more innovative ideas and formed his own new company, OSC (Oxford Synthesizer Company). Their first (and only) product was the OSCar, a three-octave, programmable, two-oscillator synthesizer using the Wasp Deluxe keyboard but adding processor-controlled facilities such as a 1500-note, 12-pattern, ten-chain monophonic sequencer and arpeggiator, plus data dump to cassette.

The OSC OSCar developed from EDP's tiny Wasp synthesizer and is now highly collectable.

The OSCar joined the very short roster of fully programmable analog monophonic synths (which also includes the Moog Source, Roland Pro Mars, PAiA Proteus, and Oberheim OB1) and is the only one regularly seen with MIDI fitted. The OSCar could also carry out harmonic synthesis and had various filter modes, including an intentional overdrive meant to simulate the richness of the Minimoog sound (examples are on the website, tracks 51–53). This was sonically very successful, the machine was taken up by Ultravox, Keith Emerson, Bronski Beat, Jean-Michel Jarre, Style Council, Vangelis, Stevie Wonder, Geoff Downes, and many others. This was despite the fact that the OSCar design had a few eccentricities of its own: the end cheeks were not wood or metal but comprised large blocks of rubber; the memorised sounds were selected from the keys themselves, forcing the musician to stop playing in order to select a new sound, as there was no keypad or individual selector button for memories; and the modulation wheel was sprung just like the pitch bend wheel, so could not be left faded up.

MIDI had been added part way through the OSCar's production run. Despite the synth playing only monophonically, this was polyphonic on output, but due to the lack of any display other than the five-octave LEDs next to the performance controls, the MIDI settings were sometimes difficult to understand. Ultimately, sales were not huge, since polyphonic synthesizers were much more in fashion, and before OSC could market a large, rack-mounted polyphonic version of the OSCar with added sampling facilities (the Advanced Sound Generator, or ASG) in 1986, the company closed down. Chris Huggett, though, went on to design for Akai.

Other British designs of the period – for instance, the Jeremy Lord Skywave, the Wem synth made by the well-known manufacturer of amplifiers and Copicat tape echo units, or the rich-sounding Freeman String Symphonizer designed by keyboardist Ken Freeman – existed in such tiny quantities as to make little or no impact. Freeman, though, was involved in another project, the SZ3540 Thunderchild, used on Jeff Wayne's *War of the Worlds* concept album in 1978:

The Jeremy Lord Skywave, typical of very short production run British synthesizers.

"One of the engineers, Nick Broome, was interested in building instruments and I asked him to design a small synth. We built about six prototypes in the end. It had a long, flat design, a bit like the Roland SH2000 preset synth, and it was good for bass parts. But we didn't have the resources to take it all the way and develop it any further in the end. I still have one in storage, but the others have gone to the great studio in the sky."

(Jeff Wayne, interview with the author, 2006)

Jeff Wayne had also used the Moog modular:

"I was one of the first two or three people in the UK to have one. Bob Moog came over personally to set it up for me and spent ages connecting it all together, lying down round the back of it until it was all connected. But then when he came to the UK mains plug he really didn't know what to do with it. He couldn't work out where to put the third wire!"

Wayne successfully used the Moog on scores of compositions, including an advertising theme for Gordon's Gin later covered by The Human League, and had a battery of other instruments available when, during the height of the punk movement, rather improbably he was recording the multi-track analog rock opera *War of the Worlds*:

"Until the Yamaha CS80, synthesizers were very unreliable in terms of keeping them in tune during the session, much less from one day to the next. So you had to live with that, and every sound you created couldn't be stored, so each one was recreated daily. I worked with my synth expert Ken Freeman and I would say 'I want to do a snowflake today'; we had notation for the music, but how do you describe on a piece of paper what the sound was? Even with a straight sound, for example the pan pipes, they had to be made from scratch."

Wayne gradually made the transition towards MIDI and eventually to soft synths:

"... but the great thing about those original analog keyboards is that they sound very rich. There is a little extra ambience around the sounds and they have something about them which gives them a quality which keeps them still in use. That's why a lot of plug-in software emulates the ARP Odyssey or the Minimoog, but with the advantage of programmability and staying perfectly in tune."

In 2006, Wayne released a remastered *War of the Worlds*, but had barely touched the analog synth sounds in the production:

"After *War of the Worlds* I worked on a lot of other productions, so I have always kept up with the new keyboard technology, but once you have them on tape, those old analog instruments still sound good, with that analog quality of fatness, and they blend well now with some more modern digital sounds."

Despite the continued popularity of American and Japanese synths, even through the punk rock period, not until Cheetah became part of the "analog revival" in 1988 was there any other significant UK synth manufacturer on the market.

GERMANY

Considering the high-profile use of synthesizers in Germany from the late 1960s by artists such as Popol Vuh, Tangerine Dream, Ash Ra Tempel,

Agitation Free, Klaus Schulze, Eberhard Schoener, Karlheinz Stockhausen, and many others, surprisingly few synthesizer design companies actually emerged from that country. Perhaps it was because some of those artists were believed to take their musical influences from outside Germany – Tangerine Dream, perhaps most influenced by Pink Floyd and Terry Riley, were widely thought to be British – that existing products from EMS, Moog, and ARP were found to be perfectly acceptable in the country for many years.

Some of the most ambitious users in Germany did at the very least want their existing instruments extensively modified and improved. This led to a small market for custom design companies, which eventually began to launch their own products. Matten GmbH (also known as Synthesizerstudio Bonn) developed sequencers for Kraftwerk and later under the name Syntec launched a six-voice, 12-VCO, five-octave synth, the Banana, using Curtis and SSM chips in a cross between Prophet and Oberheim designs. Later units had MIDI, but only about 200 were made altogether.

Tangerine Dream used custom instruments from PPG, EEH, and Projekt Electronik, as well as the more familiar Polymoog, Prophet 5, Minimoog, ARP Odyssey, Elka Rhapsody, Oberheim OBX, and Moog modular systems.

EEH (Electronic Engineering Hoffman) built small digital/analog sequencers for Klaus Schulze and others (the author now has Schulze's unit, used on albums such as *Audentity*), and test equipment designers Projekt Elektronik built larger memory units and sequencers for Tangerine Dream. PPG under Wolfgang Palm also started life in 1974 making custom modifications to Tangerine Dream's Moog modular systems and developed a short production run modular synth of their own. The 100 System was built in 1975 and used by Toto Blanke; the 200 System was built in very small quantities but the 300 System was sold to Tangerine Dream, Ruediger Lorenz, Thomas Dolby, and several others, with more than 20 different modules available.

The impact of these German custom instrumentation companies should not be underestimated. Contemporary analog sequencers tended to limit the user to patterns of eight or 16 notes, and digital sequencers had at most two or four memories. Keyboard controllers for modular synths were extremely simple and did not offer very expressive playing possibilities. When albums, such as Kraftwerk's *Trans Europe Express*, Tangerine Dream's *Ricochet*, and *Encore* (largely recorded live during concerts in Europe and the USA), and Klaus Schulze's *Live* (recorded in 1979) were released, sequence patterns could be heard that smoothly varied in length and tone, transposed in key or exactly followed a lead-line synth part, played trills between phrases, or jumped effortlessly from one pattern to another. Attempting to replicate these techniques taxed the abilities of musicians using only the widely available types of instrumentation. It wasn't until some years later that the reasons started to become clear: Kraftwerk's sequencer had been custom-built by Matten and Weichers, Tangerine Dream's Moog modulars and sequencers had been heavily customised and interfaced by Projekt Elektronik, and Klaus Schulze's multiple simultaneous Moog sequence patterns could only be controlled from a rare PPG master keyboard. During the same period in France, Jean-Michel Jarre was using a very wide range of readily available synthesizers but since he had commissioned the building of a custom MatrixSequencer and rhythm computer his distinctive sequencer patterns were also difficult for other musicians to match. In the UK, David Vorhaus was using his custom-built MANIAC sequencer with White Noise, but it wasn't until the mid-2000s that costly, short-run, MIDI-equipped sequencer products appeared that offered easier access to some of these techniques. At the time of writing there is rather a boom in more complex sequencers, such as the Schrittmacher, Sequentix Circlon, and others.

Following the success of their customisation and short-run work, PPG looked into keyboard synths from 1976 with the PPG1002, a two-oscillator, four-octave programmable design, and later the 1020 with digital waveforms and digital keyboard control. Palm was becoming increasingly interested in digital design and launched the 360 Wave Computer, coupled with a 350 Computer Sequencer; users included Thomas Dolby, sales of this unit led to a polyphonic digital design, the PPG Wave, again used early on by Tangerine Dream (not to be confused with the related, but much later, MIDI-equipped Waldorf Wave).

PPG's first mass-market keyboard was the Wave 2, a "polyphonic integrated eight-track recording system". Its wavetables were derived from

The German-built PPG Wave (seen with its optional Waveterm sampler) combined digital sounds with analog filters, and was used by Tangerine Dream, Klaus Schulze, Gary Numan, and Jan Hammer.

the 360 Wave Computer and each of 30 wavetables stored 64 different waves, giving almost 2000 tones altogether. Waves could be selected manually or under envelope control, and, depending on the table chosen and how many waves were swept through, the result could resemble an analog filter closing down – say, if the waves changed from a buzzy square to a smooth sine wave – or give much more unusual, apparently random results. For simplicity, every wavetable did have sawtooth, pulse and a string wave included, and the design offered pitch envelopes in addition to the wavetable and filter envelopes.

The Wave 2 had 30 preset and 100 programmable memories, 15 rotary controls, cassette dump, a velocity and aftertouch keyboard, and a single diagonally angled performance control; a keypad plus a numeric LCD displayed all the parameters available. But an IC-based resonant filter was included on each voice too, so as well as creating unusual digital sounds, the Wave was capable of sounding highly "analog".

The Wave 2.2 version introduced a second bank of oscillators, a second performance wheel, more memories, compatibility with the Waveterm rack-mount for sound sampling and expanded sequencer possibilities, and later MIDI. The final model Wave 2.3 versions all had MIDI, compatibility with the Waveterm B and weighted PRK Processor Keyboard, and Combi Programs for split and layered sounds.

These various PPG models saw very extensive use, with Klaus Schulze, Tangerine Dream, Frankie Goes To Hollywood, Jan Hammer (very notably on the opening theme of *Miami Vice*), Gary Numan, the experimental German synthesist, Rolf Trostel, and many others, and though they had some reliability problems, became a major resource in 1980s pop music, offering "modern" digital sounds combined with analog staples plus some strikingly unidentifiable textures.

PPG's final design, a non-keyboard unit called the Realizer, was an all-digital system intended to simulate many other synth designs with equal effectiveness. A Minimoog layout was one of the first facilities shown, as well as an FM synthesis routine, but the instrument never got onto the market, and after PPG closed, Wolfgang Palm re-emerged designing for the Waldorf company, of which much more later.

Also in Germany in the early 1980s, AVC Technik offered ready-built or kit-form modules and had over 80 different designs available. Apart from the familiar VCOs, VCFs, and VCAs, these included monophonic keyboards, a polyphonic keyboard interface, fixed filter banks, a voltage-controlled phaser, 440 Hz tuner, digital reverb, and an analog sequencer.

Around the same time, Andreas Bahrdt was designing and building instruments in extremely small quantities. In 1984 he completed a 16-voice, four-VCO per voice programmable synthesizer with FM synthesis, multiple sequencers, and hard disk storage, all run from a Hewlett-Packard computer. Referred to as "Le Bart", this was used by Lamborghini head Patrick Mimran and mixed down direct onto two albums, *Novels for the Moons* under the name Axxess and *Down to Earth* under his own name, and also used by Peter Baumann, who had left Tangerine Dream and was in the process of setting up in the USA. Baumann was also interested in E-Mu's ambitious Audity project, which used similar voice cards.

E-Mu's Audity project almost closed the company, which survived to market the Audity 2000.

BME, under R. and H.-P. Baumann (no relation to Peter), marketed a PM10 modular synthesizer in the 1970s featuring various VCOs, including a high-speed VCO that went up to 200 kHz, updating the system in 1983 to the Axiom, which featured more than 30 different modules, including a triple VCO, programmable ADSR, digital noise, sequencer and keyboard. There was also a 36-note, touch keyboard, single-oscillator synth, the BME 700.

Boehm (or Doctor Boehm), extremely well known for their organs in both kit and ready-built form, also had a modular synthesizer to offer for a short period in the early 1980s. The Boehm Soundlab design included 11 modules plus a joystick controller; modules included a VCO, VCF, dual VCA, voltage-controlled ADSR, multi-function module with noise, random, mixer, LFO and ring modulator, a 49-note keyboard, and a computer interface for control from a Commodore 64 computer. Patching used banana plugs and about 250 units were made; a book, *Die Klangwelt Des Musiksynthesizers*, by co-designer Bernd Enders took its examples directly from the Boehm Soundlab.

USA

Donald Buchla's designs, which had strayed away from the familiar facilities of the early Moogs and ARPs, were touched on in Chapter 3, but another early US designer whose instruments were very much non-standard was Serge Tcherepnin, who studied with Karlheinz Stockhausen and marketed instruments under the Serge Modular Music Systems brand.

The unusual Serge modular system, here in a flightcased three-panel version.

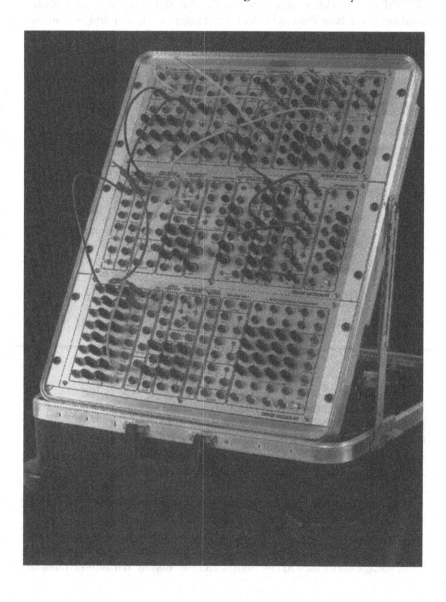

One of the major factors in the Serge design was that they had no keyboard, although sets of touchpads were sometimes available. Some facilities and terms such as VCF are recognisable, but others such as Dual Universal Slope Generator, Resonant Equalizer, Dual Transient Generator, and Analog Shift Register are less familiar.

Serge systems were relatively expensive as they use extremely high quality components. During some periods the amount of instruction given with systems was extremely limited as each user was intended to come up with personal methods of using the system. Some modules were capable of many applications – as an oscillator, a slow oscillator, an envelope generator, a frequency divider or a VCF, for instance – and systems were patched using banana plugs, which could be stacked up so as to readily combine voltages. Various types of sequencer module were also available, and the Serge, generally made in very small quantities and relaunched in 1994 by Sound Transform Systems in California, remains an unusual, expensive and challenging approach to analog synthesis. Users included Vince Clarke and Kevin Braheny.

Other more conventional modular systems manufactured in the USA also appeared from small companies formed after the various take-overs and business reorganisations at Moog Music. Aries was formed by Alan Pearce and Ron Folkman after the Norlin buy-out, and at first built vocoders under licence for Moog. The Aries modular system was intended to be compatible with Moog systems, although it didn't always fit into the same cabinets. Eventually more than 30 different modules and four different keyboards, including an eight-voice, velocity-sensitive design, were available in seven-, 11- or 14-module cases. The 300 Music System was completely modular, but the System 3 was pre-patched internally as well as offering many patching possibilities, and had an integral keyboard.

One of the major features of Polyfusion's system was an interest in quadrophonic sound, and quad panning modules were available as well as a programmable graphic equaliser and various sequencers and trigger assignment modules. Users included Vince Clarke, Hans Zimmer, and Steve Porcaro with Toto, but production stopped early in the 1980s and systems have now become extremely rare and very expensive.

Another US design company more responsible for introducing entry-level players to analog synthesis was EML (Electronic Music Laboratories). Their first product in 1968, of which only about ten were produced, is the usually referred to as the "Black Monster", produced for use in schools, apparently it was intentionally made to weigh 200 pounds so it could not be stolen. But by 1970 the company had started the Electrocomp line. The 100 was a small portable synth, the 101 of 1972 was semi-modular and offered four oscillators, the 200 was a two-VCO expander, and the 500 was a three-and-a-half-octave, two-VCO, performance-oriented synth not dissimilar to the ARP Odyssey.

After marketing the PolyBox, a tiny keyboard with an audio input and divider network intended to make monophonic synthesizers sound polyphonic, EML went further into pseudo-polyphonic synths with the Syn-Key 1550 and 2001, the latter programmable using punched cards, the marketing of which had, by 1984, finished the company off.

1975 Yamaha releases the GX1 then CS50, CS60, and CS80 synthesizers based on GX1 polyphonic technology

1975 ARP launches the top selling Omni small polyphonic keyboard

Steve Porcaro's giant
Polyfusion modular system
was called DAMIUS.

1975 Oberheim launches 4-Voice
synth in the USA built from
their SEM modules

Like EML, the oddly named PAiA was also responsible for introducing many new users to analog, largely because their early designs were usually available in affordable kit form. PAiA's founder, the late John Simonton, published a design for a drum machine in 1970, a Hz/volt-based modular design (the 2700) in 1972, and was later joined by electronic music writer Craig Anderton who had a great deal of input into the PAiA designs.

The 2720 modular, available as a complete kit, included a keyboard as well as VCO, VCA and low pass VCF modules, plus options such as glide and a band pass filter. The 4700 modular (from 1974) was the first

Devo championed EML synths (centre), as well as the split-keyboard Minimoogs, Prophet 5s, Roland CSQ sequencers, Moog vocoder, and ARP Odyssey seen here. The final item is an electric heater.

available ready built, so standards of construction are much more reliable; it featured a spring reverb, a sequencer, an analog and later a digitally scanned keyboard with arpeggiator, and various flight-case options for anyone wanting to take it on the road. The 4700C was one standard configuration featuring a single oscillator; the 4700J had three.

PAiA's P-series modular synths were innovative in using microprocessor control, which could make them polyphonic if you bought enough oscillators; the P4700C and J, in one or two cabinets respectively, were taken up by Larry "Synergy" Fast, among others. Fast also used the PAiA Gnome, designed as an absolute budget synthesizer in that it comprised a tiny box with no keyboard, just a conductive ribbon and stylus controller, for creating white noise and other sound effects. The company also marketed a string synth called Stringz'n'Things, and some organ designs such as the Oz and Organtua, interesting mainly in that they had trigger outputs to help in processing them through the synthesizer system's filters and envelopes for pseudo-polyphony.

One of PAiA's major achievements, released in 1979, was the programmable monophonic Proteus 1 (not to be confused with E-Mu's later MIDI module of the same name). An early user of the Curtis synthesizer chips that had recently been released, the Proteus had two oscillators with sync, 16 memories, 20 control knobs and an Apple computer interface. This is

1975 Neil Ardley releases the UK electronic jazz/world classic *Kaleidoscope of Rainbows*

1976 Tangerine Dream release the *Ricochet* live album featuring Moog modular synthesizers

1976 Roland launches the System 700 modular, similar to larger Moog systems

1976 Jean-Michel Jarre releases the *Oxygene* album in France using many custom instruments

1976 Ash Ra releases the *New Age of Earth* album on Virgin using ARP, EMS, and Farfisa keyboards

1977 Tangerine Dream release *Live: Encore* taken from their USA tour with Laserium laser show

one of the very few programmable monophonic synth designs but unfortunately polyphonics were much more fashionable by this time and kit building of instruments was also a shrinking field. Around this time PAiA went very quiet, continuing to make a couple of models such as a rack-mount vocoder, but the company survived through to the analog revival and continued to have several items of interest to offer, as we'll see in Chapter 6. PAiA modular systems of various kinds are quite commonly found on eBay and similar auction sites. Though their construction quality obviously varies, they can easily enough be rebuilt and despite their use of banana connections on the earlier systems and some other eccentricities, can be used to build up impressive, retro-looking, and readily portable modular systems.

To be fair to PAiA, many smaller synthesizer companies around the world were being completely closed down by the success in the 1980s of the electronic music giants. Even organ giants, Hammond and Baldwin, tried out just one monophonic synth design each – the 102200, with a 3.5 octave keyboard and a matrix of colored push-buttons to select preset sounds and performance parameters, and the Syntha-Sound, which resembled an ARP Odyssey but with preset sounds, wood panelling, and many fewer variable controllers – before getting out of the field altogether. In the USA in 1981, Gleeman launched the Pentaphonic, a three-octave synth which nevertheless offered five-note polyphonic playing with 100 programmable memories and a sequencer. Eclipsed by the success of the Yamaha DX7, fewer than 50 Pentaphonics were sold, and an even smaller number (probably fewer than 10) of the legendary plexiglass Pentaphonic Clear version.

Like Gleeman, Electro-Harmonix had just one early synthesizer product of note. Not because the company disappeared but because they were primarily a successful guitar effects manufacturer that just happened to launch a simple vocoder, a couple of drum synthesizers and trigger sequencers, and a tiny portable synth, the E-H Mini-Synthesizer. This worked off batteries, had a two-octave blue and black touch keyboard, and featured a single oscillator with interesting sub-octave and overdrive circuits, a decay slider but no release envelope, and a simple type of velocity sensitivity in the form of a piezo microphone pick-up under the keyboard, which actually translated the thump of the player's fingers into an additional control voltage for the filter. The E-H Mini-Synthesizer (which resembled EDP's Wasp, though even smaller) was shown at the NAMM trade show in 1980 and later played by Jean-Michel Jarre at the "Concerts in China", attached to a very long cable and passed by Jarre around the audience for them to play.

E-H also marketed a Micro Synthesizer for guitarists. This was an envelope follower with a resonant filter and octave divider in foot-pedal format, but remains quite useful for adding analog synth-style processing to any input sound, and has recently been reissued in both its guitar and bass variations.

Octave (later Octave-Plateau) had some success with more conventional keyboard designs, though their Cat synth introduced in 1976, and played by David Bedford and Dave Greenslade, amongst others, was so similar to an ARP Odyssey that legal action ensued. This wasn't helped

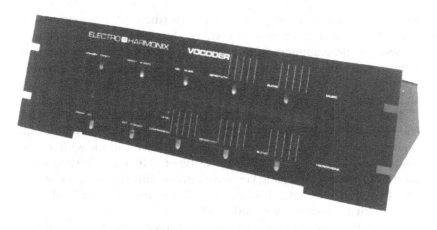

The Electro-Harmonix design was an unusually simple rackmounted vocoder.

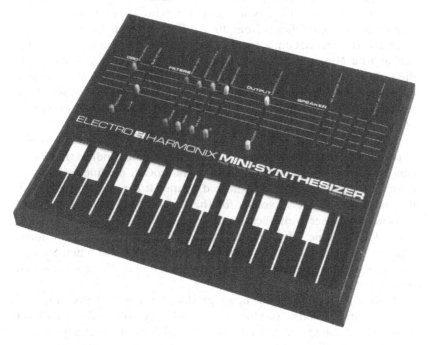

The tiny Electro-Harmonix MiniSynth is now much sought after.

by the fact that the stripped-down version, the Kitten, also very closely resembled ARP's Axxe. A redesigned Cat SRM II solved some of these problems and the company went on to produce in small quantities an early and now sought-after rack-mount programmable polyphonic, the Voyetra, extensively used by New Order (on *Blue Monday*, for instance), before going over only to software design under the Voyetra name.

E-Mu, originally comprising engineers Dave Rossum and Scott Wedge, was also designing modular systems, having started life helping on the design of the polyphonic keyboard for Tom Oberheim's multi-voice systems, and going on to help on Dave Smith's design for the polyphonic keyboard on the Prophet 5. Royalties from the Prophet 5 design helped the company develop their complete analog modular system originally introduced in 1972. Eventually, there were over 60 modules available,

1977 Oberheim launches the OB1, the first programmable monophonic synthesizer

131

1977 Korg launches the affordable Micro Preset synth played by Orchestral Manouevres in the Dark

1977 Yamaha launches the affordable smaller CS range synthesizers, including CS5 and CS10

plus monophonic and polyphonic keyboards (the 4060 microprocessor-controlled design effectively becoming the Prophet 5's keyboard), sequencers, programmers, and a tape interface. Around 100 systems were made, users including Patrick Gleeson, Herbie Hancock, Hideki Matsutake with Yellow Magic Orchestra, and Frank Zappa mainly for brass sounds.

E-Mu went on to try to develop a very large digitally controlled polyphonic analog synthesizer, the Audity, but when Prophet 5 royalties suddenly dried up, had to abandon this and quickly come up with a new product. Inspired by the launch of the Fairlight CMI sampler, they quickly got their "Emulator" sampling keyboard into production and didn't come anywhere near the analog field again until the "analog revival" a decade later, entering Eurorack after another decade.

Throughout the same period, Steiner-Parker were at work in Salt Lake City. Nyle Steiner was a trumpeter who later worked on the soundtrack of *Apocalypse Now*. In 1975, along with Dick Parker, he launched the Synthacon, a four-octave, three-VCO monophonic synth similar to a Minimoog except in having all the oscillator controls to the right rather than the left of a sloping panel. There was also a whole modular system, used by Columbia-Princeton Electronic Music Centre among others, comprising over a dozen different modules and an optional sequencer.

In 1977, Steiner-Parker launched a four-octave, single-VCO monophonic synth, the Minicon, and a small analog sound module, the Microcon (not to be confused with the later MicroCon module from Swiss manufacturer Technosaurus), followed by the SR/8, a rack of eight modules intended to act as voices for the Roland MC8 Micro-Composer sequencer or other controllers. But Steiner-Parker's best-known product, the EVI (Electronic Valve Instrument), was intended to act as an interface for Nyle Steiner's own trumpet-playing talents, and when Steiner-Parker ceased production, this design was adapted by Akai for their own wind controller.

Ensoniq, strictly speaking, never produced a fully analog synthesizer, but their first intended product was a digital synth based on a circuit design called the Q-Chip; when sound sampling began to catch on with the release of the E-Mu Emulator, a budget sampler using the same chip seemed to have more commercial potential. After the success of Ensoniq's Mirage sampler, the company was able to go back to its original idea to launch the ESQ1, a multi-timbral digital synthesizer with built-in multi-track sequencer.

The ESQ1 featured several digitally generated waveforms, including some very basic drum sounds, but as these did include sine, square, and sawtooth waves, and as the ESQ's filters did include a resonance control, the machine was very easily capable of creating some very good, apparently all-analog, sounds.

Its follow-up, the Ensoniq SQ80, added more waveforms and individual note aftertouch sensitivity on the keyboard, which is a very rarely found facility, seen only on the Yamaha CS80, the MIDI-equipped Prophet T8 and upmarket Synclavier, and a handful of others. The SQ80's keyboard design had an odd, clattering feel to it, but like the ESQ1 the synth was capable

Ensoniq's SQ80 – an all-digital synth with many analog abilities.

of routing performance controllers, such as its pitch bend wheel to filter cut-off level, so was capable of creating, using its own built-in sequencer or external sequencers, a wide range of highly animated, analog-style sequences. Though later generations of Ensoniq – the VFX, VFXSD, SQ1 and SQR rack-mount, and so on – featured more and more sampled waveforms, the Ensoniq synths were always better at doing analog than a lot of real analog designs, but the company eventually merged with E-Mu and effectively disappeared after the release of the FIZMO, of which more in Chapter 6.

1977 Korg launches the upmarket PS series polyphonic synthesizers, as played by Keith Emerson

Very short-run true analog synthesizer products existed in China, East Germany, and elsewhere too. China developed the Yinxianghechengqi, apparently a copy of the five-module Roland 100M modular system, the duophonic Polyvoks was released in Russia and resembled Korg's MS20, and the Tiracon 6V from VEB Automatisierungsanlagenbau Cottbus was the (then) East Germany's six-voice, six-VCO MIDI synth with 32 memories and a sequencer, similar to Korg's Poly 800 though with six independent filters (there was a similar, smaller but still very heavy strap-on instrument, the Junost), while the simpler Vermona synth cosmetically resembled a stretched Minimoog with some preset parameters but a much thinner sound.

There were quite a few synthesizer products from the Soviet bloc, with names such as Aelita, Altair 231, and the EM26 vocoder, but interest in their existence did not spread to the West until many years later. Many Western keyboardists later sought them out in the hope that they would offer some unusual sonic facilities and some of them became available for export, though often with high shipping costs. But most of these instruments are as disappointing as might reasonably have been expected, poorly and heavily constructed, with coarse keyboard actions and very basic facilities. Some, like the Polyvox, which resembled a Korg MS20 covered in Cyrillic legending, did have great visual appeal, but existing owners of a Minimoog, an ARP Odyssey or a Solina string synth are given nothing to worry about by these instruments. After the fall of East Germany, the old Vermona designs became more widely known and the company relaunched, offering analog studio effects and some synth modules, of which more later.

But despite the growth of analog synthesis around the world, both the major and the minor companies had all moved out of the field by the late 1980s: Moog, ARP, and Sequential had closed; Oberheim, EMS, and PAiA were almost inactive; Yamaha had gone over to FM digital synthesis, Korg and Roland to sample-based synthesis, E-Mu and Akai to sound sampling; and all the smaller players such as EDP, Crumar, Multivox, Jen, and many others had disappeared. But as we will see, all of this was about to change; the history of the "analog revival" is covered in Chapter 6.

5

Using and programming analog

Although analog synthesizers are in many ways the easiest type to use, some analog techniques that were generally appreciated 30 years ago are now not so widely understood. Recent generations have become more used to later innovations such as FM synthesis or sample-and-synthesis instrument designs, or indeed to synthesizer designs that offer only very limited editing and sound creation facilities, such as most of those conforming to the General MIDI (GM) standard.

But having a good understanding of the programming and use of analog, which is perhaps the synthesis technique closest to the physical fundamentals of sound creation, is vital to any systematic attempt to create and modify sounds using any type of synthesis system. In this chapter an understanding of the physics of sound taken from Chapter 1 and of the specific facilities available to the analog synthesizer user set out in Chapter 2 are combined into a history of and a specific guide to the programming and use of analog sounds in real-life applications.

The first principle in programming analog is to decide exactly what types of sounds are required from the instrument. The likely requirements have varied with current fashions over the years during which analog synthesis has been available: from creating entirely new classes of sounds, to imitating existing instruments, to producing rather comical sound effects, to counterpointing with something a little unusual the well-known and established sounds of other keyboard instruments, such as pianos and organs, to creating exciting and motivating dance rhythms, and far beyond.

CLASSICAL AND AVANT-GARDE PROGRAMMERS

When Walter Carlos began to work with the early Moog modular analog synthesizers, his requirements were perhaps rather surprising to some. While John Cage, Karlheinz Stockhausen, Morton Subotnick and others were interested in using synthesizers for the creation of strange new sound textures, Carlos decided to use them in the realisation of what was fundamentally rather conservative classical music. But this decision

1977 Bob Moog leaves Moog Inc., owned at the time by Norlin

was more deeply considered than it might at first appear. Carlos clearly did realise the more avant-garde possibilities of synthesized sound, having had two pieces, *Dialogues for Piano and Two Loudspeakers* (1963) and *Variations for Flute and Electronic Sound* (1964), included on the compilation album *Electronic Music* (Turnabout, 1966), alongside pieces by Ilhan Mimaroglu and others. On these tracks, Carlos had used simple electronic devices (perhaps including very early Moog modules) to counterpoint a flute and a piano, creating abstract and often pitchless noises. Later Carlos compositions, such as *Timesteps* from the soundtrack of Stanley Kubrick's movie *A Clockwork Orange*, as well as later albums, such as *Digital Moonscapes* and *Tales of Heaven and Hell*, showed Carlos's great ability in abstract sound creation very convincingly. But in the late 1960s, Carlos had realised that a combination of unusual compositional techniques and unusual sound textures was enough to put the work of Cage, Stockhausen, and Subotnick out of the field of interest of the average listener.

The more practical method of promoting synthesized sound was to use it in the creation of material that was more immediately familiar, at least in format if not in content. The first attempts to do this were through pop music. *The Zodiac* by Cosmic Sounds (released in 1967), featuring Moog work by the producer Paul Beaver, is generally regarded as the first commercial Moog album, although it mainly comprises conventional rock and pop instrumentation composed and conducted by Mort Garson, sounds more or less like any psychedelic rock album of the period, and does not even credit the Moog system as such.

Other experiments were going on with synthesizers, though still in the field of avant-garde music. Donald Erb's *Music for Instruments and Electronic Sounds*, released on Nonesuch and recorded in 1968, was partly realised at R.A. Moog Inc's studio at Trumansberg, New York, using a modular system and a "Moog polyphonic instrument" – presumably a prototype – with Paul Beaver as technical assistant and Bernard Krause as producer for one side of the album, but this simply added electronic sounds to conventional trombone, double bass, violin, piano, and percussion. By 1968, the Nonesuch label had already put out a double album, *The Nonesuch Guide To Electronic Music*, demonstrating waveforms, keyboard control techniques, filtering and delays, together with a complete piece by Beaver and Krause created on a Series III Moog with an additional X-Y controller, which generates voltages from a square touch pad. Paul Beaver died in 1975 and the album was extensively updated by Krause in 1981 as a single album, with references to digital techniques and vocoding, and including a new piece using Moog modular, Prophet 5 and Steiner synthesizers, SCI, and Roland sequencers.

Beaver and Krause, as a duo, also released *Gandharva* (Warner, 1971), partly based on live performances in a cathedral, with simple Moog sounds developed from work for Mick Jagger's scene featuring a Moog modular system in the movie *Performance*, and also *All Good Men* (Warner, 1972), which processed spoken voices and tapes, the music of Scott Joplin and other sound sources through Moog modular and Musonics Sonic V synthesizers.

Tim Souster was an avant-garde composer whose use of Minimoog and Roland synthesizers also touched on jazz styles.

Pianist, David Rosenboom, and instrument designer, Donald Buchla, started work around 1968 on a piece titled *How Much Better if Plymouth Rock had Landed on the Pilgrims, Section V,* for four voices of the Buchla 300 Series Electric Music Box, a piece astonishingly similar to Terry Riley's *A Rainbow in Curved Air* of the same period, but not actually released until ten years later on the album *Collaboration in Performance* (1750 Arch Records, 1978). Jon Appleton, who was involved with New York's Columbia-Princeton Electronic Music Centre and who had used the computer-based RCA synthesizer, released *Appleton Syntonic Menagerie* in 1969, but this classical avant-garde work used as much *musique concrete* as Moog modular sounds. The Moog was used to process tapes of human voice and snatches of big band records, Appleton in this case becoming as much an early champion of sound sampling as of analog synthesis.

So why did the analog sounds and techniques on the Walter Carlos album, *Switched On Bach*, released in 1968, have so much greater impact than any of these releases? Part of the reason is that Carlos had chosen material that was highly sympathetic to the particular limitations of the Moog system. The music of Johann Sebastian Bach, though complex, was very much dependent on the art of counterpoint, the simultaneous performance of interrelated monophonic melody and harmony lines. The monophonic Moog system could handle this well, although multi-track recording techniques (which were relatively unsophisticated at the time) still had to be developed to help create complete pieces, but this was initially far less problematic than taking on the works of composers such as Debussy, Wagner, or Beethoven, which demanded much more complex chordal harmonies and tone clusters. Having decided on the musical material to be tackled, Carlos was then faced with making the exact choice of Moog sounds and textures required. The challenge was to create something that was distinctively Moog, something new and exciting, yet

1977 Ashra releases the *Blackouts* album on Virgin and tours with laser show

which was recognisably faithful to its classical sources. This requirement divided the sound creation task into two categories: sounds that were to be analog recreations of the original acoustic instruments as far as possible, and sounds that were musically useful and reminiscent of the content of Bach's original compositions, but which at the same time could have come from no other existing or conventional instrument.

As an experienced researcher in electronic music with a good knowledge of physics and acoustics, Carlos was quite capable of considering the content of the sounds made by conventional acoustic instruments, and of thinking about how to recreate them using analog technology. Some examples of the necessary thought processes are given later, but on *Switched On Bach*, Walter Carlos was also trying to create a whole group of new and unfamiliar sounds and textures. One of the most striking sound groups on the album involves the use of envelope-controlled low pass filters. The Moog filter is particularly powerful, more so than designs released many years later, which failed to replicate its rich tone, and so even from this debut album, Carlos's arrangements were able to benefit from one of the strongest elements available to analog synthesis in subsequent years. The low pass filter sound was immediately striking because it was so unlike anything heard from conventional acoustic instruments; the closest comparison is perhaps when a trumpet or trombone player removes a mute from the instrument, creating a "wah" sound, but most conventional instruments go through very little comparable variation in tone during the course of a note. The Moog low pass filter was able to sweep a tone from smooth and dull to fully open and resonant in a fraction of a second, and do so with reproducible exactness on every note – a sound that had simply never been heard before (listen to track 29 on the website). The same kind of audible shock occurred a few years later when Keith Emerson used a Moog lead sound with a high degree of portamento on *Lucky Man*. The idea of a rock melody sound that could glide from note to note over spans of several octaves was sonically extremely novel.

One of the first Carlos pieces realised, the Allegro third movement of Bach's *Brandenburg Concerto No. 3* (the track played to great acclaim at the 1968 AES convention) did not even need to involve these relatively ostentatious innovations. The bass part is a simple, quickly decaying square wave sound, but does rise right up to the middle registers during some parts of the piece. The main melody parts are brass-like, with a slight filtered "wah" element and bright overtones (listen to track 46 on the website), while one of the counterpoint lines includes an element of white noise similar to the description of a synthesized flute later. The second (Andante) movement, not completely notated by Bach and so open to some improvisation on Carlos's part, includes some much more complex sounds that are clearly not attempting to imitate acoustic instruments at all: sounds are ring modulated to create clanging, metallic effects; oscillators are pitch modulated extremely quickly into fast, descending chimes (listen to track 58 on the website for a more extreme example of these techniques); and there are fast, percussive sounds almost like cartoon sound effects, as well as many using white noise and highly resonant filter effects sounding like water droplets. The first movement, with its distinctive fanfare-like opening, has more conservative attempts at acoustic instrument emulation as well as a few more striking filter effects; brass-like sounds are predominant, along with some using an almost human voice tone, a technique which Carlos was to develop extensively later (track 42 on the website has some simple human voice sounds using a later Korg analog synth). In this movement, another of the limitations of the Moog becomes clear; without velocity sensitivity on the keyboard it could be an instrument that very much lacked dynamics (variations in volume) in the performance. Because Carlos's arrangements are sympathetic to Bach's originals and made full use of the limited available multi-track recording techniques, these problems are overcome and the piece has a tremendous climax, but later and less sophisticated players were guilty of producing Moog performances of extreme banality.

There were some synthesists whose understanding of the potential of the Moog almost matched that of Carlos himself. In 1970, the Kama Sutra label released *A Moog Mass* by Caldera (although if you examined the back of the sleeve it was described as *Stabat Mater* by "Caldara", a confusion of both band name and title probably caused by the rush to capitalise on the success of the Carlos release). But whatever the correct name of the band, the men behind it were to have a great influence on the art of Moog programming; Malcom Cecil and Robert Margouleff later emerged as T.O.N.T.O. (The Original New Timbral Orchestra) with one of the largest Moog modular systems ever assembled and, on their albums and in collaborations with Stevie Wonder, Steve Hillage, Gil Scott-Heron, and many others, raised the creation of smooth and accessible Moog sounds to new levels. On *A Moog Mass*, Margouleff and Cecil collaborated in making the Moog create tones that were very reminiscent of human voices. Walter Carlos had also been working on this idea, some of the results appeared on the soundtrack for *A Clockwork Orange*; these techniques were more basic than the use of a vocoder, involving filtering Moog oscillator sounds in ways that reproduced the human vocal chords and the resonant cavities of the mouth. Sometimes several parameters would have to be controlled simultaneously to get a lifelike performance and Carlos developed foot pedals and other controllers to help with this. Apart from synthesized basses, harpsichords, and brass, the other interesting factor on *A Moog Mass* is the much more unashamed use of completely abstract sounds. Rapid sequencer patterns control both oscillator pitches and filter settings, repeating through deep reverb and highly resonant filters to create distant, watery sounds; oscillators are deeply pitch modulated and bent downwards to create abstract chiming effects. There are many other imaginative sounds reminiscent of the effects used by Tangerine Dream and Klaus Schulze in Germany several years later.

The magnitude of Carlos's and Malcolm and Cecil's achievements in sound synthesis can be better appreciated on listening to some of the albums released by other labels in an attempt to share in the success of *Switched On Bach*. In 1970, RCA released Hans Wurmann's *The Moog Strikes Bach*, recorded in 1969, and in addition to Bach, Wurmann took on compositions by Mozart, Chopin, Rachmaninov, and Prokofiev. But the textures of the synthesized sounds are much simpler than those of Carlos, involving much more use of wide-open oscillator sounds with little or no filtering. The arrangements are much simpler, some other keyboard

1977 Tangerine Dream tour the USA with the Laserium show, releasing the *Live: Encore* album

instruments such as the Hohner Clavinet are substituted for some of the more complex polyphonic parts. The opening tone of Bach's *Toccata and Fugue in D Minor* must use one of the most wretched substitutes for an all-stops-out church organ ever heard.

Ruth White was commissioned by EMI around 1971 to record *Short Circuits*, covering Satie, Debussy, Verdi, and others, including a clever bee sound on *Flight of the Bumble Bee*, but also with many comic sounds, and using various electronic organs as well as the Moog. White's arrangement of *The Snow Is Dancing* (or *Snowflakes Are Dancing*) was almost as convincing as the later Isao Tomita version. In 1972, RCA's quadrophonic production *Stolen Goods*, credited to Dr Teleny's Incredible Plugged-In Orchestra, had synthesizer parts by Ken Howard and Alan Blaikley more or less swamped by a conventional orchestra. In 1973, Andrew Kazdin and Thomas Z. Shephard, two CBS Records producers, turned out *Everything You Always Wanted to Hear on the Moog (but were afraid to ask for)*, taking on Ravel's *Bolero* as well as other Spanish or Spanish-influenced compositions from Chabrier, Lecuona, and Bizet. This album featured mainly silly, novelty sounds from the Moog.

In that same year, large Moog systems were still rare in the UK, but Dudley Simpson and Brian Hodgson, who both worked for the BBC's Radiophonic Workshop, released *In A Covent Garden* under the name Electrophon, using various synthesizers including the locally built EMS models, and some custom units built by Ken Gale under the name RSE. The sounds were serious and reasonably full, although not reaching the quality of those from Carlos or later Isao Tomita; with the late John Lewis who composed for some episodes of the TV show *Dr Who*, Hodgson then went on to form the duo Wavemaker, with two interesting releases, *New Atlantis* and *Where Are We Captain?*, which greatly extended the possibilities of large-capacity digital sequencers. Another Electrophon production, *Zygoat*, credited to the American composer for contemporary dance Burt Alcantara, remains, with its long sequencer build-ups and crazed, jazzy solos, one of the long deleted and most imaginative classics of early analog synthesizer music.

But in Japan, a further revolution in synthesizer programming was about to take place. Isao Tomita's Moog debut was on an album titled *Electric Samurai; Switched On Rock*, released in the USA by CBS in 1974. Conventionally for the time, this comprised covers of chart hits with a standard rock and pop backing, including tracks such as *Let It Be*, *Hey Jude* and *Bridge Over Troubled Water*, but Tomita's careful attention to Moog programming was already beginning to show. The lead sounds are full and strong, and there are also early versions of Tomita's later distinctive "human voice" sound. Tomita's *Snowflakes Are Dancing* (also known as *The Newest Sound of Debussy*), released by RCA in 1974, was the album that took analog synth programming in the classical field to new heights; the string sounds were rich and full of movement, and the overall approach was much more chordal than that of Walter Carlos. Tomita explained (interview with the author, 1999):

> "When I was writing music for orchestras I felt a kind of desperation and
> I felt there was no future for orchestral music, because I thought that any

978 Roland launches
he polyphonic Jupiter 4
ynthesizer with memories

combination of instruments I could use – horn, pianos or whatever – had already been written before. Then there was avant-garde music, but I thought there was no future for that either, because it had to be very advanced and would not have a lot of listeners. Then I met with the Moog synthesizer and I was really delighted; it seemed to be thing that I had been waiting for."

Since Tomita's instrumentation was more or less the same as that of Carlos – a large Moog modular system – what was the difference in their approach to programming? Largely, this was nothing to do with the synthesizer sound, but to a better appreciation of the ambience needed to create a recording as strong as that of a symphony orchestra (Tomita was an experienced composer for large orchestras, while Carlos was not). Tomita particularly used phasing and flanging to add movement to an analog sound. On carefully listening to any of the Tomita recordings, a great majority of the sounds are found to be processed in some way, each sound sweeping gradually during the course of a note, and any tendency to thinness in the basic wave shapes is removed. Tomita was also a master of adding reverberation, and excelled at placing sounds and moving them across the stereo spectrum:

"My basic system was to use one set of stereo speakers in front and another set behind, which is enough for Dolby Surround sound, which has front and rear directions only. But in some situations like concerts I add a third and a fourth pair of speakers, so for instance people can sit at a concert facing in any direction [Tomita played many open-air events] and will still get an effect. Recently I have been making a lot of calculations for spatial movement using the RSS [Roland Sound Space] technology."

On Tomita's later, conventionally orchestral, album, *The Tale of Genji*, you can hear synthesized sounds representing spirits apparently moving around three-dimensionally, and both stereo placement and apparent spatial movement have always been important in Tomita's analog programming.

However, as early as his breakthrough album *Snowflakes Are Dancing*, Tomita's choice to arrange Debussy pieces called for a much more lush and romantic approach than that of Walter Carlos with Bach. It was important to make the tones rich and reminiscent of acoustic instruments, which is perhaps not the task at which the Moog system most excelled. But Tomita succeeded in creating string simulations that were particularly strong and realistic, coupled with lyrical lead lines such as a distinctive portamento whistle (similar to the one on track 42 of the website), which, along with more abstract sounds using ring modulation for bell-like tones, helped produce an album quite distinct from the Carlos style. Many of Tomita's other Moog sounds are much less conventional; he developed a range of human voice-like sounds using a filter wah that have become widely associated with him, but which can sometimes seem rather comical. On *The Tomita Planets*, these were used very extensively and the relatives of composer Gustav Holst temporarily prevented the distribution of the album, describing it as "vulgar"; now available again, it actually includes some of Tomita's most ambitious uses of stereo sound placement and

1978 Tim Blake releases the *New Jerusalem* album featuring Moog, Korg, and EMS synthesizers

1978 Dave Smith's Sequential
Circuits Inc releases Prophet
5 polyphonic synthesizer

effects. By the time of the release of *Space Fantasy* in 1977, Tomita's techniques had become even more refined, although there was an increased use of non-modular instruments for chord work; a Polymoog, Mellotron, and Roland RS202 string synth as well as a Roland 700 modular system, the album ambitiously taking on pieces by Wagner and Charles Ives. See the Discography, Appendix E, for a photo of Tomita's studio.

One further outstanding example of analog synthesis of classical music of the period was *Tchaikovsky; 1812/Nutcracker Suite* by Kraft and Alexander (Decca, 1977), programmed by Jack Kraft on keyboards and engineered by Larry Alexander. Claiming to have up to 200 synthesizer tracks at some points, this uses dense, rich, and powerful orchestral sounds created on ARP 2600, Odyssey, and Pro Soloist keyboards; one of the few albums to match the density of Isao Tomita's sounds and featuring some great symphonic crashes on the "1812 Overture". This type of album was still common around 1980, when Richard Harvey (of the folk rock band Gryphon, and later a very successful composer for television and film) was the major synthesist on *Masterworks* released by TV promotional company K-Tel, which covered compositions by Strauss, Beethoven, Mozart, Ravel, Grieg, Borodin, Bizet, Rimsky-Korsakov, and Tchaikovsky using a huge arsenal of analog synths: Polyfusion 2000, Moog 15 and Roland 100M modular systems, Roland SH2, Jupiter 4 and VP330 Vocoder, Prophet 5, Polymoog, Yamaha CS50 and CS30L, Korg PE2000 and Vocoder, Oberheim 4-Voice, 2-Voice and sequencer, as well as an RMI Keyboard Computer, an advanced organ-like digital instrument.

After this time, there was something of a lack of interest in synthesized performances of the classics, but this was revived in the mid-1980s with a number of albums based on Yamaha's FM digital synthesis instruments, and later with others using the Fairlight, PPG Wave, NED Synclavier, and other digital systems. One of the most experienced keyboardists in this area has been Hans Zimmer, the Academy Award-winning composer for *Gladiator*, *The Lion King*, and scores of other Hollywood productions. Zimmer started his career working for bands such as Ultravox and The Buggles, using a Moog modular system and later a very large Roland 100M system with other modules from RSF and other manufacturers. Studying with the movie composer Stanley Myers and developing a powerful set of techniques for analog synthesis of orchestral sounds, he nevertheless moved to sampled and digital instruments as rapidly as possible, keeping some analog keyboards as "icing on the cake".

"I like the collaborative process in movies. I'm not interested in writing symphonies or having a Hans Zimmer No. 1 album. I actually like writing for films, which is all about fitting in with what everyone else is doing. I get most of my cues for how the piece should sound from the cameraman and the editor, so all I'm trying to do is an extension of how the cameraman has lit something."

(Hans Zimmer, interview with the author, July 1989)

For Ridley Scott's *Black Rain*, Zimmer purchased a little-used Yamaha CS80. Scott had probably developed a liking for its sounds from its use by Vangelis on *Blade Runner*:

"That's a fantastic synth and I use it a lot. I'm having MIDI added because the keyboard is a little short for me. My Yamaha KX88 has a nice long keyboard but I haven't come across a master keyboard I really like. More and more people seem to have decided that the best keyboard is a nasty plastic one because if you want light, fluffy string parts you need a light, fluffy keyboard. We have a Yamaha DX1 downstairs, which gives me bruised thumbs from trying to hit it hard. But I'm not a snob about instruments; I think they all have their uses. I don't have any Casios for example, but Vince Clarke has done some great things with them."

At this time, Zimmer was combining analog synths, including a rack-mounted Minimoog, Oberheim Matrix 6, Oberheim Xpander, and Roland MKS30, with digital modules, such as the Roland D550, Korg M1R, Yamaha TX816, a rack-mounted Yamaha CE20 ensemble, plus Akai and Fairlight samplers, though his Moog/Roland/PPG/Projekt Elektronik modular system remained a feature of his studio for many years, finally having a Yamaha GX1 triple keyboard as played by Keith Emerson posed in front of it. Zimmer's use of synthesizers, however, remained largely for the purpose of mocking up parts to be played by an orchestra. Later he came to rely mostly on racks of PCs running the TASCAM Gigastudio sampler to do this.

1978 Tangerine Dream release the *Cyclone* album, featuring vocals by Steve Jolliffe

Naturally, most programmers interested in the sounds of the classical orchestra eventually turned to samplers and sampling software, and the art of synthesizing the sounds of the classical orchestra using solely analog techniques has largely been lost. However, it will always remain an interesting technique. Trying to use analog sound for synthesis of classical instruments can still prove a highly educational experience. Something of a lost gem was *Wolfgang Amadeus Mozart – The Greatest Hits* released in 1991 on the Soundwings label by Stephan Kaske, the man behind the German rock/synth band Mythos, which uses a huge array of analog and digital synths for a bright, poppy sound combining some of the better elements of Carlos and Tomita. In contrast, for the fiftieth anniverary of Moog instruments and the tenth anniversary of Bob Moog's death in 2015, producer Craig Leon created *Bach to Moog* for the Sony Classic label, which integrated the Moog Modular very unobtrusively into the orchestral sounds of the Sinfonietta Cracowa.

JAZZ PROGRAMMERS

Jazz musicians also took up the early analog synthesizers with some enthusiasm, though few of them experimented in any great depth with the massively varied tone colors available. Jazz pianist and educator, Dick Hyman, released *Moog: The Electric Eclectics of Dick Hyman* on Command, a division of ABC Records, comprising compositions and jazz improvisations with synthesizer programming by R.A. Moog Inc.'s Walter Sear. On some of the pre-composed pieces, there were some interesting and complex sounds; on the improvisations, the synthesizers were poorly tuned and now sound embarrassing. There is some early use of a filtered analog drum machine too, and Hyman's influences from Greek and Indian music show clearly.

In 1972, jazz keyboardist Paul Bley and avant-garde vocalist Annette Peacock released *Revenge*, credited to The Bley-Peacock Synthesizer Show. The duo had been experimenting with processing Peacock's avant-garde lyrics through Bley's Moog and ARP 2500 synthesizers, adding distortion, ring modulation, filtering, and other effects over more or less conventional modern jazz backings. Paul Bley also worked with the jazz drummer Bruce Ditmas, who later released two albums (*Yellow*, recorded in 1976, and *Aerey Dust*, recorded in 1977), interfacing his drum kit to a Minimoog and ARP 2600 using a Moog Drum Controller. The analog sounds involved were generally percussive bass sounds, both pitched and randomly varying, and often depending on the level of the velocity signal generated from the drums (Carl Palmer from ELP had been doing something similar for some time. A good example is on *Toccata* from *Brain Salad Surgery*, released in 1973).

Avant-garde jazz keyboardist Paul Bley with an ARP 2500.

Many jazz players took to the Minimoog synthesizer in particular, notably Chick Corea (who also soloed extensively on an ARP Odyssey), Herbie Hancock, and Jan Hammer, who had emigrated to the USA from Czechoslovakia and worked with many of the early jazz/rock fusion performers. Hammer played the Fender Rhodes electric piano through an Echoplex and other effects, but in the mid-1970s started playing his Minimoog – later with an Oberheim SEM in parallel – through a guitar combo amp for lead sounds.

"It's not the same sound now, but the effect is the same simply because it's all in the playing. It's in the phrasing and the choice of notes. And you have to play through a guitar amp, though lately I've also been using distortion boxes like the Rockman. But I've recently finished a new studio in the barn outside my house which lets me turn up the volume and use big guitar amps again."

(Jan Hammer, interview with the author, May 1989)

Around this period, Hammer had been defining a whole new style of jazz/rock-influenced TV music with his scores for *Miami Vice*:

"I worked for three or four years on *Miami Vice* and I don't really separate my TV and album work, so a lot of the pieces ended up on albums. On *Miami Vice* the arrangement was unique in the world of TV scoring – I was totally free; every week I would decide where the music went, how long for and what kind of music. It was totally up to me. So if you listen to a couple of episodes you'll find the music differs very much from show to show – some of them have very Afro-Cuban roots music, some of them are very jazzy, some very much like reggae, some like acid rock. So I never knew from week to week what I would be working on, which was really great. So to build up those pieces into album tracks, when you have a 20-second piece of music which is really packed with ideas, all you have to do is re-record it and re-work the form."

Hammer had been using his analog lead sound successfully with the Mahavishnu Orchestra, often battling with John McLaughlin's lead guitar, as he later did on many albums with guitarist Jeff Beck. While Hammer moved on to using a Fairlight sampler, Linn and Kawai sampled drum machines, then Macintosh computers initially running Opcode Vision software. Stu Goldberg also gave some very expressive performances on a Minimoog, which had been extensively modified by Steiner-Parker, on The Mahavishnu Orchestra's *Inner Worlds* album, as well as on his own solo projects. Josef Zawinul of Weather Report also experimented more than most during this period with unusual sound textures. He made extensive use of two ARP 2600s, often playing both at once and sometimes inverting the keyboard voltage on one so the keyboard played its lowest pitches at the top. Some of his better analog sounds can be heard on Weather Report material, such as the track *Birdland*. Sun Ra incorporated much wilder synthesizer sounds into his jazz compositions and improvisations, but to be fair, he had already been experimenting with early electronic instruments and with processing existing conventional electric keyboards for many years. A special mention should be made of George Duke whose 1970s albums include an impressive array of sample-and-hold and other semi-random sounds alongside blistering and highly expressive lead solos.

1978 ELP on tour in USA with full orchestra, Keith Emerson playing Yamaha GX1

In the UK, Neil Ardley had been composing for large jazz ensembles and incorporated the ARP Odyssey synth on his album *Kaleidoscope of Rainbows*, released in 1975. Influenced by Balinese scales, the album used repeated phrases, creating close comparisons with contemporary works of systems music. Ardley's follow-up, *Harmony of the Spheres*, used even more analog keyboards, but an all-synthesizer album, *Intimate Vistas*, recorded with members of the experimental jazz/techno-pop band

Landscape (who had a hit with *Einstein-A-Go-Go*) and using early analog synthesized percussion was never completed (two tracks were preserved on a now rare reissue of *Kaleidoscope*).After a musical break during which he became a very successful author of reference books,Ardley returned in the 1990s with John L.Walters of Landscape, the trumpeter Ian Carr and guitarist Warren Greveson in the band Zyklus, using sophisticated MIDI technology and the rare multi-track sequencer also used by White Noise and Vangelis from which the band took its name.Ardley sadly passed away in 2004, but his music remains one of the most tonally expressive uses of the analog synth in the jazz field.

POP AND TV MUSIC PROGRAMMERS

1978 ARP launches the Avatar guitar synthesizer similar to the Odyssey keyboard

Inevitably, some record companies and some performers did not have the patience to create the complex multi-channel arrangements and sound textures of Margouleff and Cecil, Walter Carlos, or the other avant-garde and classical synthesists, and so many so-called "Moog" records were released consisting of little more than a very conventional rock or pop group backing under a simple synthesizer melody line.

To put this disappointing trend into context, in 1969 one of the most imaginative commercial electronic music albums ever recorded, White Noise's *An Electric Storm*, had been released by Island, created by Delia Derbyshire and Brian Hodgson of the BBC Radiophonic Workshop under the inspiration of the musicians and electronic engineer David Vorhaus. The tracks, including the epic *The Visitation* and the complex, poppy, and comic *Here Come The Fleas*, extensively used tape editing, filtering, and electronic processing, but used almost no conventional synthesizers as such.

> "We'd heard about the Moog synthesizers, but they were much too expensive. At the Workshop a synthesizer was a bank of 12 test oscillators like you'd get from a radio store, each tuned a semitone apart and played from big switches. So when we heard later from Peter Zinovieff with a description of his EMS VCS3 - a whole bank of voltage-controlled oscillators you could patch through a filter or anywhere else, all played from a keyboard - we were really interested. We went down to Putney to pick it up and it was his first sale, so he gave us a crate of champagne to go with it.Then we had a very quiet year - Brian was doing some work for the Royal Shakespeare Company and for the Roundhouse Theatre - so I spent that time building a 16-track tape machine. So our Kaleidophon studio became one of the first in London to offer 16-track, and then there was plenty of work."
> David Vorhaus (interview with the author, 2006)

The VCS3 and other early instruments can be heard on Virgin album projects, such as Cyrille Verdeaux's *Clearlight Symphony* and the collaborative project *Lady June's Linguistic Leprosy*, bringing David Vorhaus (who had by now bought out both Brian Hodgson, who continued to run the BBC Radiophonic Workshop and to work with the late John Lewis as the

Electronic music pioneer David Vorhaus of White Noise with dual EMS VCS3 synthesizers, MANIAC custom-built analog sequencer and Kaleidophon controller.

duo Wavemaker at Electrophon Studio, and Delia Derbyshire, who had left the BBC, more about her in Chapter 8) to the attention of Richard Branson:

"Virgin had really launched on the success of [Mike Oldfield's] *Tubular Bells*, so they were interested when I told them I had something very ambitious and instrumental but just played on synthesizers. By this time I had a second VCS3 and started to work on the *Concerto for Synthesizer*. So that album became *White Noise 2*. Then I built a multi-phasic analog sequencer, the MANIAC, which allowed you to flip patterns, change their length, make them play backwards and forwards, or even make them improvise, plus an analog drum sound bank to go with it. Those instruments are on *White Noise 3 - Re-Entry*, and by that stage programmable analog keyboards like the Prophet 5 were coming out just in time to play a couple of guest sounds on the album."

Vorhaus's musical progression closely parallels that of his erstwhile colleagues, some of whom remained at the BBC Radiophonic Workshop, which as its name suggests had been formed to create electronic soundtracks for radio programmes. No space here to give a complete history of the Workshop, which deserves a book in itself, but suffice to say that apart from the groundbreaking music for the sci-fi series *Dr Who*, the Workshop from pre-synthesizer days right through the analog era and into digital, sampled and early computerised times provided thousands of distinctive themes, stings, and electronic soundscapes for radio and TV, including various massive nature documentary series. As the Workshop moved from tape techniques to EMS, Yamaha, and Roland synths and vocoders, composers, such as Dick Mills, Malcolm Clarke, Paddy Kingsland, Elizabeth Parker, and later Richard Attree, quietly made innovations that influenced thousands of musically inclined listeners. After this early period, while the Radiophonic Workshop moved on to MIDI and Macintosh systems and eventually quietly closed down, David Vorhaus had moved over largely to the Fairlight CMI, composing with sampled sounds and digital keyboards on *White Noise 4 - Inferno*, and later recaptured the spirit of the analog experimental and tape looping days using a laptop system running NI Reaktor on *White Noise 5 - Sound Mind* and a series of live semi-improvised concerts in the UK and Europe.

Elizabeth Parker, Roland 100M, and Vocoder Plus at the BBC Radiophonic Workshop, 1980.

To return to 1969, the year of the release of the first White Noise album, in the USA Mort Garson, who had worked on Paul Beaver's *The Zodiac*, arranged and performed Moog parts on *Electronic Hair Pieces*, an album of tracks from the musical *Hair*, released on A&M. This included some rather comic sounds but also some strong tones and an extensive use of repeating sequences. In 1971, Vanguard released *The in Sound from Way Out* by Gershon Kingsley, a classical composer and Broadway arranger, with Jean-Jacques Perrey, a French-born musician and scientist interested in "creating a truly light-hearted, entertaining music through electronic instruments". The album's sleeve notes referred mainly to the duo's *musique concrete* techniques, but the album clearly had plenty of Moog content, comprising original compositions with a spacey theme, and also using many comical sounds. Later in 1971, the duo also released *Kaleidoscopic Vibrations - Spotlight on the Moog*, which included several original compositions plus covers of *One Note Samba, Moon River* and *Umbrellas of Cherbourg*, credited the Ondioline as well as the Moog, but used even more comic effects. The following year Vanguard also released *The Amazing New Electronic Pop Sound of Jean-Jacques Perrey*, which comprised entirely original compositions but still in a generally pop instrumental style. Perrey wrote:

"What I have tried to do is to bring the new electronic sonorities to popular music. In making this disc, I have used most of the electronic musical instruments available on the world market (Ondes Martenot, Ondioline, Moog synthesizer and the Allen series of keyboard instruments, etc.); I have also used magnetic tapes."

The album featured some serious sounds, but again very many comical ones.

What was claimed to be the first European synthesizer album, *The Magic of the ARP Synthesizer*, by Joop Stokkermans, was released on Phonogram Holland, again in 1971. Acknowledging Emerson, Lake & Palmer's use of the synthesizer (without pointing out that the band used Moog rather than ARP models), the sleeve noted:

"The synthesizer embraces ten octaves and while it could be assumed that a comparison with a Hammond organ is possible, nothing is further from the truth, the possibilities of the synthesizer being far greater. The unsuspecting Hammond player will find to his surprise that the instrument doesn't utter a sound when a key is depressed. Before it can be played each composition needs to be programmed into the machine, more or less like a computer. In order to make this album engineer Roddy de Hilster and pianist Joop Stokkermans did just that, spending an average of two hours per composition."

Apart from showing an apparently limited understanding of the instrument, the sleeve notes did nothing to point out that the album's cover versions of *This Guy's In Love With You* and some popular Bach and Tchaikovsky pieces were backed by a very conventional orchestra, and that the synthesizer sounds were again mostly fairly comical.

1978 Crumar in Italy launches the DS1 then DS2 synthesizers as used by Sun Ra

However comical, eventually some of these pop experiments were bound to become hits, and singles, such as *Pipeline* and 1972's *Popcorn*, both by Hot Butter - effectively the studio Moog player Stan Free plus session musicians - had a worldwide impact. *Popcorn* had been composed two years earlier by Gershon Kingsley for one of his early electronic albums, but the Hot Butter band format recording was much more successful, and the band also recorded several other cover versions of existing instrumentals, such as *Apache*, *Telstar* and *Amazing Grace*, and on their second album *Wheels* and *Tequila*. Some of the sounds used were interesting, though many were again vaguely comical, but the distinctive "popping" lead sound of *Popcorn* became so well known that it started to appear under that name as a preset sound on various small synths.

By 1972, ARP synthesizers had become as common as Moog instruments in the UK. Len Hunter and Bill Wellings were behind the band Elektrik Cokernut, which covered *Popcorn, Song Sung Blue* and other hit singles on *Smash Hits Go Moog* for the UK's budget Music For Pleasure label. Mostly, this comprised novelty Moog sounds over a conventional backing, but some tracks, such as *Morning Has Broken*, had a little more

1978 Jen in Italy launches
the SX1000, an affordable
monophonic synth

lyricism. Also in 1972, *Moog Espana* on RCA was realised by producer Andy Wiswell with Sid Bass playing Moog; despite covering tracks, such as *Lady of Spain* and *Spanish Flea*, over a conventional jazz/rock backing, the Moog sounds are relatively conservative, using rich oscillator unisons, perhaps because they were set up by Bob Moog himself at the Trumansberg studio, or so the sleeve notes claim.

Around the same time, *Switched On Rock* by The Moog Machine, produced by Norman Dolph and played by Kenny Ascher for CBS, covered *Aquarius, Hey Jude, Jumpin' Jack Flash* and other hits, and had a revealing and interesting set of sleeve notes:

"This record is virtually 100% Moog – only two instruments are live. One is the drum set; Moog drums are possible, but at this stage of the art, sound kind of mechanical . . . the second real sound we leave to the listener to spot. This is a synthesized record; all the orchestral textures, somewhere in the vicinity of 150 different varieties, come out of that funny box. Many of the sounds have no natural counterparts, so we coined neo-names to enable communication, e.g. the Gworgan, which is a Gwirped organ. Gwirping is the act of sweeping a filter with a high regeneration setting from top to bottom. It makes the sound 'Gwirp' . . . the inverse is Pwee, sweeping from bottom to top. The Pagwipe sounds like a ferocious, leaky bagpipe; the Jivehive sounds like a megaton of bees all swarming in tune. And there is the dread Moogoboe, and the Sweetswoop, a back and forth roar of harmonic sound like a jet plane flying through your head. Other parts were named by their function in the song . . . the 'Octangle', an eight-part progression . . . the 'BigBand', the 'Neoturnsolo', about half and half recognisable textures and totally new musical textures. One thing to stress to those unfamiliar with the way the machine works is that it is not a computer and it does not play or tune itself, constant touch-up of the tuning is necessary to correct its drifts over a 15-minute period."

Happily enough, the album featured many interesting and imaginative sounds.

Following the first spate of interest in analog synthesizers with the introduction of the large Moog and ARP modular systems, and then the pre-patched but still fully variable Minimoog, players had two major sonic requirements: they wanted to reproduce the imitative sounds of flutes, reeds, and brasses associated with the Walter Carlos albums; and they now also wanted easy access to the more unusual, "Moog-like" novelty sounds heard on hit singles, such as *Popcorn*. This is why a whole new generation of relatively affordable semi-preset instruments, such as the Moog Satellite, Korg M500 Micro Preset, and ARP Pro Soloist appeared with factory sounds such as flute, brass, and oboe, alongside "Popcorn", "Lunar", "Funny", "Frog", or "Telstar" (check out tracks 40–49 on the website). These semi-preset instruments introduced a whole new group of players to analog, those who were interested in using analog sounds, but who would prefer not to have to become involved in creating them from scratch. Clearly, those who were actively interested in creative analog synthesis would still prefer to go for an ARP Odyssey or a Roland SH5 or

System 100 modular synthesizer, so there were still plenty of synthesists interested in creating complex sounds completely from scratch.

1978 Korg launches the affordable MS series synths, including MS10 and MS20

In 1973, Robert Mason built a polyphonic synthesizer system, the Stardrive, for the quadrophonic album, *Intergalactic Trot*, released under the name Stardrive on Elektra, covering *Strawberry Fields Forever* as well as including several original compositions, but using a conventional rhythm section, including drummer Steve Gadd. The Stardrive looks like it was a modified ARP 2500, and certainly exhibited rock-solid tuning on the album, which makes it sound more typical of ARP than of Moog instruments. There are some good lead sounds and quadrophonic effects, and catchy compositions with a jazz/rock feel. The following year, Paddy Kingsland's album *Supercharged* was released in the UK and featured mainly ARP synths, including the tracks *Wobulator Rock* and *Money, Money* from *Cabaret*, and *Get Back* by Lennon and McCartney, amongst others. Kingsland noted:

"On this LP we have used the synthesizer side by side with conventional instruments, making its own distinctive sounds. We hope this puts the synthesizer in its rightful place as an exciting new instrument, not as a replacement for others."

Sadly, the album mostly comprised dreadful comic sounds over a conventional backing, but Kingsland, who worked extensively for the BBC Radiophonic Workshop, can be forgiven (!), as his later compositions included some outstandingly lyrical and powerful documentary soundtracks.

Around this time, many other artists also recorded Moog novelty records, Hugo Montenegro and organist Klaus Wunderlich amongst them. Wunderlich had become massively popular with his Hammond organ albums and released a Moog album titled *Sound 2000-1*, followed by one which was unashamedly comic, *Uraltedelschnulzensynthesizergags*, released on Teldec in 1974. Featuring Wunderlich in party clothing and festooned with patch cables, his Moog modular system draped with balloons on the cover, the album included arrangements of Fucik's *Entry of the Gladiators (Einzug der Gladiatoren)* and of Rhode's *Sleeping Beauty's Wedding (Dornroschens Brautfahrt)*, as well as Lincke's *Glow Worm (Gluhwurmchen-Idyll)*, retitled *Moskito-Killer*. Wunderlich wrote:

"Right from the start I should apologize to all granddads and grandmas because I didn't do this job completely with a straight face. On this LP, as on *Sound 2000-1*, I play a new kind of electronic instrument, the Moog synthesizer. With the help of a synthesizer it is possible to electronically reproduce all possible kinds of sounds and noises. You therefore hear no 'real' trumpets, saxophones or honky-tonk pianos, etc., but only the synthesizer. The only 'real' instruments are the drums and the rhythm guitar. The recordings were made using the multi-playback process because the synthesizer can only play single notes, that is, no chords."

Although many of the sounds are comic, at least Wunderlich was open about this, and there is some imaginative programming: the musical mosquito on *Moskito-Killer*, the whistling canary-like sound on *Canari-Cha-Cha*, and some excellent brass and bass sounds.

Other, much less imaginative examples of the "pop cover version" Moog album included The Carmets' *Synthesizer Electric Sound* released on JVC in 1974, featuring arrangements of *Surfin' USA* by Chuck Berry and *Slaughter on 10th Avenue* by Richard Rodgers, with Yusuke Hoguchi as Moog synthesist. This sort of album continued to contribute to a common perception of the Moog as a novelty instrument. By 1974 labels, such as Pye, were regularly commissioning Moog novelty records, such as *Plugged In Joplin*, which used just a Minimoog and EMS Synthi Aks to cover ragtime compositions, such as *The Entertainer* and *Elite Syncopations*, in very simple three- or four-part arrangements, mostly with vaguely comic sounds on almost any musical subject conceivable. In 1974, Pye also released *The Many Moogs of Killer Watts*, with synthesizers by Barry Leng, Derek Scott, and Len Beadle, using ARP 2600, EMS Aks, and Minimoog, mostly using comic noises for covers of *Telstar, The Good, the Bad and the Ugly, Amazing Grace*, and many others, and the BBC's Radiophonic Workshop was also issuing compilation albums of its work around this time. *Fourth Dimension*, released in 1973, had been entirely composed by Paddy Kingsland, but *The Radiophonic Workshop* of 1975 also featured Dick Mills, Glynis Jones, John Baker, Richard Yeoman-Clarke, Malcolm Clarke, and Roger Limb. This album included some comic sound passages and several up-tempo, pop-like themes, but also some serious-sounding music, such as *Adagio* by Dick Mills, mostly using the Radiophonic Workshop's EMS VCS3, and Synthi 100, plus an ARP Odyssey.

Happily, around this time the fashion for novelty Moog albums was more or less finished off by the rise of much more serious electro-pop styles. Following the worldwide success of Kraftwerk from around 1975, it must have been difficult to continue regarding the synthesizer purely as a novelty instrument when predominantly electronic bands were very visibly using it to create serious, innovative and commercially successful compositions.

ROCK PROGRAMMERS

Synthesizers were taken up by the rock fraternity pretty rapidly after early and high-profile use of the Moog modular systems by The Beatles and The Rolling Stones. The story of Keith Emerson's early use of the system with Emerson, Lake & Palmer was told in part in Chapter 3. Emerson's programming, although seemingly innovative at the time, was in many instances simply a case of very expressively using the most basic facilities available. Portamento (glide) on the famous *Lucky Man* solo, high filter resonance coupled with envelope sweeping on *Pictures at an Exhibition*, oscillator detuning and ring modulation on the steel drum-like solos on *Ladies and Gentlemen*, random sample-and-hold on the *Karn Evil 9* bridging sections, and so on. Tracks 29–31 on the website, featuring the Minimoog, give a good idea of the stronger sounds often used in these progressive rock music styles.

Progressive and New Wave rock have always been fertile ground for users of instruments such as the Minimoog. Examples include Dave Greenfield with The Stranglers, shown here, and Ryo Okumoto with Spock's Beard (overleaf).

Emerson's compositions often alternated between the Moog and the Hammond organ, grand piano, clavinet or church organ, and much the same selection of instruments was in evidence on Rick Wakeman's first solo album, *The Six Wives of Henry VIII*, released in 1972, though mostly stripped of the rock band backing. Wakeman had made his name in studio sessions and with The Strawbs, later replacing Tony Kaye in Yes; his first solo album particularly featured the Minimoog, with some very deeply swooping portamento sounds, but the apex of his Moog usage was really on the later live album, *Journey to the Centre of the Earth*, which used as many as five of the instruments each set to a different sound. Wakeman's Moog programming was distinctive because he made no attempt whatsoever to make the instrument imitate anything else. There are no

Rick Wakeman Minimoog attempts at flute, oboe, or human voice sounds from this period. On *Journey*, the Moog sounds vibrate, glide, wobble, and generally contrast very markedly with the symphony orchestra, which also features strongly on the album. Wakeman's ornate, classically derived playing style also seemed to bring the best out of the instrument; his trills and ornamentation appear particularly impressive played on a cutting Minimoog lead sound, while the single triggering of the filter envelope allowed a surprising degree of expression to be introduced depending on whether an earlier note was held down or released before playing a new note.

Also in the UK, Jon Lord was playing ARP synthesizers with the progressive rock/metal group Deep Purple (track 24 on the website gives some idea of the Odyssey's thinner but more precise sound compared to the Minimoog), while in the USA, other rock-oriented musicians were becoming interested in both the Moog and the ARP designs. Roger Powell, who worked as an ARP demonstrator, recorded *Cosmic Furnace* in 1973 using ARP 2500, 2600, Odyssey, and Soloist synthesizers. The album sleeve features some excellent photos of Powell's studio set-up, the playing is fluid and rather jazz-oriented, there is extensive use of sequencing, and the sounds are serious and imaginative, although distinctly thinner than those from contemporary Moog instruments. Powell went on to work with Todd Rundgren's Utopia and eventually in music software development. Larry Fast released *Electronic Realizations for Rock Orchestra* under the name Synergy in 1974; despite having only one Minimoog to use, Fast was able to multi-layer it for strong, orchestral string and brass textures. Though marketed as rock music, the album's styles included many classical and baroque influences, and Fast went on to play in Peter Gabriel's band, to program for the innovative British singer Kate Bush and others, and to release a whole series of increasingly complex albums under the name "Synergy", as well as some more avant-garde work including *Computer Experiments*, which developed the control of analog synthesizers from microcomputers such as the early Apple II.

1978 Kit-built synthesizers available from Powertran and others in the UK

Of course there were many, many albums around this period that used synthesizers as an occasional interlude rather than throughout the album. The so-called progressive rock field is rich in examples of the analog synthesizer – usually the Minimoog or ARP Odyssey – counterpointing the standard rock line-up of electric keyboards, guitar, bass, drums, and vocals. Excellent players of the period, particularly for their expressive Minimoog solos, included the late Peter Bardens with Camel on albums such as *Mirage, The Snow Goose*, and *Moon Madness*, and the more jazz-influenced Ingo Bischof with the bands Kraan, Karthago and Guru Guru in Germany. Also in Germany, Dorothea Raukes alternated between progressive rock and electro-pop styles with Streetmark, creating some outstanding Minimoog pieces, while Manfred Wieczorke with Jane also strongly featured the instrument. Hans-Jurgen Fritz with Triumvirat leaned very much towards Keith Emerson's pseudo-classical style, and after an uncertain start on albums, such as *Mediterranean Tales*, in 1972, produced some excellent performances, including some on the Yamaha GX1, which was also played by Rick Van Der Linden of the Dutch classical rock band Ekseption.

Still in the progressive rock field, Eddie Jobson, particularly with the band UK, gave some excellent performances both in the studio and live, often soloing on the Yamaha CS80, which is not generally regarded as a good soloing instrument despite its expressive touch-sensitive keyboard (though check out track 60 on the website). Jobson, who also played electric violin and had been a member of Curved Air, created some CS80 textures (such as on the track *Alaska*) so distinctive as to become the name of preset sounds on later synths, and all of his performances with UK are well worth hearing. He later went on to play with Jethro Tull and Roxy Music, and to compose TV scores in the USA.

Peter Robinson, meanwhile, spent some time with Santana and contributed some excellent synthesizer solos, particularly on stage, while Dave Sinclair created some fluid Odyssey solos with the Canterbury-based British band, Caravan, and Dave Stewart crossed boundaries between progressive rock and jazz with bands such as National Health, recording some flowing and complex synth solos in the process. Richard Wright with Pink Floyd tended to alternate between organ and Minimoog, sections of *Wish You Were Here* and *Animals* being notable examples, while Tony Banks with Genesis prominently used many analog synthesizers up to and including the Sequential Prophet 5 and ARP Quadra.

"One of my favorite keyboards was the ARP Quadra. You could layer strings, synth and bass or lead, and interface different sections with drum machines to trigger them in time, which we did from the drum machine cowbell in *Mama*. So I'd still have a reason to go back to that instrument, though the sounds weren't all that good. But it came out just before MIDI, and after that time anyone could do that quite easily. I played a Prophet 5 around the same time and I've now started playing the virtual version, as well as the Steinberg Grand, which is about the first piano sample I've come across I can really play, and Atmosphere, which has a nice range of sounds on it. It's a new area for me which I'm getting used to, but the main problem is that a lot of these things take a while to load up. When that's solved they will be easier to use, but I'm a very spontaneous person and often you quite like the first sound you get – I'm not a great one for wanting to do a lot of programming and fiddling, I want to get it to work pretty much immediately with what you've got."

(Tony Banks, interview with the author, 2004)

That liking for spontaneity is reflected in Tony's list of other favorite instruments:

"I used a Yamaha CS80 a lot on the *Duke* album, though I sold that as it got really superseded for me by other things. I still have my old ARP 2600s and other stuff; I had a big Mellotron which disappeared at some point, it went to be mended and never came back, though I was fairly pleased to see it go. But I have a couple of the smaller Mellotrons, and I have thought about getting them out again, but you think what's the point – there's always another sound available. I even shied away from sequencers when they first came in; I thought I could play everything by hand and that I didn't need them. Then when we did *Invisible Touch* back in 1986, *Land of Confusion* was the first thing where I really put in a sequenced bass, which locked in very well with the sequenced drums. We had used drum machines from about 1980 which I liked, then when I got used to it I started to realise what you can do with the computer, from early Atari stuff to what we have now. I'm moving my work completely over to the computer, but only gradually – I still have keyboards like the Roland JD800 and module versions like the JD990. I do find once you go over to modules, that's the end of experimenting with the sound – you put your sounds on a cartridge and you do tend to stick with sounds you know. But after the generation of the Roland D50, and of course the Synclavier, there were sounds so big that you hardly had to add to them at all. For example, on the last Genesis album, *Alien Afternoon* opens with this big keyboard thing just combining a JD800 sound with a Korg Trinity sound and it was swirling everywhere, a great sci-fi kind of effect."

1979 PAiA publishes Proteus 1 design, the first programmable mono synthesizer

Other rock bands contemporary with Genesis used synthesizers only very occasionally, but to great effect. The Who did so on the tracks *Won't Get Fooled Again* and *Baba O'Riley* from *Who's Next*, while another good example is Edgar Winter's album *They Only Come Out At Night* (Epic, 1972); most of the album is conventional rock but the track *Frankenstein* has many filtered, highly resonant ARP sounds and sequenced passages that remain impressive to this day.

The punk rock era from around 1977 is associated with some antipathy towards synthesizers – despite the fact that Emerson, Lake & Palmer, Jeff Wayne, and others had huge successes well after its inception – but a few bands, including The Stranglers, took a more "new wave" direction which could include some excellent keyboard work. *Go Buddy Go, No More Heroes,* and so on, were strong, punkish rock music, though embellished with intricate, baroque keyboard lines that keyboardist Dave Greenfield is happy to admit were influenced more by Rick Wakeman and Yes than by Ray Manzarek and The Doors (interview with the author, 2004):

> "I had studied music theory, but I taught myself piano and had been play-ing in a lot of bands in the UK and in Germany. The advert I saw in *Melody Maker* was for a keyboard player and saxophonist for a 'soft rock' band, but the saxophonist only lasted a couple of weeks. We were playing for a couple of years before signing up with United Artists, and I had graduated from a Vox Continental organ to a Hammond with a Hohner Cembalet electric piano. Then I was able to buy the Minimoog, which I still have, though I don't take it on tour with us any more; the organ sounds now come from a little Voce Micro B module, most of the rest from a Roland JD800 synth and from an Akai S3000 sampler, which has a lot of the sounds I've used over the years, including a choir."

In later years, Dave's stage set-up was mixed through a Yamaha O1/V automated digital mixer. His controlling keyboards – two Yamaha DX7 Mk 2s from which he doesn't use any sounds – send patch commands to reset the mixer for each song, and this sets levels, mutes, and effects for every instrument. Drummer Jet Black used an MPC2000XL sampling drum machine, also used on some songs to play very fast arpeggiated pat-terns beyond Dave's considerable playing capacity in a live setting. Finally, an Evolution controller keyboard played a Novation A-Station module for Minimoog-like sounds. The Yamaha DX7 emulator software FM7 was also used in the studio, though not necessarily for classic FM synthesis sounds, as it also includes an analog-style filter, as was G-Force Oddity, a replica-tion of the ARP Odyssey. Dave's EDP Wasp synth and Korg VC10 Vocoder used around the period of the *Meninblack* album survived into the 2000s, but were no longer taken on stage. With their 17th studio album *Giants* released in 2012, The Stranglers remained a great example of a working band taking strong analog as well as digital keyboard sounds right through from their inception in the 1970s to the present day.

While the golden era of the progressive rock field was the early 1970s, there was a second generation of such bands in the 1980s, including IQ and Marillion in the UK, and a third coming largely from Scandinavia and the USA in the 1990s. While musicians in these bands often used the latest technology, they also retained an attachment to the sounds and instru-mentation of their early heroes from ELP, Yes, and Genesis, and so showed a great interest in getting hold of genuine Hammond, Moog and similar instruments; certainly a factor driving the revival in second-hand prices of these instruments. A typical musician coming up through this period was

Jordan Rudess, currently performing solo and with Dream Theater, who originally studied at Juillard:

Jordan Rudess of Dream Theater, a typical keyboardist combining virtual analog synths (here from Novation) with digital keyboards (here from Korg and Kurzweil).

"I got assigned this Chopin piece, went away to study it, and when I came back a week later I started playing it, and the teacher took the sheet music away. She said 'You have memorized this, haven't you?' So I played that game for a year or so, but then I started hearing Yes and Genesis, and eventually I left, bought myself a Minimoog and started playing my own music."
(interview with the author, 2005)

Rudess worked for computer music companies making musical content for the Atari and Commodore 64 computers, and eventually for Kurzweil and for Korg USA. His first successful band The Dregs allowed him to go full-time and now, though he extensively uses Korg keyboards, he retains other analog favorites such as the MIDI-equipped Moog Voyager, which he played at a recent MoogFest.

"We had signed up to support Yes on some US dates so I had met Rick Wakeman, but then I got a call to perform with him at the Moog event to celebrate Robert Moog's 70th birthday and the 50th anniversary of the company in 2004. That was held at BB King's in New York for an audience of around 1000, so it was a great evening. It was like a boyhood dream to play with Rick Wakeman – we learned his track *Catherine Parr* together and it went really well."

Rudess also carries out a busy program of workshops and clinics, many in France and Italy, "because Italy is certainly the biggest market for Dream

Theater. They have always liked prog rock there, and we always get good media coverage, with people saying that progressive rock is back and Dream Theater are leading the way. I don't know if it's true, but it's very pleasing to hear." Jordan's system of recording at the time, on a G5 Macintosh running MOTU Digital Performer, ably supported his spontaneous method of composing, integrating both hardware analog and digital synths, and virtual software synths:

1979 Debut album from Yellow Magic Orchestra in Japan featuring Ryuichi Sakamoto

> "What you play spontaneously can be much more complex than what you may score out by hand, and MOTU DP works pretty well – even if you're playing freely without a click it will try to work out your tempo. And I like MOTU's sampling software Mach 5 too and gave some workshops on it at the NAMM show. I tried out their new soft synth MX4 and that looks great; I know the guy who designed it and he's very talented."

In terms of hardware, in 2005 Jordan's stage shows still included a Novation Supernova virtual analog synth and (like Nick Rhodes of Duran Duran) a Roland V-Synth:

> ". . . one of the few synths I've had to buy, but it was really worth it – it has a really unusual approach to time-stretching and synchronizing sounds, and I do things like getting my children to speak into it and really messing the sound around. For soft synths, the main one I use is Spectrasonics Atmosphere, that has some great sounds including a Theremin which really gets some attention from owners of real Theremins! I also use Spectrasonics Stylus for bass; that's on several tracks on my solo album."

In 2018 Rudess continues to innovate, and is also very active in the development of musical apps for the iPad. A contemporary of Jordan Rudess is Ryo Okumoto, the Japanese-American keyboardist with prog rock band Spock's Beard. In 2000 he was playing a complex keyboard set-up, including a Roland VK7 organ through a Leslie rotary speaker, a Kurzweil K2000 sampler including Mellotron samples, a Korg Trinity as main keyboard plus a Trinity rack module, and a Minimoog (vocalist Neal Morse also fielding an Alesis Quadrasynth while bassist Dave Meros doubled on Korg Prophecy played from a set of MIDIStep bass pedals).

> "I really like the Hammond and the Mellotron, they typify the style of Spock's Beard. There's nothing like the Mellotron. These days when you buy a new keyboard, every one has the same 128 sounds, a piano, then another piano, then an electric piano, always the same general MIDI stuff. That's so disappointing; I don't go to music stores any more. But when you have one good sound on a keyboard, that's all you need, if you know what to do with it. The most important thing is the volume or expression pedal, which most people don't know how to use. But I grew up playing the Hammond organ and listening to people like Jimmy Smith, and that's how I developed it. It's real-time control, added to the tone wheel sound generation of the Hammond, plus the percussion and the reverb unit, you get a very organic sound, more acoustic. You can move with the drawbars

from one sound to a different sound to another – it's the limitations of the sound that makes you develop these techniques."

(Ryo Okumoto, interview with the author, 2000)

That particularly applies to the Mellotron, which we haven't covered in detail as it's a tape-playing electromechanical instrument rather than an analog synth as such.

"I've had a Mellotron M400 for a while, with flute, strings and mixed choir tapes, and I have another one to go out on stage. The Mellotron lasts pretty well unless you play it too hard or hold chords for longer than eight seconds; if you do, the tape heads have to be adjusted, but apart from that, mechanically it's very strong. I used to take out a Jupiter 8 as well, and now I have the Minimoog and Trinity. But the progressive rock fans want that Mellotron sound. I tried the sampler right up against the original Mellotron, and there's a certain warmness from it that you just can't reproduce. I'd like the kids to see what the Mellotron is, so I can explain it's like a tape recorder and show the motor running. People can't believe how it works, they need to see that. And I bring a genuine Leslie rotary speaker because simulators are no good at all, they're not warm enough, they're really fake."

1979 EMS launches the unsuccessful Polysynthi in very small numbers

Ryo delivers some particularly blistering lead parts on the Minimoog, a technique also typical of Keith Emerson and Rick Wakeman.

"The Minimoog is great too, there's nothing quite like it, though I like the Prophet 5 and Jupiter 8, but the Minimoog really cuts through. In this set we do about three songs from the new album like *Thoughts Part 2*, then older tracks like *Mouth of Madness* and *The Healing Colors of Sound*, a lot of the epic tracks. And I try to get all the parts in, so you can see I'm very busy on stage, every track is very complicated with patch changes, volume changes, changing the delay, tuning the Minimoog every time before I play it, and making sure all the MIDI is sending and receiving."

Ryo Okumoto and Jordan Rudess are excellent examples of musicians influenced by the first generation of analog synth-playing prog rockers, excited to meet up and perform with them, but particulary in the case of Rudess, also willing to incorporate into his performances and recordings all three generations of analog-style instruments; genuine analog classics like the Minimoog and Voyager, virtual analog digital replications like the Novation Supernova, and computer-based software virtual analog synthesizers like MOTU MX4 and many iPad apps.

PURE SYNTHESIZER PROGRAMMERS

So far we have not looked at those groups and artists whose existence was virtually defined by the invention of the synthesizer: those whose music could not have been played on any other instrument, who were fascinated by the synthesizer as an instrument in itself, not for its applications to the classical, jazz, pop, or avant-garde fields. As most people interested in the field appreciate, this genre largely emerged from Germany in the late 1960s and early 1970s (one subsection becoming known as the "Berlin School"). Speculation as to why this should be the case when virtually no synthesizer manufacturers were based in Germany has been extensive, most theories centering on the idea that the country felt dubious about the recent martial connotations of its classical music, lacked a strong pop or rock music tradition of its own, and was ripe for the introduction of any style that owed little or nothing to the American style of rock and pop music.

Early experiments in Germany clearly did have something in common with US or British bands. It is generally accepted that in the late 1960s and early 1970s musicians, such as Edgar Froese, who formed Tangerine Dream, were influenced by groups including The Doors, The Velvet Underground, Jimi Hendrix, and Pink Floyd. Tangerine Dream's early rock performances using electric guitar, drums, bass, and saxophone would take tracks such as Pink Floyd's *Interstellar Overdrive* as a starting point and go off into long improvisations, but the early appearance of the synthesizer made an entirely new type of sound experimentation possible, and the difference between the band's first two albums, *Electronic Meditation* (1971) and *Alpha Centauri* (1972), is vast, since the latter had introduced the use of the EMS VCS3 synthesizer. By the following album, all traces of the rock format had disappeared completely: *Zeit* (1972) used the EMS synths, the large modular Moog system of Florian Fricke from the band Popol Vuh, plus some orchestral string instruments to create long, ambient textures impossible to conceive using any conventional type of instrumentation.

1979 Oberheim launches the OBX polyphonic synthesizer in the USA

True to the nature of the EMS synth in particular, Tangerine Dream were using these analog instruments to create sounds of their own and to process the sounds of more conventional instruments. Pianos, organs, voices, and flutes were all processed through the VCS3; in addition, it was used to create deep, constant drones, very abstract unpitched sounds and ring-modulated effects, bursts of white noise, and quickly repeating patterns. All of these were processed with additional studio effects to create analog sounds that matched much of the richness and texture of acoustic sounds, a discipline that was sadly neglected by some of the band's less fastidious contemporaries and later imitators, who, while often using similar instrumentation, created much less absorbing music.

Using the Mellotron (an electromechanical keyboard playing tapes of real instruments, such as strings, brass, and choirs) for chordal work, but processing this through phasers and filters as well, Tangerine Dream developed their sound using an early analog sequencer and drum machine to create a demo, *Green Desert*, for the UK-based Virgin Records, which resulted in their being able to work with two Moog modular systems at Virgin's Manor Studios in 1974. The resulting album, *Phaedra*, became a landmark in analog synthesis; the drum kit had been completely replaced by repeated bass and percussive sequences that varied constantly, and much of the melodic material was replaced by abstract sounds from the EMS Synthis. This style was developed further in 1975 on *Rubycon*, and when the band toured with similar technology (Moog modular systems,

Minimoog and EMS synths, Mellotrons and Elka Rhapsody string synthesizers) in 1975/1976 they created *Ricochet*, probably the most virtuosic display of live analog sequencing ever released. Within Tangerine Dream, Christoph Franke, who had started his career as a jazz drummer, simultaneously ran several sequence sections of the Moog systems; these could be changed in tempo, length, and tone, left to run almost unsupervised, transposed from a controlling keyboard that could also play a melody part simultaneously, or could be supplemented with the passing grace notes and trills ("ratcheting"), which eventually became a trademark of the band. Thanks to custom modifications of their instrumentation by PPG, Projekt Elektronik, and other companies, Tangerine Dream quickly became able to create analog sequencer patterns of a complexity quite inaccessible to other performers of the time.

While Tangerine Dream's sequences and textures tended to change and develop fairly rapidly during the course of a track, an ex-member of the band, Klaus Schulze (another drummer), developed a similar style, which tended to depend on even denser synthesizer textures, but extended for much longer periods of time. These reached a peak on the album *Timewind* (1975), which consists of just two half-hour pieces: one for fast analog sequences under synthesized strings, Moog lead lines and abstract EMS sounds, and one for slow, phasing chords generated on string synthesizers and organs under heavy ring-modulated sounds. Schulze went on to release a vast number of albums, keeping up with new technology as he incorporated first MIDI, then digitally sampled, then dance-oriented and most recently fully software-based instrumentation.

Klaus Schulze in concert with The Big Moog plus PPG keyboard, EMS Synthi A, ARP Odyssey, Minimoogs, Yamaha CS80, and Crumar GDS; Rainer Bloss joins in on Yamaha electric grand piano and Roland Jupiter 4.

Several other musicians in the Germany of the 1970s were working in similar styles; Michael Hoenig (who had been in the band Agitation Free and briefly toured as a member of Tangerine Dream in 1975) released one notable album, *Departure from the Northern Wasteland*, in 1977, particularly developing the style of multi-layering analog patterns in differing time signatures on the ARP sequencer, changing their length and skipping notes to give the impression of much more complex evolving patterns, while around the same time Robert Schroeder was even building many of his own analog instruments from scratch. Schroeder collaborated with Klaus Schulze, incorporated some very lyrical acoustic instruments such as guitar and cello over his synthesized music, gave some impressive large-scale concert performances, and later embraced MIDI technology for a successful series of pseudonymous and more conventionally melodic albums. Ashra, in the form of guitarist/synthesist Manuel Goettsching performing solo (the original Ash Ra Tempel group having had a more conventional psychedelic rock band line-up), released *New Age of Earth* in 1976 and *Blackouts* in 1978, both relying heavily on string sounds from a Farfisa Syntorchestra over repeated patterns from an ARP sequencer, abstract EMS Synthi sounds, ARP Odyssey, and electric guitar lead lines.

Robert Schroeder built many of his own analog modular synthesizers as well as playing PPG Wave 2 and Multimoog.

One musician who tied together many of these artists is Harald Grosskopf, a drummer and synthesist, who was in early and later line-ups of Ashra (Tempel), performed in the "krautrock" band Wallenstein, and also recorded and performed extensively with Klaus Schulze. Starting with beat groups in Berlin in the 1960s playing covers of Beatles and Kinks songs, he even joined one very early line-up of heavy rock band The Scorpions. But clearly he was unsatisfied with these conventional styles of music:

"The whole scene that became known as 'krautrock' was starting in Berlin. Bands like Tangerine Dream and Cluster as well as Klaus Schulze were based there, and there were others like Can in Cologne and Kraftwerk in Dusseldorf. When I heard some of these new kinds of music using synthesizers and improvised music, I thought that was the way ahead. In my first band Wallenstein, some of the music was getting very composed, like progressive rock, and you really had to think about what you were playing. I wanted a style where you could relax more and really feel the music, almost to play it without thinking about it."

(Harald Grosskopf, interview with the author, 2002)

Some massive jam sessions, many of them admittedly LSD fuelled, led to albums in the legendary "Cosmic Couriers" series, and out of these were born bands such as Ash Ra Tempel. Meanwhile, Klaus Schulze went off on a solo path, bought a huge modular Moog synthesizer and started experimenting with sequencers. Though also a drummer himself, Schulze preferred to call in Harald to follow the precise, metronomic beat they created, normally a very difficult task for a drummer more used to setting a beat than to following one.

This led to some wonderful collaborations on albums such as *Body Love* and X, where the constantly varying sequencer patterns of Schulze are matched by Harald's intricate percussion work, particularly on a very busy bass drum and in his spinning complex hi-hat patterns around the basic beat. "I was living at Schulze's place for a time so he would just let me know when there was a gig or a recording session and I would turn up. It was all very spontaneous." After a solo period as Ashra, Manuel Gottsching called on Harald along with guitarist Lutz Ulbrich for a new band line-up, and in 1977 they played a massive tour which included a legendary date at the Open Air Theatre in London's Regent's Park. Complete with laser show (and five-minute gaps between tracks to retune the sequencers), this caused serious problems for low-flying aircraft and probably to local residents, since the venue has never again been used for live music. An album called *Correlations*, rehearsed in an old Berlin film studio, was released later. "But I still felt that the drums were always being mixed low compared to the synthesizers and sequencers on these albums. It was like they were an accompaniment, not the driving force of the music – I wanted to record something myself where the beats were really prominent." In 1980, Harald started working with Udo Hanten, who had a synthesizer band called You, with the idea of creating a collaborative album, but after a few days Hanten handed over the keys to his apartment, saying he was sure Harald had plenty of his own ideas. Ten days later, having battled with modular synthesizers and keyboards, about which he really knew very little, Harald emerged with his first solo album, *Synthesist*:

"I was no keyboard player, but in fact none of us were. You didn't hear the sort of fast playing that came in the 1970s from Keith Emerson or Rick Wakeman. We were more interested in the sounds and the textures. I was really pleased with *Synthesist* because it was my first solo work, and though some of it sounds primitive now, there are parts I still really like."

Harald released other solo albums every few years, incorporating first Yamaha's FM synthesis technology and later sampling, and after moving out of Berlin met the techno DJ Steve Baltes:

"I had no idea that there was a whole generation of DJs and younger electronic music fans influenced by the old albums of Ashra and Tangerine

1979 Debut Human League album aims to prove that synthesizers can create pop music

1979 Moog releases the budget Prodigy synthesizer as played by Howard Jones

Dream. I knew there were a lot of people in what I call the second generation of the old school of music, but I didn't see any point in getting involved with that. But Steve and I got on immediately, we talked the same musical language, and we have been working together ever since."

Baltes gave a new techno beat to yet another line-up of Ashra, which played live in Japan and to the Sunya Beat project with guitarist Axel Heilhecker, and when Harald started solo recording again, advised him on the latest software instruments:

"We really started working on my new solo CD when we settled on a combination of Cubase VST with other plug-ins and software like Native Instruments Reaktor and Ableton Live. On the *Digital Nomad* album I felt for the first time that I could really work the way I wanted. Now I don't have space to set up a whole drum kit in my own studio, but I have a Roland hand percussion pad and can play beats on that. Any beat you play can go into Ableton Live, which is particularly well set up for taking in rhythms, time matching them and replaying them very quickly. You can pre-listen to any pattern then fade it in once it's in time, within maybe 10 seconds. That's a very spontaneous way of working. Then I can have all sorts of compressors, echoes and other effects as plug-ins for Cubase. There's nothing worse than having to stop working on a piece because you have to sort out all the cables to an effects unit, reprogram it and save the patch away. With these plug-ins I can have any sort of compression and any sort of effect I want almost immediately, and you can create music very quickly. I really feel that with this computer technology I can finally realize the type of music I have always wanted to make."

Harald Grosskopf is a good example of a musician from the analog-dominated "Berlin School" of the 1970s, who in 2018 is still working but who, like Klaus Schulze, has moved through generations of digital, sampling, and techno-oriented instrumentation to emerge with a powerful set of sonic techniques now mostly based on software. But to return to the 1970s era in which Grosskopf had his roots, in France during a similar period, Tim Blake was developing a comparable abstract synth style, performing with Gong, playing many laser-accompanied solo concerts as "Crystal Machine" and releasing two solo albums, *Crystal Machine* in 1974 and *New Jerusalem* in 1976. The former used rather simple sequences played on an EMS Synthi Aks sequencer under long, flowing Minimoog lead lines, the latter featured stronger patterns from a Roland 100 sequencer that Blake transposed rapidly as it played to create the apparent effect of much more complex sequences emerging. On *New Jerusalem*, the EMS Synthi's burbling abstract sounds are also particularly strong, while the fluid quality of Blake's Minimoog solos is outstripped only by those of his guest player, the blind French keyboardist Jean-Philippe Rykiel. Also working independently in France was Richard Pinhas, who played electric guitar through an EMS Synthi A under the influence of Robert Fripp from King Crimson. Selling his original record label funded Pinhas's purchase of a Moog modular system, which unaccompanied created many of the tracks on the early albums of his band Heldon; on later compositions, other more jazz-influenced musicians layered bass, Minimoog solos, voice, and drums over Pinhas's slowly developing sequences, and later, Pinhas's solo albums such as *Iceland* featured some coldly beautiful textures created on Polymoog, E-Mu modular system and looping guitar.

Jean-Michel Jarre surrounded by Yamaha CS60, Fender Rhodes piano, electric piano, Eminent organ, Elka Rhapsody, custom Geiss MatrixSequencer, ARP 2600, Korg PS3300 and 3200, Moog modular system, and more.

But 1976 had marked the debut of another French artist whose popularity was to eclipse of all the previous combined. Jean-Michel Jarre was, as is well known, the estranged son of film composer Maurice Jarre, he had studied avant-garde music in Paris before also entering the pop music and film soundtrack business. But his early avant-garde electronic recordings, such as the soundtrack *Les Granges Brulees*, gave little indication of the sophistication of his debut solo album *Oxygene*, which achieved a dynamic compromise between imaginative sound textures and accessible melodies that for one reason or another had been denied to earlier synthesizer artists. The album grew slowly in popularity, at first used widely as a hi-fi demonstration record, but after a huge free open-air concert in Paris at the Place de la Concorde, it became a worldwide success. Jarre was clearly a technology junkie, happy to talk about his choice of instruments and the way in which they had been adapted and modified, and to drag his entire studio set-up out for live appearances. His instrumentation was highly varied: Moog and ARP synthesizers, an Eminent organ that had rich string sounds (basically those of the Solina string synth), particularly when processed through a phaser, a Mellotron, and an EMS VCS3 for abstract sounds, among others. He also solved some of the limitations of these instruments, commissioning the construction of a custom sequencer (the "MatrixSequencer") much more powerful and expressive than those currently available, and adopting some more unusual instruments, such as the RMI computer keyboard (an early digital instrument), plus a rare, easily-interfaceable Eko drum machine built in Italy and also used by Manuel Goettsching of Ashra:

1979 Debut album by Michael Garrison, *In the Regions of Sunreturn*

"I was at the Music Research Centre in Paris working with Pierre Schaeffer when I recorded my first electronic music piece. It was for a TV ballet, all made on a big bank of 48 oscillators, but it was not a synthesizer at all, you had no envelopes, just a big matrix board and I used an LFO technique to make rhythms. I tuned the different oscillators, so I had a sort of sequence and I had to change the speed of the LFOs live, but it was still one of the first sequencers. I asked a drummer to come in, and we did two versions, one with both of us live, then another where I did it on tape and he played over it. It was an eight-minute piece and we made a single and probably sold 54 copies – now I have to look for copies myself because the record company has destroyed everything.

So that time was even before I started to use the EMS Aks synth – the first one I got was in 1966, and it was very exciting to have everything all in one box. I worked with an Aks and a VCS3, and I did all the music for the French Opera House ballet, the first official thing I did on stage, which was quite ironic because most of the time it's at the end of your life that you are asked to do that sort of work. I used four Revox tape machines manually synced, so it was eight tracks all mixed onto a fifth machine, and my patch bay was built into an old shoe box, but it worked quite well and I still have it.

That wasn't very commercial music, and at the beginning *Oxygene* was not considered commercial – record companies refused to release it because there was no singing and no drums, but from a distance I think what *Oxygene* had was the fact that I didn't consider the machines as machines, as so many German groups were doing. They think that the machines can make music, you even see Tangerine Dream leaving the stage as they are playing, but I always considered synthesizers like ordinary instruments, I never wanted to make a specific statement about electronic music being better or worse, just that it used the instruments of our generation. I could have created music in other contexts with an orchestra or a rock band, but I think it's this attitude that had the album received by audiences in a different way. I think the German groups wanted to make quite heavy statements about the technology, their music was very linear, and what annoyed me a bit about their work is that you start with a very good mood and then nothing happens, so you have no vertical structure: when you've heard the first five minutes you've heard the first 20. On *Oxygene* you have a more symphonic structure like a set of songs, and that's why people liked it. You can expand on structures but you can't invent a new one, it's like saying the day has 24 hours but mine will have 26 – that's why an ideal movie is always between 90 and 110 minutes long: if it's less than an hour it's too short, if it's more than two hours it's too long. If you make a movie seven hours long it won't work, it's a question of the rhythm inside people, a question of structure, and on *Oxygene* I was interested in coming back to a very ordinary structure."

(Jean-Michel Jarre, interview with the author, 1987)

Jarre's further album releases, such as *Equinoxe*, and later open-air concerts in Houston, Lyon, London and Paris, increased his popularity, but at a fairly early stage he began to use sampled and digital sounds on albums, such as *Magnetic Fields*, and even more notably on *Zoolook*, which combined analog synthesis with ethnic and vocal music:

"I've always been involved in ethnic music, though I thought the way a lot of people have been using ethnic music was a little superficial. Sometimes it works, like the Brian Eno stuff, it worked the first time, but for me what was more interesting was not making a particular statement about recording in Africa or in China, but taking some sounds and having exactly the same attitude as when you were in front of a Moog 55 or a modular system, replacing the oscillators with a bank of actors or people, treating them through the Fairlight or the EMS synth, and establishing an orchestration only using voices. Apart from sampling I used a lot of different rhythm machines like the LinnDrum 1 and 2, also a lot of different analog synths like the ARP 2600, the Moog 55, the EMS Aks, the Oberheim OBX, the Prophet 5, and a bit of DX7 and DX9. But I'm not particularly a fan of the FM synths; the DX7 is a good instrument for stage but for creating sounds it's too complicated. I've been through a lot of different instruments but I think there's a trap with the musical instrument industry putting out a new instrument every two months. I have the feeling that instruments of the first generation are sometimes the best; on the second generation the company has a name but they don't use the best components. A lot of people tend to think that all electronic instruments can sound the same because they are electronic, but even from one Minimoog to another you can hear a difference. The last purchase I got last year was an old Moog 55 owned by Robert Moog, and that had a lot of modifications, a more balanced power supply and oscillators which move a lot less in terms of pitch. Synthesizers with discrete components have a special sound, a wider sound you can never have with the Japanese synthesizers. They're too precise – what gives a wide sound is the imperfections of the instrument, the fact that the oscillators are moving gives that wide sound, and a DX7 can't compete with that. On *Zoolook* I really made a big effort to have the same dynamic as when you are using acoustic instruments, because synthesizer sounds are not as rich in harmonics as acoustic sounds, so I tried to replace that using a lot of live feedback or regeneration. I'm using live ambience, even sending sounds to the bathrooms of the studio, for instance, for a very bright and short delay, a very acoustic ambient with a lot of high frequencies, then you have to delay the original sounds to get them back into sync with the reverberant sound."

1980 Teisco (Kawai) synths launched in Japan with the monophonic S60F

Although Jarre used a Moog modular system, early on he hadn't used the more affordable Minimoog:

"I never used the Minimoog all that much because I started on the EMS synth and I was a freak with that, and also the ARP – it was also a question of money, but I was so happy with the VCS3, you can have incredible sounds with that because of the matrix board, which I think should be used much more. When you are forced to edit sounds with just one knob that's a bad idea, because if you have ten fingers you should have more control. On the EMS matrix board you can send a lot of controls to the same parameter instead of just one leading to another, so if you want to make a new connection you have to destroy the existing one. And I still have Michel Geiss's matrix sequencer which he built for me years ago; it's a huge matrix and a very flexible instrument."

Having integrated sampling and later MIDI technology into his music, Jarre later became an enthusiastic supporter of the "analog revival" on albums, such as *Oxygene 7–13* (1997), by that time using his revived original analog instruments alongside their newer digital imitations, while by the release of *Metamorphoses* (2000) he had learned how to reproduce these effects and many more almost exclusively using a Pro Tools computer system with many plug-in software synthesizers. At the time of writing Jarre is celebrating his fiftieth year in music, more about that in Chapter 8.

A spate of releases influenced equally by the early style of Jarre and by the developing "New Age" movement for relaxing, meditational music also began in the USA around the late 1970s. Musicians, such as Steve Roach, Kevin Braheny, Robert Rich, Michael Stearns, and Richard Burmer, who prematurely passed away in September 2006, helped these originally European styles to survive, albeit with a generally less experimental attitude, through a period in which the European players such as Jarre himself had generally moved into new musical territory involving digital and sampling instruments. The USA's Michael Garrison, for instance, released throughout this period a series of almost interchangeable albums, such as *In the Regions of Sunreturn* and *Prisms*, which had an extremely simple, though compelling, approach to analog synthesis, comprising short tracks layering together very simple analog bass sequences, melodic lead lines, extreme abstract sounds and strictly 4/4 time signature analog electronic percussion. When Garrison passed away in 2004, his studio set-up was sold intact and preserved by an enthusiast.

1980 Tangerine Dream East Germany shows, release of *Staatsgrenze West* and *Quichotte*

This rarely seen photo of Michael Garrison was shot by Ross Chandler of chandlerphoto.com in Bend, Oregon, scanned from a print and kindly supplied by Craig Padilla, and processed by Chris Jenkins. Seen clockwise; Yamaha YC45D organ, unidentified, early model ARP Axxe, EML Electrocomp modular system, string ensembles, ARP2600, Syntar by George Mattson, and Minimoog. The Syntar was a portable analog monosynth with a guitar-style neck based on Mattson's earlier modules. See the Discography section for a closeup picture.

While some of the players who had created fascinating early albums using analog synthesis quickly disappeared, there were other interesting musicians, such as Zanov, Bernard Szajner, who used a Laser Harp and produced the richly textured *Visions of Dune*, Philippe Guerre, and Besombes/Rizet, for instance, only seeming to have produced albums during this very short early period. A few of the musicians mentioned in this section, including Jean-Michel Jarre, did help to bridge the gap between the first period of analog synthesizer music in Europe from perhaps 1972 to 1982, and its revival almost ten years later.

TECHNO-POP PROGRAMMERS

Another genre that could not have existed without the invention of the synthesizer was techno-pop. Again, an early source was Germany; the group Kraftwerk initially comprised two classically trained musicians, Ralf Huetter and Florian Schneider, both influenced by Stockhausen and by avant-garde music. Their early albums (*Kraftwerk* and *Kraftwerk 2*) filtered and processed flute, guitars, electronic oscillators, voices, and other instruments, but the duo wanted more efficient control over sound generation and invested early on in EMS and Minimoog synthesizers, in a custom vocoder first heard on the album *Ralf and Florian*, and in having electronic triggers built to control their percussion sounds.

Kraftwerk's early influences were strictly analog, as seen here in 1973.

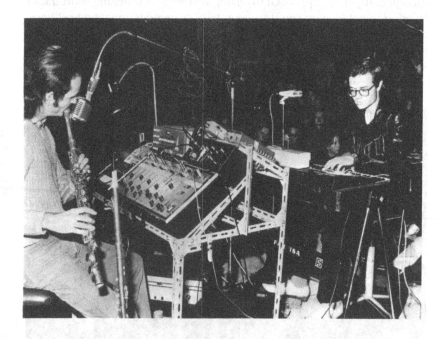

The band's commercial breakthrough album was their fourth, *Autobahn* (1974), which layered flute, guitar, and voices over repeated synthesized bass sequences and electronic percussion sounds. Lead lines on Minimoog, ARP Odyssey, and later a Micromoog, simulated passing vehicles, car horns, and air brakes, and there was a general avoidance of the more abstract, bubbling sounds created on EMS synths by the other German and French synthesizer artists. Kraftwerk did make extensive use of custom-built vocoders and models from EMS and other manufacturers, and also had custom controllers built, including a powerful sequencer by Matten & Weichers of Bonn, developing their sequencing and electronic percussion powers on the albums *Radio Activity*, *Trans Europe Express*, *Man Machine* and *Computer World*. Kraftwerk's creation on these albums of analog-synthesized percussion sounds – a bleep, chirp, or "thwip" instead of a conventional snare drum or cymbal, largely the responsibility of Karl Bartos, who continues the band's 1970s and 1980s tradition as a solo artist – has become a universal staple of pop music. But there was an extremely long delay before the band's next album, *Electric Café*, by which time they were concentrating as much on sampling technology as on analog synthesis and percussion.

By the 1980s, Kraftwerk had become masters of digital as well as analog sound creation.

Bartos and fellow percussionist, Wolfgang Fluer, left during this period, and much of Bartos's subsequent output from the album *Esperanto* (released under the name Elektric Music) gave the impression that his contribution to the Kraftwerk sound had been much more significant than previously imagined. For instance, Bartos had made up the set of sampled phonemes (parts of speech) that allowed Kraftwerk to play almost any vocal part simply from a preset bank in a sampler:

> "That's one technique for vocals. I use samples of individual phonemes assigned in groups on the Akai samplers. Then I can play any word from the keyboard, and pitch it using a vocoder. Another way is to use a speech synthesizer designed for blind people. You place it on a book and it will read the text."

> (Karl Bartos, interview with the author, 2004)

Bartos by that time was using an Access Virus keyboard for vocoder parts and stored many other analog sounds, particular Minimoog textures, in his Akai samplers:

> "I use Logic, but only as a recorder, not for running soft synths. All my sounds are in Akai S3000 sample format, and sometimes I sample another instrument like the Minimoog. With the Akai samplers I have learned the system, so now I don't have to think about it – when you are working long hours in the studio, that's important. And they are very cheap now, so I would prefer to have a whole bank of Akai samplers than to try to make my computer play samples and soft synths. You would really need four or five computers to achieve the same capacity and to avoid any playback delays, so I just don't think computers have reached that stage yet."

On a later album, *Communication*, Bartos continued a very strong concentration on the three-minute song structure, without lengthy instrumentals:

> "There are no really long instrumental parts. There was too much in the 1990s of that idea of a sequencer fading in after six minutes, playing for a long time and fading out again. You know a lot of the world today is about commercials, and people now expect all their important information to come in short sections."

There was also a clear return influence in some songs from the British techno-pop band New Order:

> "After releasing *Esperanto* I was in Manchester and met people like Bernard Sumner [of New Order] and Johnny Marr [of The Smiths]; Johnny had some nice guitars and I found myself playing the acoustic guitar and singing more. Bands like The Beatles and others on Radio Luxembourg were my first influences, but I had always been in Kraftwerk, so there was no chance to use that kind of influence. It was there at some level in my brain, I wanted to get it out, and someone offered to pay to make an album [released under the name Electric Music – note the 'k' disappearing from the original project name]. If I had done it earlier in my career, I wouldn't have had to do it at that time, but I got that sound out of my system. Then we had some shows to play and I wrote one new song, *15 Minutes of Fame*, which has this kind of New Order influence, which took me back towards the [techno-pop] style of the *Communication* album."

1980 Moog releases the Liberation strap-on guitar-style synthesizer

When the original Kraftwerk duo re-emerged around 2000 and eventually released the laptop-based album, *Tour de France Soundtracks*, the contrast with the strong, analog-based sound maintained only by Bartos could not have been clearer. But the original success of Kraftwerk from 1974 had set off a huge burst of techno-pop activity, mainly, it has to be acknowledged, in the UK. Gary Numan, John Foxx, Orchestral Manoeuvres in the Dark, Devo in the USA, and slightly less obviously Ultravox, Visage, New Order, Duran Duran, and Spandau Ballet, all emerged from this stimulus. Numan found a Minimoog in the studio while recording a guitar-oriented punk album and used it for simple lead and bass sounds, the surprise success of his singles, *Are Friends Electric?* and *Cars*, leading to a series of albums largely using Minimoog, ARP Odyssey, and the basic string setting of the Polymoog, superseded later on in his career by digital and sampled sounds from a PPG Wave system. John Foxx, on the other hand, only recorded one significant analog synth album before moving on to other interests; *Metamatic* made great use of a Roland CR78 drum machine (some analog drum sounds from the similar but slightly later Roland TR808 are on tracks 32 and 33 of the website), as well as Roland, Moog, and ARP synths. Many years later, Foxx created abstract ambient albums and also had a low-key return to techno-pop.

Orchestral Manouevres in the Dark created some quite experimental work early on, but became best known for the simple melodic synthe-sizer lead lines between verses of their songs, at first played on a Korg Micro Preset (check tracks 40–42 on the website), while Basildon-based Depeche Mode started their career using various affordable monophonic synths – Moog Prodigy, Korg, and Teisco included – and after landing a good recording deal were able to multi-track many parts, even using an ARP 2600 under the supervision of Mute Records head, Daniel Miller, to create most of their percussion sounds using analog synthesis. While Depeche Mode quickly moved on to digital sound sampling and more use of guitar, ex-member Vince Clarke, with Yazoo and Erasure, stuck largely to analog sound synthesis, and built up one of the world's most impressive collections of analog and modular synthesizers, including Serge, Buchla,

Yazoo, featuring Vince Clarke's Roland Jupiter 4 and TR808 and Sequential's Pro One.

and E-Mu models, often sequenced from a MIDI-equipped BBC Micro Computer, all set in a purpose-built circular studio. Sadly, following structural problems in the studio building the collection was broken up, and Clarke has more recently moved towards using a laptop system based on Apple Logic software.

After early avant-garde experimental recordings, The Human League developed a hybrid style on their debut album for Virgin Records, *Reproduction* (1979), which featured analog sequencing on a Roland 100 system but processed many of the synth sounds through distortion units for guitar-like effects. The band's lack of musical training led to their spontaneously creating rather unusual song structures, intermingling

Martyn Ware of The Human League/Heaven 17 and his original Roland 101/102 analog synth.

Electro-pop pioneer Gary Numan in 2012, when punk collided with the analog synthesizer, he never looked back.

synthesized percussion and bass sequencer sounds at will, processing voices in unusual ways, and generally stretching the possibilities of small monophonic analog synthesizers to the maximum. The follow-up album, *Travelogue*, in 1980, took a slightly more pop-oriented approach, but still includes some brilliant songs and startling sonic innovations. When the band split in two, vocalist Phil Oakey continued The Human League as a slightly more conventional but still electronically-based group that had worldwide success and continues in a slightly more low-profile manner to this day, while Martyn Ware and Ian Craig Marsh formed Heaven 17, added soul singers and conventional instrumentation to the mix, had some notable chart success and again survive to this day, though Martyn Ware spends much more time using an Apple Logic system sometimes in collaboration with Vince Clarke, creating soundscapes for major installation projects worldwide. Martyn Ware outlined the early days of The Human League and the reasons for the band's division:

1980 EDP launches with the affordable Wasp synthesizer in the UK

"We didn't really have a conventional musicianly approach to the band, none of us was trained and when we talked to other musicians at the time a lot of them were envious of us in that respect, because we could approach music without any preconceptions. We were approaching as much from a timbre point of view as a tune, though the irony is that we were all obsessed with pop music, so we loved the melodic aspects, and that's what appealed on a general level to a lot of people. But the thing that made it sound so weird was that we would have been equally happy doing completely abstract stuff, though we were determined to make electronic music a popular medium.

I did most of the keyboard playing but I had no idea what I was doing in terms of chords or keys, I just did whatever sounded great. But we wanted to use some conventional musicians combined with electronic music, and the original template of The Human League became a bit restricting. Phil Oakey wanted to cling onto that and with *Dare* he made a fantastic album, but if we wanted the sound of a great bass player, which you still couldn't reproduce with a synthesizer, we couldn't have it. So with Heaven 17 we had that great bass part on *Fascist Groove Thang*, for instance, and worked with drummers like Simon Phillips, putting electronics against that, and that's what we found exciting. Going into Heaven 17 I didn't exactly study more conventional musical structures, but we did start working with a lot of great musicians. It worked, and I still think it was the right thing to do at the time, but the downside of that is that it does force you into slightly more conventional musical structures and you do start learning certain techniques, however free you want to keep your mind, and inevitably you fall into certain routines. The normal dynamic of a creative group in those days would usually be that you'd peak after a couple or three albums and then you'd become a bit repetitive, and that's what happened with us really."

(Martyn Ware, interview with the author, 2002)

The Human League's use of analog synths, such as the Roland Pro Mars and the System 100, for which Ian Craig Marsh built a huge metal framework to rack up modules and sequencers, was innovative in its use of effects such as guitar distortion, but by the early 2000s Martyn was getting the last use out of his analog, MIDI and sampling synthesizers:

> "I think a lot of my old keyboards will be going in the bin now except for the Roland System 100, which is there for sentimental reasons. I stopped using the Roland S750 sampler, though I still have an XV5080 which reads its disks, but I mostly use the EXS24 sampler within Logic; I still use the Roland JP8000 and the Quasimidi Sirius, but now I have all the soft synths that come with Logic Pro, which works fine on my Apple iMac laptop. It's what I have been aiming towards for a long time. I have my little M-Audio Oxygen keyboard and powered speakers in a backpack, and Vince Clarke has the same set-up so we're completely compatible. It's ironic, because he has moved from the hugest possible studio full of gear to the smallest you could imagine. I'm also running the Arturia Moog Modular V software, as well as Absynth 2, and Spectrasonics Stylus and Atmosphere. If I need some older synth plug-ins I can also go back to Mac OS9."

1980 Jean-Michel Jarre performs in China, later releasing *The China Concerts*

Martyn Ware is a good example of a musician who embraced analog synthesis, adopted new technologies as they became available, but welcomed the return of analog imitations when they became available in software. By 2006, Duran Duran's Nick Rhodes had travelled the same course but ended up considerably more committed to the larger analog keyboards. Starting life around the same time as The Human League in 1979, Duran Duran took a slightly more conventional approach to pop music, which fitted into the "New Romantic" style alongside Spandau Ballet, but which still relied heavily on the synthesizer parts played by the self-taught Nick Rhodes:

> "I started very simply – my first synthesizer was a Wasp, which had a touch keyboard and only played one note at a time. That was lucky, because if it played two notes I wouldn't have known what to do at that stage! I also had a Stylophone 350S – the large model with a double stylus – a little Kay rhythm unit, and a reel-to-reel tape machine on which I'd record anything from traffic noises to sirens to church bells and play them back in the songs; this was in the original line-up even before Simon [LeBon, vocalist] was involved. By the time we got to the 1980 lineup I bought the Crumar string synth, which was what I played the pads and melodies on, and before the first album I also got a Jupiter 4 and then a Prophet 5, maybe an SH2, and sequencers like the Roland CSQ100. So I really had enough in the battalion to make varied sounds, but we'd still just be getting in a room and jamming together until we found what worked. And often that would come from a sequencer – there's something about the way we play that seems to make that work, to have something giving us the tempo and perhaps a key, and then we'll develop things over that. But I think I was one of the prototypes for the synthesizer generation – I wasn't classically trained, so there are no boundaries I can't cross and no notes that won't go together. Since then I've learned a lot more, but I don't let that stop me doing what I want. What makes the music special is energy, style and guts. I don't think it's technical playing capability. If I needed something played like Chopin I could get someone else to do it, but the sound, the melodic structure, the unique arrangement and a touch of surrealism are more important."
>
> (Nick Rhodes, interview with the author, October 1990)

During the 1990s the band was playing with a more experimental line-up, featuring guitarists Warren Cuccurullo and drummer Sterling Campbell:

"Warren and I like to have ambient technology wars, looping pieces of sounds and putting them into delays, letting them build up, like two wizards hurling spells at each other. On one track from *Liberty* [the band's seventh album from 1990], Sterling copied a pattern from a tiny rhythm box playing live, my Roland D50 synth was playing some chopping, dissonant chords and the whole song came about just from those sounds."

By that time Rhodes was also playing an Ensoniq VFX and a Roland D50, but explained:

"I've also sampled some sounds from my old Roland Jupiter 8 into an Akai S1000 sampler. Of all the keyboards I've ever used, the Jupiter 8 is the one I have total command over. And my old friend the Crumar Performer, which was on all the older albums, also got a look in; it took a while for my roadie even to find it, but it's still working well."

1980 Classic line-up of Duran Duran forms with Nick Rhodes on synthesizers

While Rhodes got further into MIDI and computer technology during the late 1980s and 1990s, working extensively with the Fairlight CMI then a Pro Tools sequencing set-up, by 2000, when the band reformed with its original line-up he had settled on a good mixture of digital and classic analog synths both for stage and studio work, insisting that the classic analog models remained his preference for controllability and ease of use. On stage an Apple G4 Powerbook stored some sound effects, but the major keyboards were: the Alesis A6 Andromeda, a large pure analog synth; a Jupiter 8 retrofitted with MIDI; a Roland V-Synth, which features digital, analog-style and time-stretched sounds; and a Kurzweil K2000 for samples.

"A lot of the old stuff I use is so old and fragile now it's OK for a one-off somewhere in London, but sticking it on ships or in the back of a truck is no longer feasible. So the only bit of true vintage analog I take with me is a Roland Jupiter 8, and I take a back-up for that, plus I'm looking for another one in good condition. For a keyboard that old it has held up remarkably well; there's just something about it – whenever I sample it onto anything else it never sounds the same, it has a warmth when the oscillators are warmed up which is just not like anything else. I used it on every album we've ever made, from sequences to pads to sounds to melodic parts. I've taken the precaution of backing up the sounds as close as I can get them in the Roland V-Synth in case the thing goes down in the middle of a show, and I also have the Alesis Andromeda, on which I've got a lot of my old analog sounds replicated to a fair degree. Especially things like the old Crumar string synth, I've got as close to that as I've ever been. I really do love the Andromeda, of all the new analog synths that came out. I tried all the virtual analogs; the Access Virus I quite like for some of its sounds, also I used a little of the Clavia Nord Lead on the *Pop Trash* album, particularly on a song called *Hallucinating Elvis*, but it wasn't really me, whereas the Andromeda fits into my general sound bank. While we've been a very technological band, I like to keep a lot of the organic

1980 Moog launches the
Memorymoog programmable
polyphonic synthesizer

stuff in there too – I think the reason our sound works is the balance between the warm analog synths and the edge of the guitar, so it's still a very human sound even though it uses a lot of technology."

(Nick Rhodes, interview with the author, 2004)

Returning to the early period of Duran Duran in the late 1970s, in Japan, Yellow Magic Orchestra had also been feeling the influence of Kraftwerk. On their first album, released in 1979, Ryuichi Sakamoto, Yukihiro Takahashi, and Haruomi Hosono created techno-pop influenced by Japanese styles and musical scales on tracks such as *Firecracker* and *Yellow Magic*. YMO owned and used almost every synthesizer in production, from Moog and E-Mu modular systems to Roland Micro Composers, Polymoogs, and Prophet 5s, and engineer Hideki Matsutake went on to become an expert sequencer programmer, under the name Logic System, releasing several excellent, largely instrumental, techno-pop albums of his own, such as *Venus*.

This techno-pop style has persisted in one form or another since its inception and has generated many successful singles and albums for bands from all over the world, such as Men Without Hats and Trans X (Canada), Telex (Belgium), Yazoo, Heaven 17, and Cabaret Voltaire (UK), and Yello (Switzerland), as well as more "underground" success early on for artists such as Fad Gadget, Thomas Leer, and Robert Rental and The Normal. Later, the techno-pop style diversified into techno-dance, created by artists such as Grandmaster Flash and Mantronix (one early example, *Planet Rock* by Afrika Bambaata, was simply lifted from Kraftwerk's *Trans Europe Express*), into industrial dance from bands such as Startled Insects and Attrition, and into Electronic Body Music (EBM) coming from bands such as Nine Inch Nails, Test Department, and Front Line Assembly, a style which continues to extensively use analog sound, but which is also very much involved with sound sampling and drum looping techniques.

PROGRAMMING FOR ORCHESTRAL IMITATION

The programming techniques used for the classical, avant-garde, rock, pure synthesizer, and techno-pop styles of analog synthesis differ greatly from one area to another and call for differing levels of understanding and playing ability.

Clearly, the creation of authentic-sounding classical instrument sounds calls for a good understanding of the acoustic principles that apply to the creation of the sounds by the original instruments, matched with an understanding of the appropriate performance techniques and equally with an understanding of the principles of orchestral arrangement. It would be difficult to create a good impression of an acoustic instrument without having a clear understanding of how its distinctive sound comes about acoustically, and pointless to take a good emulation of an acoustic sound but play it outside its normal range, or with a playing technique atypical of the actual instrument. Equally, it would be pointless to take a well-programmed, well-performed instrumental imitation but obscure its fine detail in your arrangement behind other instruments that shared its frequency range or tone.

1980 Sequential Circuits launches the dual keyboard Prophet 10 synthesizer

Take, for example, the flute: obviously the sound is monophonic and so was a good prospect early on for successful analog synthesis. The basic tone is very smooth and lacking in high harmonics, so the oscillator sound of choice would most likely be a sine wave. The envelope is easy enough to set up; since a flute begins to sound almost immediately when the player blows into or across its mouthpiece, the envelope would have almost no attack (fade-in) time, a sustain level which remains at full as long as the player is blowing (or the synthesizer player is holding down a key), and a release (fade-out) time which is almost zero (that is, the flute stops sounding as soon as the player stops blowing, or the synthesizer player stops holding down a key – listen to track 8 on the website for an illustration of different envelope shapes).

1981 ARP closes, leaving CBS to develop the Chroma synthesizer

This would create a very simple and unsophisticated impression of a flute sound, but even from the earliest examples of this approach to synthesis as heard on the *Switched On Bach* album, programmers, such as Walter Carlos, would then have examined other aspects of the sound of the genuine instrument in more detail, realising, for instance, that the start of a flute note is accompanied by a slight peak in volume as the column of air in the flute first begins to oscillate. This would call for an adjustment to the synthesizer's volume envelope; the sustain level would be decreased a little and a fast decay chosen to bring the volume down to this slightly reduced sustain level.

This would give us a more realistic flute effect with a slight volume peak at the onset of each note, but when recorded through a microphone, a flute sound is also often accompanied by some element of the player's breath as he or she blows hard enough to bring the column of air in the flute into oscillation. Breath is well simulated using a little white noise, but this would have to occur only at the start of each note; to achieve this, a white noise source passed through a separate voltage-controlled amplifier controlled by a second envelope shaper triggered from the same keyboard, and mixed with the sine wave oscillator sound, could be added. The second envelope would be set to a very fast attack and decay with no sustain level, since the white noise is not intended to continue throughout the course of the flute note, only at its start.

We now have a flute with a distinctive "chiff" of breath and a volume peak at the start, followed by a fairly plain sustain period. While some instruments, such as the violin, benefit from having the option of vibrato (a slow regular change in pitch), the expression that can be added to a flute sound is usually better represented with tremolo (a slow regular variation in volume) or with filter modulation (a slow regular variation in tone). But sine waves do not vary much on being passed through a low pass filter because they already mostly lack the high harmonics that low pass filters are designed to cut out, so to add some expression to flute sound, a combination of tremolo perhaps with some filter modulation would be called for. This would involve the patching of a slow-running oscillator (LFO) to the voltage-controlled amplifier – remember that the early Moog systems, for instance, had none of these options ready patched, so every decision of this kind involved reaching for another patch cable or two – and then perhaps dividing the LFO's output to simultaneously modulate a low pass filter as well. The player could then manually fade up the filter modulation and tremolo level after the start of each flute note, or have that depth automatically controlled by the level of yet another voltage-controlled envelope. Walter Carlos also had various foot-pedal and other control interfaces built in order to help create this kind of expressive performance, so it can be seen that the creation of even a relatively simple sound, such as a flute, could involve patching together many individual modules and controllers. Check out tracks 31 and 41 on the website for two impressions of an analog flute sound.

1981 Roland launches the Jupiter 8 polyphonic programmable synthesizer

Other instruments, such as the oboe (probably starting off with a square wave rather than a sine wave, and using vibrato more than tremolo), can be emulated in a similar way (listen to tracks 40–49 on the website for analog impressions of oboe, bassoon, trumpet, recorder, horn, and other wind instruments). Thinking about the nature of the original instrument is also very important in terms of choosing the performance controllers to be added to analog programs.

Let's take a detailed look at a typical project in analog synthesis. Some years ago, the author set about recording a folk music piece arranged for pipes, whistle, fiddle, bass guitar, accordion, bodhran (goatskin drum), and bones, using only a Sequential Prophet 600, perhaps not the most obvious instrument for such a project. For a start, the percussive sounds of the bodhran and bones would probably call for a white noise element, and the Prophet 600 has none. But pipe, whistle, and similar sounds are invariably found on preset analog synths, so seemed a good starting point.

The melody part was intended to replicate a Uillean (Union or Elbow) pipe, the Irish version of the bagpipe using an airbag pumped with the elbow and having a much more cutting, nasal sound than the Scottish bagpipe, well-known players including Davey Spillane (on the soundtrack to *Rob Roy*) and Paddy Moloney with The Chieftains. The original Uillean pipe sound is distinctively cutting yet nasal; the Prophet's square wave gives the best impression of this, set to a thin pulse width to increase the nasal quality. The filter resonance was boosted to give a cutting edge; the filter cut-off position was very carefully tuned to enhance the nasal quality without losing the cutting edge of the sound. In the case of this instrument the volume envelope is easy to set; the pipe is monophonic, has no release time and has an extremely rapid onset once the wind pressure in its air bag reaches a certain level. Because the player is not blowing into the bag or pipe there is no mouth vibrato or tremolo possible as such, but one method of creating expression is to rapidly play a note one tone or semitone higher than the one sustained at the time, by bringing a spare finger down over the next adjacent fingerhole in a trill. This is difficult to do rapidly and evenly by actually playing the Prophet's keyboard, so an alternative is to make the effect available from the modulation wheel.

While sine or triangle shapes would normally be selected to create a vibrato, the solution here to creating a Uillean pipe trill is to choose a square wave, setting it so that full-depth modulation raises the pitch exactly a whole tone or semitone. You need to bring in the vibrato smartly, because half-depth vibrato would not sound a whole- or halfnote trill but something in between notes. There is also a more difficult playing technique for bending notes on the Uillean pipes, but careful use of the synthesizer pitch bender can emulate this effect quite well too. The resulting pipe sound was saved for future projects and can still be heard on the author's CD album, *Mexico Rising*.

1981 Yamaha launches the large CS70M programmable polyphonic synthesizer

Instruments such as the violin, which have a more complex tone than the flute, oboe, or pipes, and a wider variety of expressive possibilities too, can accordingly be much more demanding to synthesize. Firstly, the one or two oscillators of the typical small monophonic analog synthesizer can be insufficient to capture even the basic tone of a violin. On an instrument, such as the original Minimoog, with three oscillators available (providing the player is willing to lose the possibility of vibrato), a more complex tone can be achieved by selecting different wave shapes and mixing the oscillators carefully to differing levels. For instance, two oscillators set to a triangle wave can emulate the main body of the sound, while a third set to a square wave an octave higher can be mixed in to give a more edgy overtone. Some portamento (glide) can, for instance on the Minimoog, be quickly switched in and out using the control switch to the left of the keyboard, or using a footswitch; a patch with portamento on every note will be much less convincing than having the ability to bring the effect in and out at will.

If the imitated violin is intended to be playing smoothly (legato), some slow attack can be added to the envelope, but this is not so easy to quickly switch in and out in order to alternate with more rapid (staccato) notes. So many other techniques are also available to the solo violin player, including bouncing and scraping the bow, that solo violin can become an extremely difficult task for any small analog synthesizer to emulate; even Isao Tomita generally does not try to emulate solo violin, despite the brilliance of his emulation of ensemble violins. Perhaps the most successful ever analog emulation of the solo violin is Steve Jolliffe's performance using the Lyricon breath-controlled synthesizer on Tangerine Dream's *Cyclone* album from 1978; unfortunately, the expressive possibilities afforded by the Lyricon wind controller are not so easily available to the keyboard player.

Brass players will also insist, quite rightly, that the full expressiveness of their instruments is difficult to emulate, though reproducing a very basic brass sound is among the easier of the analog tasks (The Human League give a wonderful example on their track *Being Boiled*). The sawtooth wave generally gives a good impression of brass instruments; a little filter resonance helps to add a cutting edge. Adjusting the volume and filter envelope attack times can vary the sound from a sudden, blasting effect to a swelling, slowly building brass crescendo. Chorus, if available, helps build the impression of a whole section of trumpets, horns or saxophones playing simultaneously rather than a solo instrument. On the Minimoog, setting two oscillators to the modified sawtooth wave, with the filter cut-off at "0" (this is the mid setting), about 50% resonance and a filter envelope attack at about 600ms gives a good synthesized trumpet; if played an octave or two lower, a deep tuba-like sound is produced. Brass players can produce a harsh vibrato effect using their lips. This can be emulated by setting the slow oscillator about twice as fast as normal (to perhaps 15 Hz) and routing it through the modulation wheel to filter modulation rather than to vibrato. Some semi-preset analog synthesizers refer to this effect as "growl"; check out tracks 44 and 46 on the website.

The human voice, as discussed in the sections on Walter Carlos and Isao Tomita, can be synthesized using fairly simple analog techniques, though the effect is sometimes a little comic. Synthesizers with a single filter, however, can only give a very basic impression of the human voice; those with two filters that can be placed in series, such as the Synton Syrinx, Crumar Spirit, Korg 800DV, and modular systems, can more easily simulate the combined effect on a vibrating column of air of the human throat, mouth and lips. A combination of low pass and band pass filters, each given some variation during the course of the note to emulate the creation of vowel sounds, can often be successful; check out tracks 42, 61, and 62 on the website.

If a modular system with plenty of oscillators, envelopes, and filters is available, the possibilities quickly become more extensive. More ambitious programmers may like to try using an oscilloscope to display the waveform of an acoustic instrument; an attempt can then be made to match this using as many analog modules as necessary. Again, a simple static imitation of the wave shape is not going to give a very convincing emulation; the player will have to think about how the wave shape may change over the course of a note, how the original instrument is played, staying within its typical frequency range and getting techniques, such as pitch bending, tremolo, vibrato, and staccato or legato, playing correctly for that particular instrument. Although, in the 1970s, Rick Wakeman generally performed with orchestras and kept his synthesizers for non-imitative playing, he more recently created some extremely successful orchestral imitations on keyboards, and explained:

"The important thing is to make sure you write within the range of the instrument; a lot of people don't, and it sounds completely unconvincing. And you mustn't fall into the other trap of playing on a keyboard a part which is technically impossible, say, on a clarinet. Although the general public wouldn't immediately realize that was happening, they would be hearing something they'd never heard before, so it would sound unconvincing."

(Rick Wakeman, interview with the author, 1999)

Orchestral arrangement is a subject beyond the scope of this book, but a few basic principles can help enormously in a successful job of imitative orchestral synthesis. All of the instruments of the orchestra – the various strings, brasses, winds, percussion, and so on – have their typical frequency ranges, and the secret of successful orchestral arrangement is to keep each frequency range relatively uncluttered. In programming each sound, careful filtering can help; high-pitched instruments, such a flutes, piccolos, and oboes, should have high pass filters applied to ensure that their sounds do not have much unwanted low-frequency content. Bass instruments (orchestral basses, tubas, and so on) should not be too "bright" (that is, should not have too much high-frequency content) or they will interfere with the perception of the flutes, piccolos, and oboes, whenever they are playing, so low pass filters should be used on these instruments. The mid-range instruments (human voices, violins, horns, and so on) should be filtered in both ways to keep their effect centred around the mid frequencies.

If polyphonic synthesizers are being used it's also important not to play any given instrumental sound over a very wide keyboard range at any given time, in other words, avoid the temptation to play a huge chord with both hands a couple of octaves apart using, for instance, a single violin sound. The correct procedure, for clarity and authenticity, would be to play the violin sound in a limited register, perhaps no more than an octave wide, and then to separately play another, lower part more specifically programmed for violas, then if necessary a third part programmed for orchestral basses. In the orchestra, strings are usually divided into first and second violin sections, these can be distinguished in your arrangement by giving these slightly different tones and equally by giving them differing stereo positioning. Rick Wakeman again:

1982 Datanomics gives up control of EMS and new product development halts

> "When I do violins in that way I use a sample off the Kurzweil K2500R edited a little bit, and lots from the Technics WSA, which isn't such a well-known instrument. But I combine those, and using various instruments I add analog strings underneath to give it a little phasing element, which lets you get pretty close to a full string section. Then on the second violins I change the analog sound a little, and re-edit fractionally or double up one of the samples. Clarinets are interesting too, because I double them with a Minimoog just on a plain square wave with all the filters open, and again that gives you the slight phasing which you can hear from a clarinet section. These days I get my oboes from samples – the little Korg X5DR does a nice oboe, and the old Korg DW6000 or DW8000 synths have a sound which, put up an octave, does a perfect oboe, or at least does the same job as the Minimoog would."

It's useful also to study albums by Isao Tomita for good examples in programming string sections and applying reverb and phasing to them without making the overall sound too "muddy".

Be aware also that the orchestra will normally only have a small complement of some instruments: perhaps two oboes, three or four French horns, and one or two bassoons. Certainly, do not go over the top with percussive sounds, such as timpanis, or vary them over a range much wider than a real timpani set can span, which will inevitably result in a less than convincing overall sound. Rick Wakeman again:

> "For timpanis now, I use samples from the Kurzweil K2500 or Korg 01/W – but you must be careful not to put too many timpanis together because it starts to sound very full, and there's generally only one timpani in an orchestra."

1982 Synton in the Netherlands launches the Syrinx monophonic synthesizer

PROGRAMMING ROCK, POP AND ELECTRIC SOUNDS

The basic tones of electric instruments such as lead, rhythm, and bass electric guitars can be very easy to reproduce on any small analog synth, but almost as important as the tone is the use of pitch bend and other techniques to convincingly simulate a guitarist's string bending and vibrato. Passing almost any synthesizer waveform through some distortion can give a passable lead electric guitar sound but, again, the important imitative factor here is to practice bending notes as a guitarist would, Jan Hammer (who actually played his synthesizers through a guitar combo amplifier) being the best keyboard artist to study for these purposes. A guitarist can create vibrato by moving the string from side to side with the finger, this is not always well imitated by the fixed, regular vibrato created by a synthesizer's slow oscillator, which usually sounds instantly artificial. Keep in mind that a guitarist generally has to bend *up* to a desired note (it's more difficult to start on a desired note and bend *down* on the guitar, unlike the synthesizer keyboard), so the following technique has to be learned on the keyboard: bend the pitch down while playing nothing, then play the desired final note, then bend up to reach it. Doing it any other way will instantly give away the artificiality of your carefully programmed lead guitar sound, even if only on a subconscious level. The first two albums by The Human League, *Reproduction* and *Travelogue*, on Virgin Records, are particularly rich in examples of adding distortion to synthesizers to give an electric guitar-like effect; their Roland monophonic instruments and small string and brass synthesizers are regularly overdriven, either using a "fuzz box" or at the mixing desk, not so much to produce lead guitar solos as Jan Hammer did, but to create huge crashing backing chords that give the music a great deal of added energy.

Some other electric instruments such as the Hohner Clavinet can be reproduced in an unambitious way on any analog synthesizer because their basic tones (though produced acoustically by plucking a metal tine) are somewhat simpler than those of traditional acoustic instruments such as the violin. But it's important to look into the performance aspects of these instruments too; the clavinet was adopted as a substitute for the rhythm guitar in certain styles of music so a choppy, rhythmic style of playing is the best one to adopt for a convincing emulation (listen to early Stevie Wonder and George Duke albums for inspiration). The clavinet was not particularly sensitive to velocity, so a convincing imitation of it should not be too velocity sensitive either; it was, though, frequently played through a phaser, and a phased clavinet imitation may immediately start to sound a little more convincing than a "straight" one.

Electro-acoustic pianos such as the Fender Rhodes and Wurlitzer were also popular targets for imitation by early polyphonic analog synthesizers, although it has to be admitted that none of them, up to and including the Polymoog, was particularly successful. The synthesized electric piano sound really took off with the introduction of the digital Yamaha DX7, which had a Fender Rhodes/Hohner Pianet imitation as one of its basic

1982 Introduction of MIDI on the Sequential Prophet 600 and the Yamaha DX7

presets, and it must be said that this was more convincing than anything analog synthesis had offered up until that time.

The creation of drum sounds using analog techniques is perhaps a subject for a whole book in itself. Suffice it to say that once it was realised that a little white noise shaped by an envelope could create a basic snare drum or cymbal sound, and that a slightly bending oscillator sound could imitate a tom-tom, composers in every field from the avant-garde to electro-pop wanted to investigate the idea of synthesized percussion. Unfortunately, the synthesizer very early on had become keyboard oriented, and with the exception of the fact that Moog introduced a drum controller, prospective percussionists were very much left out of the scene. By 1974, Kraftwerk had developed a set of trigger pads, which at first were used to play analog percussion sounds, and which later actually played melody parts as well, controlling small synthesizers like the Minimoog and Micromoog. In the UK, Depeche Mode, under the supervision of producer Daniel Miller, started to create whole kits of analog sounds, generated by synthesizers such as the ARP 2600, often under the control of a sequencer such as the Roland MC8 Micro Composer. Eventually, some companies started to create stand-alone analog electronic percussion kits; the Pearl SY1 SynCussion, comprising two independent circuits in a single casing, or the Star Instruments Synare and the Pollard SynDrum come to mind. Sounds from the very flexible JHS SD1 Pro Rhythm Drum Synth are on the website.

Analog percussion really took off with the launch of the SDS5 from the British company Simmons (ironically, Dave Simmons was a keyboard player), and these modules were able to merge in any combination white noise plus a pitched oscillator with a variable amount of bend to create electronic tom, bass drum, snare drum, high-hat, and simple cymbal sounds (a really convincingly metallic crash cymbal sound was beyond the capabilities of simple analog circuits, and was offered in the form of a digitally sampled sound later on). The Simmons sound was fashionable for a while, but "real" drummers probably preferred the sound and feel

1982 Yamaha launches the tiny CS01 non-MIDI monophonic synthesizer

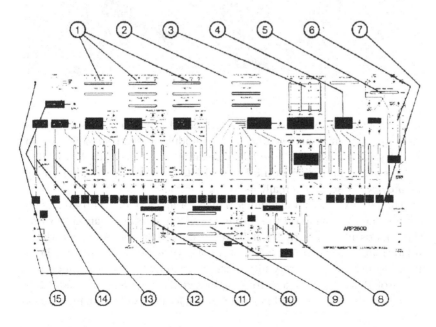

Mastering each synthesizer parameter – here diagrammed for an ARP 2600 – is vital.

of an acoustic kit, while non-drummers could synthesize similar sounds
easily enough on a small synthesizer or modular system, so analog drum
kits (despite the release of models from amongst others Tama, Pearl, MPC,
and Ultimate Percussion in the UK, and kits from the Amdek division of
Roland) never became a really essential purchase.

The history of drum machines, which for an early period used analog
sounds before moving on to sampling technology, is again another story.
The sounds in very early "beat boxes" from Hammond, Ace Tone (later
Roland), and other manufacturers, were of course analog, comprising little
more than short bursts of white noise or a pitched oscillator, sometimes
put through a ring modulator for a slight metallic edge. Roland's CR68 and
CR78 drum machines, used by Phil Collins, John Foxx, and many others,

Roland's TR808 became the classic analog drum machine, now re-born in the *Boutique* series.

A powerful early studio for experimental analog sounds – Moog modular, Roland TR808 drum machine, CSQ sequencer, RS09 organ/strings and guitar synthesizer, Korg Vocoder and Octave Catstick controller, 8-channel multitrack recorder.

used just this technology. When the Roland TR808 appeared it was seen as just a slight improvement on the CR78, with a less preset-based approach to pattern programming, a better selection of audio outputs for external processing of the sound, and improved interfacing and triggering abilities to lock its performance into that of its contemporary sequencer, keyboard, and other products. The TR808 certainly did not sound particularly realistic; its bass drum could be strong and its snare extremely "tight", but again its cymbals were not very convincing, and although it was taken up by many techno-pop bands, such as Yellow Magic Orchestra and Cabaret Voltaire, it did not put many human drummers out of work. The next generation of drum machines, including the Roland TR909, started to introduce sampled sound technology and this quickly led to fully digital machines, such as the LinnDrum and Sequential DrumTrax. As far as the drum machine designers were concerned, they had at last achieved realism. Most of the early analog sound drum machines started to appear on the secondhand market at extremely low prices. No one suspected that these neglected analog instruments would soon become one of the driving forces behind a huge revival of interest in analog synthesis.

1982 Elka launches the Synthex programmable polyphonic in Italy, later fitting it with MIDI

PROGRAMMING ABSTRACT SOUNDS

Giving guidelines about the programming of abstract sounds for avant-garde and pure synthesizer music may be a little redundant, since even a small analog modular system or a couple of interfaceable keyboard synthesizers plus some effects will offer almost endless possibilities, and the user's imagination and originality are what will set him or her apart from other synthesizer users. However, there are a couple of standard techniques that are useful to master in creating some of the entertaining effects used in the past by well-known synthesizer artists. For example, the distinctive whirling, abstract sounds created by Jean-Michel Jarre, and by Tim Blake in his solo work and with Gong, generally came from a particular set-up on the EMS VCS3 or Synthi A. Although simple, this can be surprisingly difficult to match on other instruments. For instance, the

Moog's early Prodigy – too basic for generating complex abstract sounds.

Minimoog, Moog Prodigy, Roland SH101, and their modern reproductions, along with many other small synths, though highly variable, are often found incapable of generating this kind of effect unaided.

The secret here lies in modulating not the oscillators but the resonating filter, and in having two slow oscillators available that are variable both in terms of speed and depth. When a low pass analog filter is set to the highest possible resonance setting and begins to "whistle" (though not all designs do so) it is producing more or less a pure sine wave. This tends to sound much more pure and cutting than the sine wave from an audio oscillator; it is certainly generally louder, and care needs to be taken when setting a filter to resonate that the volume does not suddenly become damaging to the speakers or amplifier system. Once the filter is oscillating, any audio oscillators still operating can be turned down or switched off. A slow oscillator is then patched to modulate the filter; on a Minimoog this is just a matter of pushing the Filter Mod switch. The modulation wheel or other depth control then controls the depth with which the oscillating filter is swept in pitch.

Unfortunately, a single slow oscillator sweeping the filter's pitch repeatedly up and down does not sound very interesting, even when the LFO shape is changed – say, from sine or triangle to sawtooth or square – or when its speed is altered. Simply modulated sounds of this kind tend to be the mark of the rather unimaginative synthesist. This type of sound only starts to become interesting when a second LFO is also available and is also applied to the filter, ideally with a different shape and at a different speed from the first. At this point some very dramatic, "cosmic" sounds are created and the degree of variability of these will depend on how easily the performer can alter the speed, shape, and modulation depth of the two LFOs. Unfortunately, on the small synthesizers previously mentioned, this set-up is very difficult to achieve because there is simply no second slow oscillator available (on the Minimoog either oscillator 1 or 2 can be switched to a slow setting but neither is immediately routable to be used as an additional modulation source).

The portable Moog Sonic 6, however, does have two LFOs, which as well as having independent speed controls can also have their speed controlled automatically by the level of the envelope, so that they slow down during the course of a note (check out track 56 on the website for an epic demonstration of these techniques), while some of the early Roland synths also have a second LFO, or in the case of the SH3A a sample-and-hold section that can act more or less as a spare modulation LFO. Any other such synths would also be able to create the type of sounds in which we are interested, but not always with a great degree of spontaneous control available. The difference on the EMS VCS3 or Synthi A is that several parameters can be controlled together using the instrument's joystick; both the speed and the modulation depth of one oscillator to the filter can be controlled by the upward and downward movement of the joystick, and both the speed and modulation depth of a second oscillator to the filter can be controlled by the left-to-right movement of the joystick, or these parameters could be inverted, or differently combined (see tracks 48 and 50 of the website, and "Ten great early analog sounds" at the end of this chapter).

1982 Roland launches the TB303 Bassline, which later becomes the basis of the "analog revival"

1983 Teisco synths rebranded as Kawai in Japan with the SX210 polyphonic model

For those with access to only a single LFO (such as purchasers of their MS10 synth), Korg manufactured a "modulation controller", the MS04, a foot-pedal with a built-in LFO, the output depth of which could easily be controlled, while some other companies offered modifications, for instance adding a second independent LFO to the Minimoog (standard now on the succeeding designs). If there is an external control voltage input to a synthesizer's filter, even if it is designed mainly for an optional foot-pedal to produce "wah-wah" effects, it can generally be used to insert an additional control voltage from the LFO of a second synthesizer; if this is not normally accessible from your second synth, it is almost always possible to "tap off" an LFO's control voltage output and add a new socket (usually on the rear panel) to make it accessible. Going for Eurorack or another modular system makes this all much easier to achieve.

A simpler technique that also creates an entertaining sound is to quickly modulate the resonating filter with the random voltage output from a sample-and-hold section (again not available on the original Minimoog, a stand-alone Moog 1125 Sample-and-hold unit was available but these are now very rare) or from an LFO with a random setting (such as on the Korg MS20). This effect is particularly easy to set up on the original ARP Odyssey and its modern Korg reissues and creates a high-pitched "tinkling" sound like a wind chime that Klaus Schulze used extensively on his early albums; on the Odyssey, the sample-and-hold section can also be set to glide from one level to another rather than stepping sharply from one level to another, which gives an even more abstract sound, and the Odyssey's filter can also be modulated in many other ways (examples of sample-and-hold on the website are on tracks 27 and 39).

Modular systems, or patchable systems such as the Roland 100 or original Korg MS series, are obviously rich sources of extremely abstract sounds. By their very nature, modular systems are also particularly suitable for use with external effects units, since sounds can be patched from one part of the system to an effects unit, then back to the modular system for further processing. For instance, an oscillator sound could be echoed before being filtered, or equally easily echoed after being filtered, creating quite different end results. Modular systems sometimes include voltage-controllable effects modules (such as an analog "bucket brigade" delay, or on the early Delta Music Research system a voltage-controlled flanger) and it is also often possible to integrate external effects pedals and other units more closely into an analog synthesizer system using specialised output modules, more about that in Chapter 8. Older pedal-type envelope followers, phasers, and flangers often incorporated a voltage-controlled filter or an analog delay line with voltage-controlled clock speed, it's possible to "tap in" to these and apply to them the output voltage sent from a synthesizer or modular system's LFO (electrical safety precautions should be observed when attempting any modifications to achieve this). The output voltage of an LFO set to Random, when applied to a phaser, produces an interesting sample-and-hold phasing effect that can be stepped in time with another repetitive effect; if there is one thing more interesting than an imaginative use of effects units, it is the ability to time the effects exactly to coincide with what is going on with regard to another aspect of the sound.

1983 Sequential launches the weighted keyboard Prophet T8 MIDI synthesizer

Since it is unimaginative to regard the analog sequencer only as a source of repeated patterns of pitches or notes, effects can also be repetitively "sequenced". The analog sequencer is probably one of the most under-used modules available to the synthesist; mostly, sequencers have been used for simply creating repeated eight- or 16-note beats, but more imaginative users tend to use them to create repeatedly varying patterns within the span of a single note. Again, the early Isao Tomita albums include some good examples; Tomita is rarely interested in setting up a repeated pattern of notes with the sequencer (the nature of the classical pieces on which he chose to work would rarely call for this), but on close listening, many of his sounds can be heard to contain within them repeated pitch, filter, or effects changes created using the modular system's sequencer.

Tomita was also the master of ring modulation; the ring modulator module takes two input sounds and outputs new pitches having the sum and difference of their frequencies. These new pitches are generally not musically related to the input tones so the result is often dissonant and can be described as metallic or bell-like. In the creation of very abstract effects, as found very early on by Karlheinz Stockhausen and others, the ring modulator is extremely useful, although it is not an effect that finds very many applications in conventional music (listen to track 58 on the website). Similar effects to ring modulation can also be created using FM (frequency modulation) synthesis. If this term seems familiar from its appearance on the decidedly non-analog range of Yamaha DX digital synthesizers, analog users may be surprised to learn that exactly the same principle can be put to work using their analog instruments. See the Dixon and Stein interview in Chapter 8.

1983 Roland launches the JX3P polyphonic synthesizer with MIDI and optional programmer

Try starting by adding vibrato modulation – at about 7 Hz – to an oscillator. If this modulation is speeded up, perhaps to 15 Hz, it no longer sounds like a convincing vibrato, as track 3 on the website demonstrates. On some synthesizers, such as the Sequential Prophet 600, the slow oscillator does not go much faster than this; on other synthesizers, such as the modern SE Boomstar, and on modular systems, the slow oscillators can go much faster, or can be substituted by a faster-running audio oscillator. As the modulation speed goes up, the basic pitch of the oscillator being modulated appears to waver and a new pitch soon appears that is more closely related to the speed of the slow oscillator. With even faster modulation, other tones will appear as well; the end result is something like ring modulation and can produce similar metallic and bell-like effects. These can also become extremely voice-like at times; a single oscillator with some very fast modulation going on can take on a much more interesting tone, which (in a much more precisely controlled way) is exactly what is taking place in the Yamaha brand of digital FM synthesis.

1983 SIEL launches the MIDI-equipped Opera 6 polyphonic synthesizer in Italy

The Sequential Prophet synthesizers can also usually route the voltage output of one oscillator to the pitch of the other, or to the cut-off point of the filter. On the Prophet models this is referred to as Poly-Modulation, but shows just the same FM principle in action (track 64 on the website demonstrates this). Preset 55 on many Prophet synths has a good example of this: a falling, bell-like tone notably used by Lyle Mays on Pat Metheny's *As Falls Wichita, So Falls Wichita Falls*. Unfortunately, this type of effect was denied to analog models with only one bank of oscillators, such as the simpler Sequential Six-Trak and Max.

Another way of creating interesting overtones – sometimes musical, sometimes rather abstract – is the use of sync, the coupling of two oscillators so that the pitch of one restricts the pitch of the other by resetting its cycle. This is not a very interesting technique in itself because it simply removes the "beating" between slightly detuned oscillators, which usually acts to make the combined sound richer, but when an attempt is made to alter the pitch of the "slave" oscillator – either manually, or using an LFO, or the level of an envelope – the slave oscillator is not able to change its pitch freely and new harmonics are generated that create a distinctively cutting, metallic sound. Controlled by a pitch bender, this effect can be useful in creating very sharp, "screaming" lead or bass lines, an effect denied to single-oscillator synthesizers but which could be particularly well created by the Moog Prodigy and Oberheim OB1 (examples on the website include tracks 23, 24, 38, 51, 52, 68, and 79). Adding the option of sync between oscillators 1 and 2 was also a common modification for the original Minimoog.

Apart from oscillators and resonating filters, the other common analog sound source is white noise. Many white noise effects are very obvious: wind (particularly with a little high resonance added), sea or waves (usually with some filter modulation happening), percussive effects, and so on (check out tracks 16, 20, 42, and 69 on the website). Less obviously, white noise can be used as a modulation source and, on many synthesizers, the white noise source is routed through a sample-and-hold circuit clocked by an LFO, so random effects at a controllable speed can be created. This is a very common technique – heard early on in use by Keith Emerson on *Karn Evil 9* thanks to the Moog modular system – but unusually on the original Minimoog, modulation is generated by a mixture of the output from the third oscillator and the white noise source and there is no sample-and-hold circuit available. The effect of using white noise to directly modulate an oscillator is not so striking, leading more to a slight wavering of the pitch than anything else, but the technique is worth looking into for the creation of more subtle abstract sounds. Don't be afraid to experiment in this way with the interconnection of any analog modules; most systems are designed so that no output can possibly damage any input. Different combinations of modules can often produce surprising and pleasing results. Piling up modules will generally increase the complexity and interest of patches, for example, an LFO can be set to produce random sounds but a voltage-controlled LFO with its own speed modulated by another LFO can produce random sounds at random periods, which could be even more interesting.

The interfacing possibilities offered by modular systems in particular, and to some extent by small synthesizers as well, can be used to extend their function towards other instruments too. An analog filter, phaser or ring modulator module can be used to process a keyboard such as a digital synthesizer with no resonant filter of its own, or an older electric instrument such as the Hohner Clavinet, electric piano or organ. The master of this technique was Irmin Schmidt of the German experimental rock group, Can, who played a Farfisa organ with the band and wanted to extend its sonic possibilities:

1984 Oberheim releases the MIDI-equipped Xpander keyboardless synthesizer

"I saw things like the EMS Synthi 100, which was huge, you couldn't work with it on stage, and even with the Minimoog people used to sit with their headphones on and when they finally found the sound and got it in tune, the band had gone on to something else. I wanted something which I could use immediately, just put a knob down and the organ sounds different."

(Irmin Schmidt, interview with the author, 1999)

Schmidt's solution was to commission from a Swiss engineer an analog processing module, which came to be called the Alpha 77 (picture in Chapter 2 on Ring Modulators):

"I wanted it to contain ring modulators, filters, a couple of audio oscillators, and it also had a huge tape loop with about 12 playback heads for echoes. I had the Farfisa [Pro Duo] organ and the Farfisa electric piano, and they went through the ring modulators which had one output, or the filters which had another output, and the whole box had another output, and I could put the ring modulator on one or on both, or reverb, or filter – it was just one switch, and there it was, I could just play. It wasn't that rich at creating sounds, but at that time the sounds available were not that rich anyway."

Schmidt was also fond of using overdrive effects:

"One of the very technical sounds which everybody thought was a very tricky complicated thing was very simple. The whole Alpha 77 went through a Farfisa pre-amplifier which had what they called 'Sphaero-Sound', which was an electronic imitation of the Leslie [rotating speaker], which by itself, when it worked as they designed, was awful. But when you overloaded the pre-amp so it distorted horribly it made quite a nice sound, especially when the 'SphaeroSound' was on, so it started sort of phasing. Then if you played on the organ very high and very low with a registration which made the high end and the bottom end very strong, the distortion became totally crazy. The Farfisa also had a glissando or pitch bend pedal which went down a whole octave, and it was really handy, because you could do very strange things – it sounded like a whole pack of Hell's Angels on their motorbikes!"

Journalist Duncan Fallowell described Schmidt's sound at the time as "like the engine of a flying saucer". Some good examples of the speed with which Schmidt could call up different analog-based effects on the Alpha 77 and Farfisa organ combination are on *Can Live 1971-1977* (SPOONCD42/43) released by Spoon Records in Germany and by Mute Records elsewhere.

Despite playing instruments, such as the Polymoog, on later albums, Irmin Schmidt never became a fan of polyphonic synthesizers:

1984 Crumar launches MIDI synthesizers under the brand name Bit (or Unique in the USA)

"I liked them for certain things, but it's only ten years ago that they started becoming real instruments – now, of course, you can use the very old ones as instruments, because an ARP or a Moog is now just one register among many others, rather than having to try to do everything with it."

1984 Korg adds MIDI with the
Poly 61M model, then the
portable Poly 800

On the early Can albums, *Ege Bamyasi*, *Tago Mago*, *Soon Over Babaluma*, *Future Days*, and *Landed*, Schmidt's keyboard sounds excel almost anything else of their period and are undeniably unique, despite the fact that Schmidt's analog "synthesizer" was purely a processor rather than an instrument in its own right, a very good example of the way in which imaginative use of what was effectively a modular system could create outstanding results. Filters, ring modulators, phasers, and other modules can also usefully be applied to guitars, voices and other instruments. A modulated VCA or VCF can create tremolo effects on guitar or wind instruments; running a human voice through a repeated envelope can "chop" it into short sections, speeding up this process can create unusual modulations and overtones on the voice. A voice processed through a ring modulator, with a carefully chosen oscillator pitch at the other audio input, can be a simple substitute for a vocoder to help create distorted "robot" voices.

One problem with processing polyphonic sound sources through single synthesizer filters is that they will lose their individual tonal response for each note. In other words, a whole chord is processed at the same time (the same applies to some of the more inexpensive early analog synths, such as the Korg Poly 800, Moog Opus 3, SIEL DK80, and most old string and brass synthesizers, which only have one filter to offer regardless of how many notes are played).

While it is easy enough to pass any external sound through a filter that is static or being swept by an LFO, it is difficult to get a synthesizer's envelope-related functions to function as well because the source instrument is usually not generating any trigger that might set them off. The only useful value that the input sound has is its own volume level, so if the user wants to use envelope processing, it is necessary to use the input signal's volume in some way to derive a trigger. A very few synthesizers, such as the original Korg MS20 and Sequential Pro One, have this Envelope Trigger facility built in. As well as deriving a trigger from the volume of the input signal, the MS20 can also detect the pitch of any monophonic input sound and use this to determine the pitch of the synthesizer oscillator and other functions (pitch-to-voltage conversion).

The same envelope triggering process is taking place in the effects pedal known as the Envelope Follower. This looks at the incoming volume, and when it is high enough, starts to sweep the cut-off point of a variable resonance filter downwards under the controller of an envelope with variable length. The typical application is to turn a bass guitar sound into a twangy, synth-like funk bass. A milder version of the effect is the guitar auto-wah pedal, and the Electro-Harmonix MicroSynthesizer (and

1984 Roland begins to separate
modules from keyboards with
the MKS30 MIDI module

old EMS Synthi Hi-Fli), which is a guitar or bass processor, works on similar principles, in addition to mixing in frequency-divided versions of the input sound.

The Sequential Pro One can also detect a sound at its filter input socket and not only use this to trigger the envelopes, but also to step along its internal sequencer, so the synthesizer and its sequencer become reliably synchronised to almost any incoming signal, such as a drum beat on tape. The EDP Spider sequencer also had a tape synchronisation click input, with such a widely variable sensitivity range that it could be triggered from almost any instrument output or recorded music, and since the Spider had analog control voltage and trigger outputs in addition to its digital output to the Wasp synthesizer, this was one way of setting off an analog synthesizer envelope from an external sound. The author processed a Welson string synthesizer through a Moog Sonic 6 through a Spider sequencer in this way on the CD album *Analog Archives*. Some designs of audio noise gate, which are common as an item of studio equipment, may also be able to generate a trigger for a synthesizer or modular system whenever they open in response to an incoming sound or, if not, their output can sometimes be amplified up to the required level (often five volts for a reliable trigger, which is very high for a line level audio signal). Korg did make an interface device, the MS03 Signal Processor, which helped handle this sort of task and was similar to the external audio input sections of the MS20 synthesizer packaged in a smaller casing; these are now very rare but Eurorack and other modules exist to do a similar job.

Obviously, the creation of very complex, abstract sounds using analog synthesizers and modular systems will lead to a problem of reproducibility. If a recording cannot be finished in one session, or if the piece is supposed to be played live, it can be nearly impossible to go back and recreate the sound exactly as it was. Tiny changes in LFO speeds or oscillator tuning can completely alter a complex abstract sound, writing down these settings in note form can be very time-consuming and still not completely reliable. Many small synthesizers, such as the original Minimoog, ARP Odyssey, and Korg MS20, came with a "patch sheet", a blank diagram of the front panel that could be photocopied and filled in to represent any given sound, and for modular systems of a customised layout, the user could create his or her own blank patch sheets, which as well as control levels would also have to indicate the interconnection of patch cables. But this method is also far from perfectly reliable, this was one major reason for the initial decline in interest in the modular system and the move towards programmable synthesizers.

Sequential did, early on, make a programming unit, the Model 700, intended to provide programmable envelope and LFOs for modular systems, but this was only a partial solution and that unit is now very rare; early Oberheim synthesizers could memorise at least some of the major parameter settings, and early polyphonic synthesizers, such as the Yamaha CS80 provided "memories" not digitally, but in the form of tiny reproductions of the front-panel controls using miniature preset sliders. But no system of memorising patches was entirely successful until the Sequential Prophet 5 appeared complete with 40 comprehensive digital memories which could be backed up to tape.

This ease of storing and reproducing, for instance, a basic brass, string, bass, or lead line patch, if anything, made it less likely that users would continue to look into the creation of very complex sounds; in any case, instruments such as the Prophet did not offer the complex interfacing capabilities of a modular synthesizer. Ease of use and ready programmability inevitably won out in commercial terms and so the imaginative creation of abstract sounds using modular synthesizers became much less common, only picking up again after the so-called "analog revival".

A typical clarinet imitative
patch, here for a Korg PS3300.

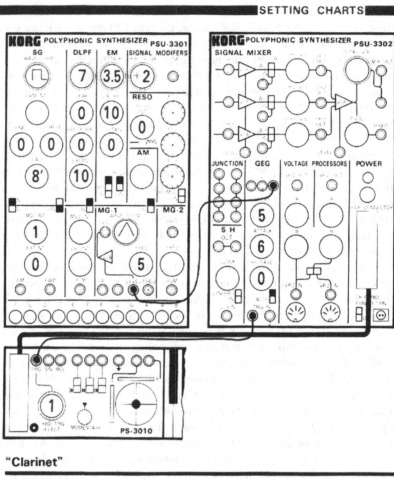

"Clarinet"

クラリネット＝ディレイ・ビブラートは、任意に　　　　Adjust tremolo and vibrato as you like.
設定してください。

1984 Casio goes into pro
and semi-pro synthesizers
with digital MIDI-equipped
instruments

One solution to the problem of reproducing complex abstract sounds, of course, is to create a complex sound on a modular system and associated effects units, then sample it on a hardware sampler, such as those from Akai or E-Mu, or into a software sampling package. Even a master of analog synthesis, such as Isao Tomita, for instance, has said:

"I always used the Moog modular system but I also started using a Synclavier, because one disadvantage of the Moog synthesizer is that it had no memory. On the Synclavier the hard-disk sampling system was very good, so I put the sounds made by the Moog into it. Now there are much better sampling systems as well."

(interview with the author, 1999)

The digital sampling technique can very dependably recreate the complex sounds of modular systems, but of course the sampled sound has become static; regardless of what facilities may be provided within the sampler itself, the individual elements of the modular system sound become more or less fixed. A lot of the appeal of complex sounds created by modular analog systems lies in the way they are modulated, developing and interacting in different ways depending on what input levels and playing techniques are used; once that sound is sampled, most of this variability is lost. After some twenty years during which polyphonic, MIDI and memory-equipped synthesizers, and samplers dominated, this is perhaps why non-programmable, fully variable analog synthesizers and modular systems started to become popular again. Players simply recognised that they would have to give up the luxury of complete reproducibility, or even tuning reliability, in order to regain some of the flexibility and excitement that the earliest generations of synthesizers had to offer.

TEN GREAT EARLY ANALOG SOUNDS

1. JEAN-MICHEL JARRE, *OXYGENE PART 2*: SPACE SOUNDS (ON *OXYGENE*)

An EMS VCS3, or Synthi A with filter resonating, two oscillators set to slow controlling its frequency, and their speed and depth controlled from the joystick. Add plenty of echo and joystick movements creating various abstract sounds, no envelope control needed.

On other synthesizers, similar sounds can be created if you can find two LFO sources and turn off any audio oscillators. Reproducing joystick-style control is more difficult, it depends on how many controls you can adjust with each hand.

2. KEITH EMERSON, *LUCKY MAN*: CLOSING SOLO (ON *EMERSON, LAKE & PALMER*)

Added as an afterthought to Greg Lake's acoustic ballad and virtually on the first take, this solo uses a fairly cutting sawtooth setting on a pair of oscillators with about 30% portamento so that, if played too quickly, notes do not quite reach their final pitch. Volume and filter envelopes release fairly quickly, but sustain is full up so they don't do so unless the note is released.

Dramatic low notes, which add a marked portamento, alternate with faster high notes. Towards the end of the solo, the filter resonance goes up and the envelopes are allowed to decay, closing down with a pronounced whistle. Later, Emerson played the low notes from a set of MIDI bass pedals so the dramatic portamento swoop is now as much as four or five octaves.

1985 Akai develops synthesizer lines with analog models before moving on to samplers

3. RICHARD WRIGHT, *ON THE RUN*: SEQUENCE (ON PINK FLOYD – *DARK SIDE OF THE MOON*)

One of the first analog sequences heard in rock music, credited to an EMS VCS3, though probably recorded using an EMS Synthi A with a KS keyboard, which includes a 256-note digital sequencer.

A simple eight-note sequence is played fast on an organ-like setting. The filter is slowly opened to make the sound brighter and more aggressive. Later, some LFO modulation is applied to the filter to make it "wobble". Simple for almost any analog synthesizer with a built-in sequencer, such as the Roland SH101, Sequential Pro One, Moog Mother 32, or Malekko Manther.

4. KRAFTWERK, *AUTOBAHN*: AUTOMOBILE SOUNDS (ON *AUTOBAHN*)

Industrial sounds on *Autobahn* were mainly created on an ARP Odyssey. The passage eight minutes in starts with white noise effects; adding some phasing gives an impression of movement, and repeating the envelope of the white noise sound simulates air brakes or other machine effects.

The sounds then become more pitched. Two oscillators are slightly detuned to create a motor noise; the Odyssey's pitch bender (which on early models was just a rotary or slider control) is pulled all the way down to simulate the Doppler shift effect of passing vehicles. At the same time, the filter is slightly opened to create a "growl", a change in tone which simulates the vehicle coming closer. At the same time, sounds pan from one side to the other in stereo. As this effect would require a third hand, it was probably added at the mixing stage.

5. RICK WAKEMAN, *CATHERINE OF ARAGON*: SYNTH BASS (ON *THE SIX WIVES OF HENRY VIII*)

1986 Roland launches the Super JX10, their last truly analog early keyboard synth

An archetypal Minimoog sound well reproduced by most analog synths. Two detuned square wave oscillators give a full, rich effect. Amplifier and filter envelope release times are medium and the filter envelope depth high. Some portamento is added, filter resonance is set to high and filter cut-off to medium.

The sound responds well to playing style; notes close together played quickly sound plain, while notes far apart played slowly show glide and the filter cut-off begins to close down, the slight whistling effect of the high filter resonance setting becoming obvious. This setting is a good way of getting expression from an analog synthesizer that has no aftertouch or other expression controls.

6. PETER BARDENS, *SONG WITHIN A SONG*: LEAD SOLO (ON CAMEL – *MOON MADNESS*)

A great demonstration of how to combine different expressive techniques, in this case on a Minimoog, Bardens' solo starts with a long glide using a slow portamento setting. The end of each bar is marked with modulation, not of the oscillators with triangle vibrato as normal, but of the filter with a rising sawtooth wave. A subtle and unusual staccato modulation effect.

Since the Minimoog has low note priority, when there is some portamento, bends can be selectively created by playing a lower note while a higher one is still held. The speed of playing determines how far the bend reaches. This technique is combined with the modulation effect before the end of the solo.

7. THE HUMAN LEAGUE, *THE BLACK HIT OF SPACE*: FUZZ GUITAR (ON *TRAVELOGUE*)

When a lot of distortion is applied to an analog synthesizer sound, it does not matter very much what the original sound is like. The excellent "Velo-Crunch" guitar sound on the Roland JD800 is very simple when the built-in distortion, reverb, and delay are all switched off.

On this sound, the source is probably a couple of oscillators of a Roland 100, through a highly resonant filter that can occasionally be heard. There's a long decay but little release time; the sound is played in a staccato, stabbing manner, the distortion created with a guitar fuzz box or just by overdriving the mixer input channel. A good example of why analog sounds don't have to be "clean" all the time. Many Eurorack modules now offer this overdrive effect.

8. ISAO TOMITA, *ARABESQUE NO. 1*: WHISTLE MELODY (ON *SNOWFLAKES ARE DANCING*)

Tomita's Moog modular human whistle sound was copied in the presets of many smaller instruments. It's similar to Keith Emerson's solo sound on the previously described *Lucky Man*, though without the filter effects and with softer oscillator waveforms, typically sine waves. Played high up the keyboard, portamento is again in use, so notes that are further apart have more glide between them. Vibrato must be subtle and brought in by hand or with an automatic delay.

1987 Release of *Acid Tracks* album, defining the Chicago/ Detroit dance music sound

9. CYBOTRON, *ACID TRACKS*, APHEX TWIN, ETC.: TB303 BASS (ON VARIOUS ALBUMS)

The Roland Bassline (TB303) sound is harder to imitate than it may appear, since the TB303's filter and envelopes were so unusual. To create TB303-like bass lines, only use one oscillator, set filter resonance and filter envelope depth high, choose a 12 dB/octave rather than a 24dB/octave setting if available (the actual TB303 design lay somewhere in between at around 18 dB/octave cut-off), and experiment with bringing glide in and out between some notes. Making the filter and envelope also respond to velocity is important too, as the TB303's "Accent" control responds in a complex and unconventional way.

1988 Closure of Sequential Inc., with the R&D department going to Yamaha

10. KEITH EMERSON, *KARN EVIL 9*: SAMPLE-AND-HOLD (ON EMERSON, LAKE & PALMER – *BRAIN SALAD SURGERY*)

The dramatic sample-and-hold sound linking Sides 1 and 2 of ELP's album is not difficult to achieve. A source of random voltages – an LFO if it has a random setting, a white noise source running the sample-and-hold section on an ARP Odyssey or modular system, or a sequencer playing a long random pattern – modulates a filter through which oscillators play a fixed drone. Speed, depth of effect and filter resonance can all be varied. If the modulation needs to be synchronised to another instrument, the random source simply needs to be externally clocked in time with it.

6

The analog revival,
1980s–2000s

As we saw in Chapter 4, by the late 1980s the production of analog synthesizers had almost come to a halt: Moog, ARP, and Sequential had closed down; Oberheim, EMS, and PAiA were almost inactive; Yamaha had gone over to FM digital synthesis, Korg and Roland to sample-based synthesis, E-Mu and Akai to sound sampling; and all the smaller players, such as EDP, Crumar, Multivox, and Jen (amongst many others) had disappeared.

The sound of popular music in the mid to late 1980s was dominated by digital sampling; sampled drums, such as those from the Linn Drum machine, sampled sequences from the Fairlight and Ensoniq Mirage, and eventually even sampled vocal lines thanks to the Emulator (used by Paul Hardcastle on *19*) and the upmarket Synclavier. Percussion sounds were often either sampled or played on an electronic kit, such as those from Simmons, but the analog elements of these products were also rapidly being superseded by digital design.

1988 Release of 808 State LP *New Build*, spreading the analog revival sound to the UK

The technology was moving so quickly that products released only a few years earlier were regarded as old, very much out of date and of very little value. A Minimoog which had sold for £1000 in the 1970s would fetch less than a quarter of that; analog drum machines, such as Roland's TR808 were perceived as having been completely superseded by the more authentic-sounding sampled Linn Drum or Sequential Drumtrax, and were sold off cheaply; and some products that had not caught on even when analog was the height of available technology, such as the Roland TB303 Bassline, were practically being given away.

It was the extreme affordability of instruments such as these – which were being found in pawn shops, junk stores, and second-hand advertisements even more commonly than in music stores – that led to their being taken up by a whole new generation of musicians, mostly involved with the club scene in the USA. Many of these were DJs rather than keyboard players and so were interested in instruments that would to some extent play themselves rather than demanding too much virtuosity in performance; so one obvious resource was the drum machine, and various models were in plentiful supply, though comparable instruments that would play a more melodic part were few and far between. Using a large MIDI keyboard, a polyphonic sequencer and a stack of MIDI modules seemed much too musicianly for these DJ-oriented players. What was needed was a unit that could be programmed almost at random by non-keyboard players, but that could still produce interesting melodic results.

Much of this type of activity occurred in the Chicago and Detroit club scenes in the USA, where in 1985 Juan Atkins founded Metroplex Records. His music, originally as part of the duo Cybotron (not to be confused with the earlier Australian synthesizer band of the same name) featured repetitive drum machines, short sampled sounds and vocals, analog bass lines, and the distinctive gliding patterns of Roland's TB303 Bassline and SH101 synthesizer (some examples are on the CD *Classics*, issued under the name Model 500, R&S Records RS931CD). In 1987, the dance single that came to define this whole style, *Acid Tracks*, was released, and strongly featured the Roland Bassline, an instrument by this time five years old and out of production for three of those, which, given the speed of development of contemporary technology made it practically ancient history. But there had been no obvious replacement instrument since that time, the TB303 having accompanied the release of the TR606 drum machine but with no parallel release to accompany the later TR707 or TR909. Although originally designed to play bass parts – which it did in a straightforward, not very exciting or authentic way – the 303 was used on *Acid Tracks* and similar pieces, more for a mid-frequency, repetitive melodic part. Notes, accents, and portamento could all be programmed independently on the 303. Although the programming system was complex, relatively good results could be had by entering values for all of these almost at random until something useful emerged, then skipping during playback from one memorised pattern to another, as was possible on the contemporary drum machines.

But something else made the 303's sound particularly compelling, possibly because the Roland designers had not put too much effort into perfecting it, because the built-in sound was only intended as a "scratch" one for programming the final sound to be generated from a larger synth or modular system connected to the TB303's CV and Gate outputs. The TB303 filter is unusual – perhaps an 18 dB per octave cut-off rather than the more common 12 or 24 – and the Accent facility does more than simply increasing the volume on a particular note, apparently affecting the filter setting, resonance, and envelope shape as well (some examples of the TR808, of the MC202, which has more or less the same sound as the SH101, and of the TB303 are on the website, tracks 32–37).

In the UK, bands such as 808 State were also taking up the now very affordable TB303. The band's debut album, *New Build* (originally released on vinyl in 1988, later on CD on Rephlex CAT080CD), using TR808 drum machines, Basslines, SH101 synths, and other affordable analog equipment, was apparently recorded on heavily edited second-hand tapes, and went against the contemporary fashion for melodic pop music. It was extremely repetitive, had no vocals and sounded unashamedly technological; the Detroit and Chicago scenes were also developing, and a long series of album releases under the title *Street Sounds Electro* helped to popularise these styles. Artists including Kevin Saunderson (who had formed Cybotron with Juan Atkins) and Derrick May (who had attended the same school as the duo and launched his own Transmat label) popularised the style. May's music from 1987 to 1997 was later compiled on a double CD, *Innovator*, which strongly features busy high-hat patterns, analog string pads, and much use of analog sequence lines alongside short samples and percussion sounds. Analog sound quickly began to get back into mainstream music too. In 1990 The Beloved had a crossover dance hit with *The Sun Rising* (from the album *Happiness*), which featured sliding analog sequencer sounds alongside digital sound sampling and digital piano tracks more influenced by contemporary Italian club music styles.

These dance-oriented analog styles also spun off a more contemplative type of electronic music for "chilling out" after a long club session. In 1992 the Warp Records label in the UK released *Higher Intelligence*, a compilation of various artists described as "electronic listening music". These tracks showed influences from the first generation of analog technology fairly clearly; Richie Hawtin, Alex Paterson (later of The Orb), Dice Man, and Musicology all quoted their influences as including Kraftwerk, while Autechre mentioned Tangerine Dream, and other artists credited Juan Atkins, 808 State, and Brian Eno. This music was full of gliding SH101 and TB303 sounds, 808 drums, short analog sequences and highly resonant filter sounds. Richard James had been working in the UK as The Aphex Twin and under various other names; his *Selected Ambient Works 85-92* (R&S Records AMB3922) also remains a good example of the use of analog sounds for dance and much more abstract styles.

Both the dance and ambient scenes were still extensively using analog sounds well into the next millennium; the use of analog, particularly for bass lines, by artists, such as The Shamen, Ace of Bass, No Doubt, Madonna (particularly when produced by William Orbit and later by Mirwais and Stuart Price), Air, Sash, AFT, and many others in the dance field is well known, while as an example of a more ambient style, the Manchester-based label SCAM Records in 1998 had released *Soup* by Bola (SCALD2), a set of slowly rhythmic ambient sequences and arpeggios, not particularly dance oriented but again more suitable for "chilling out".

To meet the needs of artists in these fields there had been some move in the USA in particular towards modernising older analog instruments and bringing them into the MIDI era. Studio Electronics, for instance, had begun rack-mounting Minimoogs, Prophet 5s, and Oberheim SEMs, while Groove Electronics in the UK offered to rack-mount pairs of Wasp synthesizers and add MIDI, calling the result The Stinger. But the supply of original instruments available for conversion was strictly limited and, in any case, enthusiasts for the newly emerging dance and ambient styles (sometimes with more money to spend than the innovators of the style in Chicago and Detroit) found themselves unable to reproduce the required sounds on any more modern instruments. In their enthusiasm they started pushing up the price of the original TR808, TB303, and other instruments so that they doubled, then multiplied by five times, then by ten times. At the height of its popularity the TB303 was selling for over £1000 (US $1500 at that time) and the TR808 drum machine – and later, as fashions changed a little, the TR909 – for around the same price.

Obviously, these inflated price levels began to drive a demand for a more affordable substitute for the TB303 sound, because while this period also marked the beginning of an increase in prices for other analog synths, such as the Minimoog and ARP Odyssey, these were not found to reproduce

1990 Cheetah expands its computer games hardware business, adding the MS6 analog synthesizer module

the 303 sound particularly well either, quite apart from lacking its built-in sequencing capability. So why did the major synthesizer manufacturers not respond to this huge demand? The reasons are simple enough: all of the US and UK companies were out of business apart from E-Mu and Ensoniq, which were fully involved with digital sound sampling technology, Oberheim being at best dormant; the major Japanese companies Roland, Yamaha, and Korg were all fully committed to digital synthesizer design. Although the idea must have been put to them many times, Roland did not wish to face the prospect of re-tooling for a new production run of TB303s and so, on the whole, the major manufacturers were extremely slow in responding to the rebirth of analog.

1990 The Beloved hit single *The Sun Rising* revives the use of analog sounds

Three or four years into the analog revival, the major companies still showed little sign of responding; a thriving business had developed in modifying and improving TB303s through companies such as Devil Fish in Australia. But this business was again limited by the number of old models available for conversion, it became clear that a new production run of some kind was needed. Unfortunately, there were few small manufacturers set up for such a commitment despite the fact that analog circuits (even if MIDI controlled) were much simpler to design than the currently fashionable and more complex digital instruments. Few companies would believe that the analog revival could last, and so it took an astonishingly long time for any new product to actually reach the market.

One product that did appear relatively early on came from the British manufacturer Cheetah, better known for its computer joysticks and other accessories. Deciding to launch a musical instrument division in 1988, the company created the MS6, a single-unit, 19-inch mounting MIDI six-voice polyphonic, rich-sounding multi-timbral synth module using a similar set of chips to those in the Oberheim Matrix 1000. This appeared to be less a response to any analog market revival than an admission that partly analog design was initially more straightforward than digital, as the company's subsequent musical products – before they returned within four years solely to computer accessory manufacture following the failure of the ambitious Zeus 24 synthesizer design – were entirely digital. A planned MS6 Mk 2 with effects and multiple audio outputs never emerged.

Amongst the major manufacturers, Roland did make one relatively early response to the analog revival, at least in the sense of trying to reintroduce analog-style controls, with the 1990 launch of the JD800 keyboard (track 79 on the website), which had scores of rotary and slider controls and multi-waveform digital oscillators that also included all the basic analog waveforms, coupled with excellent analog-style resonant filters and multi-effects. The JD800, available second-hand at the time of writing at moderately low prices, remains a wonderfully powerful, sophisticated, and versatile instrument, capable of everything from sweeping and spacey pads to cutting lead lines and heavy basses, thanks in part to a versatile set of built-in effects coupled with generous split and layering options. The JD800 could be heard in use by Chris Franke on the soundtrack to *Babylon 5*, by Klaus Schulze, over a decade later by Dave Greenfield with The Stranglers, by Ron Mael with Sparks, and by many others who appreciated its firm semi-weighted keyboard and general versatility. With the added stability of digital design and MIDI control, the author heartily recommends the Roland JD800 as possibly the finest available overall purchase of all second-hand analog-style keyboard synthesizers. A later JD990 module version, which could also be edited from a JD800's control panel and had even more flexible effects, seemed to differ very subtly in sound, but the keyboard had already been judged expensive and a poor seller at launch so Roland's next entries into the analog revivalist field, some years later, were to take a quite different approach.

In late 1993, Tom Oberheim also decided to return to analog design with the Marion Systems MSR2 (Marion Systems Rackmount 2), a single-unit, 19-inch module designed to hold two circuit boards, of which the first to appear was the ASM, an analog-style eight-voice, 400-memory design featuring digitally stabilised "High Definition" analog oscillators and an audio input to the filters. But further planned modules (probably to include sampling, effects, and digital wavetable synthesis modules) failed to appear. A simplified version of the MSR2, the Pro Synth, featuring just one permanently installed analog module and released in 1995, was not successful either. The sound of the synth's HDOs was rather sterile – when tuned apart, very little typical analog "beating" resulted – though rarely seen second-hand MSR2s and Pro Synths could become collectable in the future, Tom Oberheim having returned with products much more in line with his early designs.

It was to be five years after Cheetah's launch of the MS6 before another significant British instrument appeared, this time from the instrument retailers Music Control, working under the name Control Synthesis. Launched in 1993, the Deep Bass Nine was a single-unit, 19-inch rackmount with a single, simple analog oscillator (as on the Roland TB303 itself), a small number of front-panel controls, and an audio input as well as MIDI input and CV/Gate input and output. MIDI modulation signals could control the filter, while MIDI portamento on/off signals switched the glide facility, so despite having no internal sequencer the Deep Bass Nine could reproduce all the TB303's distinctive sound features quite easily. There was no full envelope, just the Accent, Decay, and Envelope Modulation controls found on the original TB303, but the Deep Bass Nine was inexpensive, relatively reliable and generally accepted as imitating the 303 sound very accurately; on recordings they were practically indistinguishable, and consequently second-hand prices for the 303 began to drop just a little.

Several other companies now entered the market. Synthology launched the Clone 3, which had MIDI but was just two-thirds of a rack unit wide, while in Germany the Braintec Transistor Bass 3 (TB3) was launched in 1994 offering a second VCO with oscillator synchronisation, and CV and Gate out (though not in). Another German company, MAM (Music And More) of Erlangen, launched the MB30, which was very similar to the Deep Bass Nine, this was later re-badged and marketed in the UK under the name FAT 383 Freebass. Also in Germany, in 1994 the kit design company, Doepfer, launched the MS404, a more general-purpose synthesizer again in a single-unit, 19-inch rack-mount format with MIDI, CV and Gate In/Out, and filter input, well capable of reproducing TB303 sounds but with a full envelope shared between the VCA and filter and a much wider

1990 Roland launches the JD800 synthesizer with analog-style controls

range of standard analog synthesizer possibilities; meanwhile, PAiA in the USA launched the Fatman, a more powerful-sounding two-VCO MIDI synth in kit form with rack-mount or desktop casing options.

Roland's TB303 Bassline and one of its first imitators, the Deep Bass Nine.

Also in 1994, the British company, Novation, which had launched with a small add-on MIDI keyboard for the Yamaha QY10 sequencer, showed an updated version with its own built-in sounds, the Bass Station. This two-DCO synth with a resonant filter and just seven memories could produce Roland Bassline imitations but was also capable of much more powerful sounds and was used by Massive Attack, the space rock band Ozric Tentacles, and many others. A year later, the company rack-mounted their design; the Bass Station Rack had 99 memories, oscillator synchronisation, CV and Gate inputs and outputs to drive Hz/volt (Korg type), as well as 1V per octave (Roland-type) synths, front-panel controls that all transmit parameter changes through MIDI, and an audio input. A later Super Bass Station version added a sub-oscillator, ring modulator, white noise, analog chorus, and distortion. Also in the UK, Exclusively Analog, which specialised in repairing and customising modules, launched their own genuinely analog design, the five-unit-high, three-VCO Aviator, which was semi-modular and used by Vince Clarke, The Human League, and others. But the production run was very limited, and a smaller two-VCO, 3U-high Mini-Aviator was made in even smaller quantities and only for a very short period.

1991 Chris Franke plays his only post-Tangerine Dream solo show at London's Astoria Theatre

Meanwhile, in Germany in 1995, a new company, Access, launched a MIDI programmer for the Oberheim Matrix 1000 (as well as one for the locally produced Waldorf MicroWave), a small editing panel in the spirit of the earlier Roland PG model programmers, which once again emphasised the power that could be had from a fully editable and programmable analog synthesizer. Doepfer also disregarded the fashion for polyphony, programmability, and MIDI to launch the A100, a 19-inch mounting system of compact analog modules in the Eurorack instrumentation standard, with individual modules smaller even than those of the Roland System 100M, though similarly using mini-jack patch cables. This system started something of a trend for small modular system designs, within a couple of years two UK companies were also making roughly A100-compatible modular systems. The Concussor from Analog Solutions specialised at first in percussion sound modules, including TR808, TR909, and CR78 bass, snare, high-hat, and conga modules, an eight-step trigger controller, a master clock, and a Fill module to automatically switch sequences every four or eight bars. Within a couple of years the system also offered sequencers and more comprehensive synthesizer modules, such as the 2-VCO (a dual oscillator) and the SY1 (a filter with envelope generator), which could be combined in a compact plastic casing to form a mini modular system. The RS Integrator from the similarly named Analog Systems also offered all the standard modules – oscillators, filters, and so on – plus some more unusual ones, such as a digital scale programmer that converts voltage inputs to pre-defined musical scales. Some users in Germany and particularly in the UK started to build up very impressive systems, often based on a mixture of modules from these manufacturers (despite the fact that neither their rack-mount holes nor their power cables were designed in exactly the same way), though to build up a physically large and impressive system like the old Moogs demanded the inclusion of hundreds of modules. These systems comprised the start of the Eurorack success story which now encompasses more than 300 manufacturers, much more about that in Chapter 8.

1991 Oberheim shows an early version of the OBMX module from Don Buchla

An early example showing the birth of the now massively popular Eurorack modular format.

The fashion for simpler TB303 clones also continued, however, Syntecno of the Netherlands showed the T303 (TeeBee 303), another one-unit, 19-inch rack-mount, and one of the few with a very basic internal sequencer, while Will Systems showed the MAB303, a half-rack clone with an LCD readout, manufactured in Korea.

Another fashion that began around this time was for stand-alone analog filters, designed to add an analog feel to digital synthesizers or to drum loops and other sound sources. Mutronics showed the Mutator, based on Curtis chips, while Peavey launched a MIDI-programmable analog filter in their Spectrum range – the Spectrum Synth module, which already existed, was a powerful but difficult to program digital imitator of analog sounds – and MAM launched the Resonator apparently based on the Korg PS3100 synth's three-band resonant filter section with a three-band envelope follower and dual LFO for filter sweeps, later badged by the Turnkey store in London as the FAT Resinator. Juan Atkins, Steve Hillage (System 7), and others started in the same year to use the Sherman filter bank, a dual rack-mount analog filter with MIDI manufactured in Belgium. Designed by a guitarist, Herman Gillis, the Sherman was intended to exploit the possibilities of overdriven sounds combined with analog filtering (some example sounds are on tracks 70 and 71 of the website), and became (along with a later upmarket Quad Filter Bank) popular with those looking for more extreme filtered sounds. In 1995, Sherman also built about ten prototype units of a design called Chaosbank (one went into use with The Chemical Brothers), which used analog technology to create even more extreme sounds, explaining:

1992 Warp Records compilation album *Higher Intelligence* released

"The Chaosbank is a kind of oscillator bank; there are three rows of oscillators, with functions like frequency, and when you link the oscillators to each other with banana jacks, you get weird-sounding 'chaos' because one oscillator can influence another in a random way. When you play with frequencies and other functions, the sound varies from simple notes to very complex paths – so it's not really a musical instrument, but more some kind of philosophy or concept. The making of the chaos is so interesting that it can keep you busy for hours and hours, it's never boring. Herman's philosophy is like the Zen Buddhists (!). The sound quality is good – the oscillators are warm sounding, but depending on the temperature they may sound off key, though for chaos, this doesn't matter."

Elsewhere in 1995, another major development came from the unexpected direction of Clavia, the Scandinavian manufacturer of the "ddrum" sampled percussion system. Clavia had decided to attempt a complete emulation of analog synthesis using digital circuitry, resulting in the launch of the Nord Lead, a compact, four-octave, four-voice (later expandable to 12) programmable synth with an arpeggiator and polyphonic portamento but no effects, aftertouch, or audio input, and a very limited three-LED numeric display. Despite its apparently limited specification, this could turn out good imitations of Prophet and other standard analog sounds, and became popular with Jean-Michel Jarre, Mike Oldfield, Prince, and others. Modelling analog sound with digital circuitry had the advantages of greater tuning stability, but had possible limitations in that any analog filter overdrive effect was, by definition, artificially simulated, and that the variation of important parameters, such as filter cut-off, could possibly be limited to a relatively low resolution. Implementing such parameters with perhaps only 128 possible digital levels could lead to a "stepped" sound when editing, but the Nord Lead did not suffer too badly from this (less so than the much earlier and ostensibly genuinely analog Sequential Prophet 600). Later a Nord Lead 2 version with 16 voices was launched, as well as a rack-mount version, the Nord Rack. A related product, the Nord Modular, simulated a complete analog modular system with similar digitally generated voices, giving access to many more editing parameters using a small, two-octave keyboard with just 18 knobs, all programmable, plus a PC (or later a Macintosh) running sophisticated editing software. The software simulated more than 70 different modules and sequencers, and the Nord Modular also became available in a keyboardless rack version, in a Micro Modular version with almost no front-panel controls and with just four voices rather than eight, and later, after this was sadly discontinued, in a totally revised G2 version available as a five-octave G2X keyboard, or a controller-free rack-mount module intended for editing from a computer. Clavia persists to this day with newer Nord Lead models and Nord Stage models more oriented towards producing high quality organ and piano sounds.

1993 Mark Jenkins performance in Brazil and release of debut CD *Space Dreams*

Waldorf's Wave, the largest of the company's digital synthesizers to feature analog-style filters.

Waldorf were the successors to PPG in Germany and had successfully marketed the MicroWave module with digital wavetable sounds and analog filters, the MicroWave 2 with an all-digital design, the MicroWave XT with an extended set of front-panel controls, and the huge and costly Wave keyboard synthesizer, which resembled a giant polyphonic Minimoog with a sloping control panel. They now launched a much more affordable instrument as a return to the analog tradition: the Pulse was a three-DCO, two-unit-high monophonic rack-mount with 100 memories, an analog filter, and an arpeggiator. The Pulse had a simple three-digit LED display, 16 modulation sources, oscillator cross-modulation and synchronisation, plus a random patch creation facility, MIDI memory dump, and controls that all transmit data via MIDI. Some sounds are on the website on track 68. The unit had no power switch or dedicated volume control, but the Pulse Plus model marketed from 1996 did add an audio input, CV and Gate In and Out. This unit became a popular source of strong analog bass, lead, and sequence parts with musicians including the German techno DJ Steve Baltes, who worked with the revived Ashra. More on Waldorf in Chapter 8.

By this time the demand for simple, single-oscillator imitations of the TB303 had started to diminish, most new designs tried to offer more analog power and flexibility in a compact and accessible form. In 1996, Spectral Audio of Switzerland launched the Pro Tone, a two-VCO rackmount, but despite having an audio input, resonant high pass as well as low pass filters and an eye-catching bright red colour, this had MIDI In only and a single envelope, was non-programmable, and was generally seen as something of a compromise instrument (some sounds are on the website, track 67). The same year, the Orgon Systems Energiser, launched in the UK, was praised for its nine-mode filter, but this rather expensive, box-like, single-VCO, non-programmable design lacked MIDI and was built only in very small quantities, along with an equally rare small modular system.

As mentioned earlier, several companies had marketed rack-mount MIDI conversions of the Minimoog – in the form of the SEI Mini Remote in the USA and the MIDIMuck in Germany – while Studio Electronics in the USA had more or less run out of original Minimoogs to convert around 1994 and had started to market their own SE1, a programmable MIDI three-unit rack-mount synthesizer inspired by their earlier rack-mount "MIDIMoog" conversions. In 1996, the company also launched the ATC1, a simplified two-unit-high version with touch membrane switches like the Moog Source rather than mechanical controls. This so-called "Analog Tone Chameleon" offered different filter designs on plug-in cartridges intended to imitate the TB303, Minimoog, ARP 2600, or Oberheim SEM, with an optional switching box available to mount all four filter cartridges. Later, the company launched an updated version of the SE1, the SE1X, as well as the Omega 2, 5, and 8, upmarket and rather expensive two-, five- or eight-voice multi-timbral analog rack-mount designs with two VCOs, two VCFs, and three LFOs per voice. SE continues to offer this choice-of-filters approach in the current Boomstar desktop semi-modular synth line.

Also in 1996, Technosaurus of Switzerland previewed the Selector, a powerful and professionally constructed modular analog system. The system developed slowly, launching with basic oscillators, filters, and envelopes, but later introducing more unusual modules, such as wave shapers and resonator banks. The Selector abandoned the idea of miniaturising modules through the use of mini-jack connectors, sticking rather to full-size, professional quarter-inch patching, and so in its final layout and cabinet size more closely resembled the classic large modular systems from Moog, Roland, Polyfusion, and E-Mu. The circuitry also established the highest possible standards, offering what were at the time claimed to be the fastest envelope speeds ever found on a modular system. The Selector system is featured in the module examples in

Chapter 2 and on tracks 2–18 of the website, though the Technosaurus company is now long gone.

Readers may wonder where, in all this activity, were the major instrument manufacturers, such as Roland, Yamaha, and Korg in Japan, and Ensoniq and E-Mu in the USA. The fact is that all these companies remained extremely slow to catch on to the "analog revival", believing perhaps that it could not last and would threaten to leave them with large unsold stocks of any new product, or that the potential market was too small to even justify the design of new product in this market area. The extreme reluctance of these mass-market companies to enter the area probably kept the price of original analog synths artificially high, despite the successful introduction of powerful virtual analog instruments from some of the smaller and newer companies.

One alternative to offering actual analog sound creation was to offer ready-made samples of analog sounds, an approach which tied in much more closely with the large manufacturers' preferred use of digital technology. E-Mu in particular had been very successful with their Proteus modules, which used digitised samples of real instruments. It was only a short step from these to the design of a module full of imitated analog sounds, the Vintage Keys, and later a simplified version, Classic Keys. These modules offered sample-based sounds of analog synths as well as of Mellotrons, electric pianos, and other classic instruments, and certainly offered a wide range of sounds, though never quite matching the degree of control available through having the actual original instrument at hand. To be fair, E-Mu's later Morpheus and Audity 2000 modules did have very versatile, multiple-mode controllable filters that could also offer many analog-like effects, while their Orbit (and later Orbit 3) module was full of short analog sound samples and also added a system for simulating and filtering sampled drum loops. Akai's tiny half-rack SG01V module was also packed with 256 sampled analog keyboard and percussion sounds, and had multi-mode resonant filters, two LFOs, and three EGs per voice, but with just five front-panel buttons, these facilities were difficult to access. Roland's MVS1 rack-mount had originally been produced as a slot-in "Vintage Synth" ROM circuit board (SRJV80–04) for the JD990 module and JV series of modules and keyboards, and again featured plenty of sounds, but with no easy access to real analog-style control.

In 1996, some ten years after the analog revival had begun, Roland finally responded in a concerted manner with the MC303 Groovebox, which basically added analog-like drum patterns, effects and a filter cut-off/resonance controller to a standard Roland General MIDI synthesizer chip. Also launched at the time was the XP10 keyboard synth, a lightweight five-octave model that took a parallel approach by offering just two control sliders, usually assigned to filter cut-off and resonance, coupled with many sampled analog sounds of Mellotrons, TB303s, Oberheim synths and so on added to the basics of the standard General MIDI sound set. The later JX305 Groovesynth simply added a keyboard to the MC303, while the MC505 extended the "Groovebox" idea with facilities, such as an on-board mixer and infrared "D-Beam" controller, that responded to the player's hand gestures, while the Roland EG101 added a keyboard and speakers.

1993 The Control Synthesis Deep Bass Nine is one of the first of the TB303 imitators

The Roland JP8000 synth, launched in 1997, took a different approach from that of the Grooveboxes, having much more in common with the "virtual analog" design of the Clavia Nord Lead (and consciously aiming to reflect Roland's earlier analog successes, such as the Jupiter 8, although the rights to that particular name had been temporarily lost). A very simple selection of analog waveforms was simulated using highly stable digital technology. Once again plenty of panel controls were offered as well as an arpeggiator that sends notes over MIDI and a dual effects unit; the rackmount JP8080 version also added a vocoder setting to the effects unit. But despite offering a pattern recording sequencer called RPS, polyphonic portamento, keyboard splits and layers, a ribbon controller, white noise, additional waveforms, such as the chorus-like "Super Saw", and a compact four-octave keyboard, the JP8000 ultimately seemed artificially limited in its abilities, as the much earlier JD800 had been significantly more powerful. For instance, the JP8000 had only limited modulation routing options, no aftertouch, and just a single filter with selectable low pass, high pass or band pass modes.

Roland finally entered the analog/drum loop market with the MC303 Groovebox.

Around the same time, Kurzweil were showing the digital K2000 and later the K2500 keyboard synthesizers, which also included analog-style routines amongst their imitative algorithms. These were used by Richard Wright with Pink Floyd (notably on the "Pulse" concert, finally released on DVD in 2006) to replace his Minimoog and earlier analog keyboards. Around the same time, Kawai showed the K5000W and K5000S, respectively a workstation with a sequencer and a synthesizer with a small control panel of 20 knobs. These were both powerful digital harmonic synthesizers with many added analog-like facilities.

1994 Novation launches the Bass Station virtual analog synth in the UK

Yamaha's response to the analog revival had at first also been extremely limited. Since the launch of FM synthesis the company had been strongly committed to digital technology, though finally adding filters to some of their FM synthesizers and offering analog emulations on their "virtual modelling" synthesizers, such as the VL1 keyboard and VL70M module. Later Yamaha instruments, such as the EX5 workstation and EX5R module, combined sampling, sequencing, virtual modelling, and a version of analog sound synthesis, while the AN1X (released in 1997) quite convincingly combined virtual modelling with some analog-like parameter control, including a ribbon controller and recordable knob movements. But Yamaha's main response to the analog revival was simply to make filter cutoff and resonance and envelope parameters more accessible on dance-oriented keyboard synths, such as the CS1X and later CS2X, adding an arpeggiator and other rhythm-oriented facilities. The same technique was also applied on the affordable DJX, a domestic keyboard with built-in speakers and limited sampling facilities aimed at DJs and dance music enthusiasts (Martyn Ware of Heaven 17 enjoyed using one for many years). On the later CS6X keyboard and CS6R rack-mount, Yamaha took an alternative approach in offering optional sound cards for different synthesis routines; an analog board was one early option, the current MX, MOX and Montage series of synthesizers and workstations have many analog imitative facilities, and there is more on the small analog oriented ReFace keyboards in Chapter 8.

1995 Launch of Access in Germany with their Oberheim Matrix 1000 Programmer

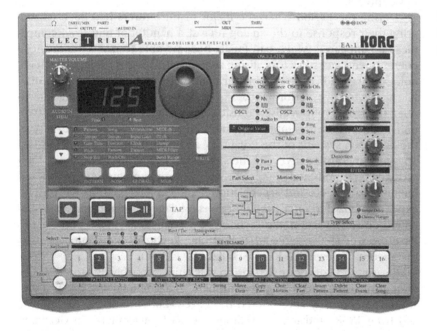

Korg's Electribe EA1 was an early modelled analog desktop synth.

Having licensed some of Yamaha's early FM technology for their 707 and DS8 keyboard synths, Korg had continued to market many keyboard synthesizers and workstations lacking either extensive front-panel controls or properly resonant filters, though eventually offering an analog emulation routine on the Prophecy solo synthesizer (which could also be installed in the Trinity workstation in the form of an additional voice card). Some filter and envelope controllers were reintroduced on smaller keyboards, such as the N5, but these still lacked proper fully resonant filters. The Z1 "multi-oscillator synthesizer", amongst other imitative algorithms, did offer 12- or 18-voice analog emulation, including two multi-mode filters in series or parallel, five envelopes and four LFOs per voice, a programmable polyphonic arpeggiator and an X-Y touchpad performance controller, while the Trinity workstation and the later Triton sampling workstation at least did have more powerfully resonant filters. However, 1999 saw the launch of two interesting analog-oriented products, the Korg Electribe EA1 synth and ER1 analog beat box, a small dance-oriented sequencer/synthesizer and drum machine respectively, the EA1 having a layout very similar to the Roland Bassline, with filter cut-off, resonance, decay, and envelope depth controls, and a rather similar sequencer facility. Launched simultaneously was the KAOSS Pad, a DJ-oriented multi-effects unit and sampler in the form of an illuminated desktop X-Y controller that responds to finger pressure, this included some very exciting and highly controllable analog-style filter effects. Later, Korg updated the Electribe models with units including one with valve audio technology, and developed the KAOSS pad with Version 2 and 3 models, as well as launching a KAOSS Entrancer, which created video as well as audio effects. Korg's much more substantial re-entry into analog sound with new workstations and ARP re-issues is covered in Chapter 8.

1995 Doepfer launches the A100, first of the small Eurorack format modular systems

During the period when the major manufacturers had been considering their response to the analog revival, a handful of mid-sized companies had appeared which had finally become able to make the financial commitment necessary to realising analog-style designs using digital technology. Quasimidi launched in Germany with a floppy disk-based MIDI file player for live performance, the Style Drive, but were soon able to offer their first fully realised musical instrument, the Quasar. This two-unit high, 19-inch mounting module appeared to combine the best of the operating systems of the Roland JV and E-Mu Proteus modules; it had 16 simultaneously active channels, was packed with 1100 short, sampled analog sounds and 300 performance or multi memories, and offered built-in dual effects and arpeggiation. This was followed up with a TRE expansion board to offer even more analog dance-oriented sounds, a simplified one-unit version, the Technox, and finally a keyboard version, the Raven, as well as a versatile MIDI controller keyboard and sequencer used by Tangerine Dream and Klaus Schulze, the Cyber 6.

Quasimidi's corporate expansion was rapid, probably due to their constant advertising and strong commitment to a single field of musical creation. The company later offered the Rave-O-Lution 309, the Polymorph, and from 1998 (and initially selling at 1998 Deutschmarks in Germany before the conversion to the Euro) the Sirius, a dance workstation with analog sounds, a sequencer, effects, and a simple vocoder. On the Sirius, knob movements could be recorded as part of user presets, and the step-time sequencer has separate tracks for bass drum, snare drum, high-hat, percussion, and three synthesizer parts. The vocoder was not highly intelligible, but the short four-octave keyboard helped make the Sirius a compact and powerful analog-like instrument. The Polymorph, on the other hand, was a rack-mount module offering 16-voice polyphony or four separate monophonic synthesizer parts with simultaneous independent sequencing, rather inspired by the way Moog modular systems were used by Tangerine Dream or Klaus Schulze and by the early analog-heavy albums of Kraftwerk, although also including many sampled waveforms for Mellotron and other sounds, while the Rave-O-Lution 309 (with a code number intended to reflect the heritage of Roland's TB303 bass synth and TR909 drum machine) offered monophonic analog synth and analog drum sounds, a built-in sequencer, and expansion cards for additional sounds and sequence memory, and was later offered in a black, Klaus Schulze signature limited "309 Groove-X" edition with the expanded synth and drum sounds included.

In all this frenzy of revived interest in analog, there was intense speculation as to what had happened to the Moog trade name. Bob Moog himself had left the company named after him long before its closure and was running his own new company Big Briar. Apart from reviving the building of Theremins, Big Briar had launched a series of pedals that superficially resembled guitar effects under the name Moogerfooger. These were all analog based, eventually coming to include a familiar low pass filter, an analog delay, a ring modulator, a phaser and later a stepped filter, the MURF, but all these sported control voltage interfacing, creating speculation that they could be built up into a modular synthesizer system, perhaps once again under the Moog name. But in 1998, a Wales-based company launched a convincing Minimoog copy with extensive added MIDI facilities (though not including MIDI transmission of front-panel parameter changes), claiming to have obtained the UK rights to the Moog name, though by the following year, with just a handful of units on the market, the company seemed to have disappeared. There were also two US companies promising new Moog analog modules or Minimoog copies, and again a very small number of replica Moog modules did reach the market, but behind the scenes Bob Moog had been working to regain rights to the Moog brand in order to launch a new keyboard referred to as the "Performance Synthesizer". This eventually appeared as the Voyager, of which more later.

1995 Clavia launches Nord series of virtual analog polyphonic synthesizers

Back to late 1998, when Ensoniq showed the FIZMO, a 24-voice, four-part multi-timbral "transwave synthesizer" with arpeggiator, effects, and many analog-style possibilities, and the following year a rack version in very limited quantities. These both had limited displays and so were difficult to program, seemed to be cheaply constructed, and so were rapidly discontinued after Ensoniq's merger with E-Mu. Having been sold off at extremely low prices, the FIZMO and particularly its rare module version are currently on the second user market at extremely high prices, but unfortunately suffered from an underspecified voltage regulator transistor in the power supply, so most units quickly died.

Around this time, at least two companies decided to try to work around the limitations of the General MIDI synthesizer standard by offering controller boxes, perhaps inspired by the Roland PG series and by the Matrix 1000 and MicroWave editors by Access. The Phat Boy from Keyfax Hardware was a simple editing panel for GM and XG synths offering fast access to filter cut-off and resonance, envelope, effects level, and other parameters, while the Control Freak from Kenton Electronics – who were also responsible for the design of the Phat Boy – was a more general purpose programmable editor for any MIDI parameter, which could act as a MIDI volume mixer as well as carrying out all manner of editing tasks. In 1999, a Phat Boy Version 2 with added sets of control functions in software, and a Control Freak Pro with 16 sliders, were also announced, and though both of these were a little ahead of their time, they marked the start of an interest in control surfaces that became much more relevant when software synthesis became more common, as we will see.

Meanwhile, Access, a German company launched with the Matrix 1000 and Waldorf MicroWave editor boxes, had been showing a fully equipped keyboardless desktop or rack-mount synthesizer, the Virus. This used powerful digital electronics to offer multi-timbral analog-like facilities with 256 memories and 128 Multi programs, the designers unashamedly comparing its 12-voice sound to that of the Prophet 5 and Memorymoog. The instrument was later updated as the Virus b, and a five-octave keyboard version, the Virus kb, was shown in 1999. The Virus appeared to offer the ideal compromise between simplicity and power; although all the most important parameters, such as filter cut-off, resonance, and envelope parameters have their own dedicated frontpanel controls, more detailed parameters, such as those for effects levels, were present though less prominently accessible. The Virus did have two definable knobs that could create several types of modulation simultaneously, and a "relative value" mode for all controls, which means they the can always start from their current programmed value, giving consistently smooth edits, while a system called Adaptive Parameter Smoothing avoided stepping or "zipper noise" during parameter editing. All sound editing changes were transmitted through MIDI for recording into a sequencer as part of a pattern, and an "Expert Mode" was included for editing the more complex functions. The Virus featured two multimode filters per voice, useable in series or in parallel and with programmable balance. Apart from the standard analog waveforms, there were 62 digital waveforms, dynamically mixable to facilitate PPG-like digital sounds, triple stereo audio outputs, and a stereo or dual mono audio input for processing external sounds both through filter and envelopes and through the effects, which included delays and a vocoder. Some example sounds and factory patterns are on the website, track 74.

1996 Waldorf launches the Pulse, a small true analog monophonic MIDI module

The Access Virus kb was a powerful early polyphonic virtual analog synth.

The Virus gives every appearance of being a powerful analog synthesizer, but in this case appearances are deceptive because its voices are

entirely derived through software programming, to such an extent that the Version 3 software announced in late 1999 not only instantly increased the number and variety of effects, but also increased the polyphony from 12 to 24 voices with no hardware change necessary. With Version 3 software installed in the b or kb version, the Virus offered 82 simultaneous effects (five individual effects on each of 16 Multi parts, plus 32-band vocoder and stereo delay), a new Retro Phaser effect comprising a six-stage, 24-pole stereo phaser, a distortion mode with bit reduction and sample rate variable for various saturation types from "Light" to "Hard", an analog bass boost effect, the chorus/flanger available in full stereo for all 16 Multi parts, a ring modulator to process the Virus sound against an external audio signal, again in stereo, FM (frequency modulation) for the oscillators with external input signals or noise also in stereo, an envelope follower allowing external audio signals to modulate the Virus filter or any other parameters chosen using the modulation matrix (which is useful for processing sample loops) and, most surprisingly, four oscillators per voice – three main oscillators plus one sub-oscillator (though polyphony was reduced by up to six voices when using the third main oscillator per voice).

Virus software updates including sound sets from artists, such as Rob Papen, became downloadable, "B" and "C" revisions, both in keyboard and module form, finding use with musicians, such as Karl Bartos of Kraftwerk, who particularly liked their vocoder effects, and the Virus C also became available in a more affordable, single-unit rack-mount version with very limited controls, while the Virus Indigo was a three-octave keyboard version with a striking colour scheme. Access still markets the later Virus TI keyboard and compact Polar, and their desktop versions.

1996 Roland launches MC303 Groovebox, first in a long line of analog imitators

Novation's Nova was intended as a virtual analog synth, effects unit, and vocoder.

Around the same time that Access launched the original Virus, British manufacturers Novation launched the Supernova, a very much extended

20-note polyphonic eight-part multi-timbral version of the Bass Station, expandable through hardware up to 40 voices and in a 3U rack-mount format. The Supernova featured three oscillators per voice, three envelopes per voice with two separate sets of controls, multi-timbral playing with independent effects on each part, multiple audio outputs and eight independent arpeggiators, though only a single programmable mode filter per voice. In 1999, a simplified 12-voice desktop or rack-mount version, the Nova, was shown, and this added audio inputs that could allow external sounds to be processed through the Nova's effects or through its new vocoder section. Effects included delays, reverbs, distortion, EQ, flanging, and phasing, and as well as an arpeggiator and complex built-in demo pieces, the Nova featured short demo patterns suitable to help assess bass, lead sounds, string, effects or other types of sounds (examples are on the website, track 75). It also had some powerful facilities not usually associated with analog synthesis routines; oscillators (because they are virtual) could be set to a Sync mode without tying up a second set of oscillators to do so and losing the possibility of tuning them apart, and the Sync mode could be altered to create even more extreme distorted and metallic sounds. Filter modes could quickly be set to 12, 18, or 24 dB, and the Nova was marketed in almost equal measures as an effects processor, synthesizer and vocoder.

Also during 1999, MAM showed several analog drum synthesizers including the DRM1 (eight-channel), DRM2 (single channel), and ACM2 (handclap), which emulated classic Simmons, TR606/808/909, and CR68/78 sounds. Launched simultaneously, MAM's SQ16 was a paperback-sized step-time MIDI sequencer for creating TB303 and TR808/909 type patterns, with a DIN 24 socket to synchronise to such older Roland gear. MAM was developing the stand-alone analog filter market too; the Warp 9 MIDI Space Filter had four filter modes similar to those on the EDP Wasp synthesizer, as well as envelope controls, 32 memories, and a MIDI-synchronisable LFO.

Another German company, JOMOX, had some success with an analog sound drum machine, the XBase 09, and in 1999 launched a rack-mount derivative, the AirBase99. This came in 1U format with many variable parameters for its analog bass, snare, and other drum sounds, and had 1024 set-up memories and MIDI. JOMOX also previewed an eight-part multi-timbral, eight-voice desktop analog synth, the SunSyn, intended to have a multi-mode filter, 350 memories, 40 knobs, 42 buttons, and a card slot for new waves and sounds. This was very slow to reach the market but reached a Version 2 before being discontinued some time after 2006.

During 1999, Technosaurus of Switzerland continued to develop the Selector modular system, previewing the 24dB/octave "VCF1" module, but also showing two small complementary instruments: the MicroCon and Cyclodon. The MicroCon was a miniature keyboardless analog synthesizer module with a single VCO plus sub-oscillator, and a filter, LFO, EG and VCA. The unit was small enough to sit on top of many other keyboards, such as the Yamaha AN1X (a MicroCon 2 version with MIDI followed), while the matching Cyclodon was a 16-step analog sequencer in a similar compact casing, and the matching Effexon (introduced in early

1997 Jean-Michel Jarre releases *Oxygene 7-13* using both analog and digital technology

2000) offered an additional LFO, ring modulator, overdrive circuit and two parametric equalisers.

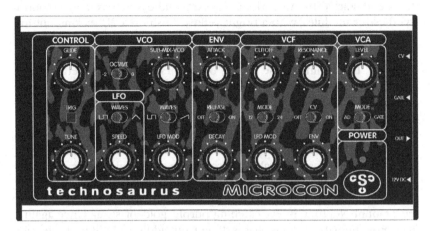

The tiny MicroCon analog mono synth from Technosaurus, now highly collectable.

A relaunch from the analog past came in the form of the Electro-Harmonix Guitar and Bass Micro Synthesizers, pedal-sized floor units mixing straight guitar with Octave Up, Down and Square Wave sounds through an analog filter, much favoured by The Chemical Brothers. All sorts of synthesizer and auto-wah effects were available (at the time of writing in 2018, both units remain in production) and can conveniently be added to digital keyboards, drum loops, or other sources as well as guitars. The fashion for stand-alone filters continued through 1999 with a new release, the Filter Factory from Electrix, which also showed the Warp Factory Vocoder and the Mo-FX for time-based effects. Waldorf, who had also been marketing stand-alone analog filters in the form of the 4-Pole and X-Pole, each derived from the MicroWave design, updated that instrument to the MicroWave XT, with many more physical controllers and an XTK keyboard version, and also showed a new keyboard, the Waldorf Q Synth. This was a digitally-based analog imitator, which, at first, appeared to be a new design, but later looked more like a simplified version of the very powerful and expensive Wave keyboard at perhaps a third of the cost. However, initial versions had non-functioning sequencer software and other problems, such as the lack of a dedicated control for LFO depth to the oscillator or filter. In early 2000, Waldorf showed a five-unit-high Q Rack version, with 16 voices expandable to 32 voices, 300 sounds, and a sequencer, and previewed a cut-down Micro Q, only two units high and with fewer controls, but still offering a powerful sound.

Simpler, but in many cases very powerful analog modules, had continued to appear around the world. The FutureRetro 777 was launched in the USA in early 1999 and, initially at least, sold only directly to users. The 777 was unusual in having genuine analog oscillators with a built-in 16-step sequencer, with CV and Gate In and Out as well as MIDI, but in having no memories or LFO and limited envelopes. Very much dedicated to creating bass sequences for the dance music market, the 777's monophonic two-oscillator sound was apparently strong (the sequencer

1997 Roland launches the JP8000, a versatile virtual analog keyboard synthesizer

became available separately later as the Mobius). In the UK, 1999 also saw the launch of the DSTEC OS1 Original Syn, with the involvement of part of the design team behind the Deep Bass Nine. This instrument continued the trend started by Waldorf and Spectral Audio for extremely colourful synthesizer modules, with a bright yellow finish on a 2U rack-mount format, and featured programmable memories, a simple three-digit display and two digitally stabilised analog oscillators (some sounds are on the website). If anything, these analog oscillators had slightly better sound quality than obvious competitors, such as the Waldorf Pulse, although the Original Syn had a simpler arpeggiator and much more limited modulation options, and made very little impact on the market.

In Switzerland, Spectral Audio had built, in limited quantity (around 75 units), the SynTrack, a 1U-high MIDI 99-memory monophonic synth with a 100-waveform digital oscillator, Moog-style analog filter, distortion circuit and optional audio input; then to add to the profusion of brightly colored analog modules available, followed up their Pro Tone with the 2U rack-mount Neptune in eye-catching purple, gold, or silver. Once again a non-programmable two-oscillator design, but adding a built-in "fuzzer" for distortion effects.

Red Sound Systems, a UK company formed by ex-Novation staff, had considerable success with their DJ equipment lines and in late 1999 announced the DarkStar, an eight-voice polyphonic, five-part multi-timbral analog-emulating synthesizer with white noise and audio inputs, in a desktop or rack-mount format (some sounds are on the website, track 73). Resembling the company's DJ products, it featured a joystick in the centre of the control panel, which could mix the balance of the oscillators or the ring modulator, or control the filter cut-off and resonance. Initially criticised only for its rather awkward method of sound selection, the DarkStar featured optional EPROM conversion kits (initially to offer monosynth and vocoder facilities) and though lacking their effects and arpeggiator sections, was launched at around half the cost of the competing Access Virus or Novation Nova. Later, the company showed the EleVAta (the emphasised VA standing for Virtual Analog), which was an expanded version in a rack that swivelled for desktop or mounted use, but this was not sonically very powerful and had little impact.

The NAMM 2000 in the USA and Frankfurt Musik Messe in Germany showed a continued interest from manufacturers in virtual analog designs. Novation launched keyboard versions of both the Nova and the Supernova. The Supernova II came with a five-octave, semi-weighted keyboard with aftertouch and in standard (24-voice), Pro (36-voice), or Pro-X (48-voice) versions, with 12- and 24-voice expansion cards available for the standard and Pro versions. The Nova II came with a four-octave, semi-weighted keyboard in a standard (12-voice), IIX (20-voice) or IIXL (36-voice) version with 12- and 24-voice expansion cards available for the X and XL versions. Each had a similar complement of controls to its rackmount version, with the exception of the envelope controls, which were sliders rather than rotaries.

Oberheim showed the virtual analog OB12, again derived from their collaboration with the Italian design company Viscount rather than from Tom Oberheim himself. However, this was a very striking and powerful

1997 PropellorHeads releases Re-Birth software, imitating the TB303 and TR606

instrument, with 12 voices, two oscillators per voice, four-part multi-timbral playing and a compact four-octave keyboard. It has a large number of knobs and faders, plus a ribbon controller, an arpeggiator, a motion recorder to record knob and fader movements, graphical representations of each parameter edited, including envelope shapes and waveform shapes on a large LCD display, and an S/PDIF digital audio output. A few OB12s remain on the second-hand market and can be an excellent purchase despite a tendency for the LCD to die.

For the millennium, Korg showed the MS2000, a three-and-a-half-octave (F to C), four-voice polyphonic virtual analog synthesizer inspired by the analog MonoPoly design. The synth had four dual oscillator voices with a powerful unison mode, the DWGS (digital waveform generation system) digitally sampled waveform mode of the DW8000 and EX8000 synths, a multi-mode resonant filter, ring modulation, white noise, modulation of various parameters by velocity, and keyboard tracking, external audio inputs, four-voice vocoder mode, a step-time sequencer featuring 16 patch select/sequencer step buttons, each coupled with its own sound edit control, and a six-mode arpeggiator. A 6U-high MS2000R rack-mount/desktop version was also available, initial pricing making it slightly more affordable than the Access Virus or Novation Nova. The MS2000, despite its four-note polyphonic limitation, seemed popular as a stage instrument and was seen in many concerts and festivals around the world, the ease of switching to its vocoder mode boosting interest in this facility.

Korg's MS2000 came in keyboard or rack-mount/desktop versions.

1998 Quasimidi launches the Sirius synth/vocoder/sequencer at 1998 Deutschmarks

E-Mu had also launched new lines for the millennium. The Proteus 2000, which was a general-purpose rock and pop module, updated the earlier Proteus/Morpheus module design with internal slots for optional expansions, but the Audity 2000 referred specifically back to the never released Audity analog synth from the early days of E-Mu. Sharing the many filter types of the Morpheus, the Audity had specific waveforms from the earlier synth, as well as many routing, patching, and effects options. The E-Mu lines became a popular resource for analog-type sounds; the Orbit V.2 was superseded by the XL1 Xtreme Lead, offering a similar selection of techno and electronic sounds with an improved Beats mode for drum loops, 16 simultaneous arpeggiators and an expansion slot for additional sound boards, such as the "Techno Synth Construction Yard". The Vintage Keys and Classic Keys modules were superseded by the Vintage Pro in the same series (as well as by desktop "Command Station" versions and a rack-mount Proteus 2500 with additional controllers and a sequencer), with its basic internal chip of analog synth and electromechanical keyboard sounds, plus the possibility of adding the Techno Synth board, an XL-1 board, a Mo'Phatt board for dance rhythms, or any boards up to a total of four including classical, rock, and percussion sounds; the Vintage Pro in its turn became briefly a popular resource with players such as Paul Shaffer in the USA.

Roland updated the MC505 Groove Box with an MC307 version with 64 voices, 240 rhythm patterns, and Turntable Emulation controls for adjusting a pattern's pitch and tempo, while Sherman launched their QMF (Quad Modular Filterbank) in a limited edition of 100 units. Bob Moog's Big Briar showed the first prototypes of the Minimoog-like MIDI monophonic "Performance Synthesizer", which was to be launched later as the Voyager. Also from the USA, Peavey launched the Paradox with an impressive remote control panel, bristling with knobs and complete with a pitch bend strip and a motion-sensitive aerial controller, the synthesizer itself was a minimalist one-unit rack-mount with a minimum of editing possibilities and a strong monophonic two-VCO layout, but never reached the market, and even the prototype was apparently disassembled.

1998 Ensoniq shows the FIZMO virtual analog synth but soon merges with E-Mu

In early 2000, Germany's Touched By Sound had announced the Mephisto, described as "*die Harley fur den Soundpuristen*", a six-voice, five-octave analog synth with three oscillators, two envelopes, two LFOs, and two filters per voice, ring modulator, sample-and-hold, 100 memories and 36 motorised controls, which rotated to show the current setting whenever a new sound is chosen. But the instrument was expensive and completed at most in very small quantities.

Nord had a similar design idea, updating the Nord Lead to the Nord Lead 3 (and later the Nord 4), with a circle of LEDs around each control, which again show the current setting of every parameter each time a new sound is chosen. The simplified Nord Electro keyboard included virtual models just of the sounds of electromechanical instruments specifically the Hammond organ, the Hohner Clavinet, and the Fender Rhodes and Wurlitzer electric pianos, and is still marketed in a revised Mk 5 version at the time of writing, while the Nord Stage, in compact, 76- and 88-note versions, was introduced to add back a slightly simplified virtual analog synthesizer section.

In the USA, Alesis, previously only known for digitally based synthesizers and effects, announced the Andromeda, again a "genuine analog" synth. This had 16-voice, 16-part multi-timbrality, two oscillators with sub-oscillators and five waveforms per voice, three LFOs and three seven-stage envelopes per voice, 32 multi-mode filters, "hard" and "soft" oscillator sync, 16 mono audio outputs plus stereo outputs, ribbon controller, 72 knobs and 144 buttons arranged in a futuristic front-panel layout styled in Germany, an arpeggiator, sequencer, effects and analog distortion. Andromeda looked like being one of the most powerful analog synths available. At $3000 it was quite an expensive proposition but was taken up by Klaus Schulze and others.

In contrast, in early 2000 Seekers of Japan, which had marketed a successful MIDI controller and vocoder, had previewed the design of the SMS1000, a 200-memory, three-VCO monophonic design with oscillator sync, low and high pass filters, and a large LED matrix that impressively displays the current wave shape, but it's not clear if this reached the market at all, while US valve effects specialists Metasonix showed the PT1 Phattytron, a rack-mount MIDI module using an unconventional selection of valves in both its oscillator and filter circuits. With a minimum of front-panel controls and highly unusual filter response, the Phattytron – with an LFO, VCA, dual envelopes, audio input, programmable modulation routing and a single oscillator, or with two oscillators to special order – created a warm and, if required, highly overdriven sound quite different from anything that could be expected either from a TB303 or from a Minimoog, and Metasonix also offered a valve distortion processing module, the TS21 Hellfire. The Phattytron's sound is featured on the demo website (track 72).

1999 E-Mu launches the Audity 2000 module inspired by their unreleased Audity synth

The Alesis A6 Andromeda was a genuine analog synth which set new standards in the field.

From around 1999, the manufacturers involved in the analog revival had been given second thoughts by the rapid appearance of simulated analog sound software. This should not have been a surprising development; virtual analog synthesizers, such as those from Clavia and Access, were basically microcomputers running a specialised set of instructions, though usually accompanied by dedicated DSP (digital signal processing) chips. If a standard home computer's own microprocessor became efficient enough to run similar sound creation software and to handle the DSP functions as well, then any computer could be turned into a professional musical instrument.

Obviously, the use of computers to create some sort of electronic music – from the giant RCA computer of the 1950s programmed from punched tape to the domestic Commodore 64, Apple 2 and Atari ST used to run MIDI sequencing software or to create simple bleepy noises from a built-in sound chip – was by that time commonplace, but around this time it became possible for the first time for an affordable computer to create sounds closely comparable to those of a professional synthesizer. The relatively simple analog sounds were among the first targets for emulation, so from this time the analog enthusiast was faced with a choice of searching out classic analog, or buying newly manufactured virtual instruments, or going the software route. This obviously had an impact on hardware development, though it turned out not to be as detrimental as might have been expected. A look at expansion in analog simulation software is given in Chapter 7.

The introduction of analog simulation software did have an initial impact, though, on the second-hand value of some classic instruments. While the Roland TB303 Bassline had reached a second-hand value of over £1000/$1500, this had been only slightly dented by the availability of hardware imitators at perhaps a few hundred pounds. But when convincing TB303 simulation software appeared in the form of PropellorHeads Re-Birth, musicians (particularly those already owning a computer) really started to wonder if there was any longer a need to buy classic hardware instruments at very high prices. After this time, there was nothing as simple as a hardware TB303 emulator marketed, (though that has now changed with the launch of the Roland Boutique and similar lines) and designers of both genuine analog and of virtual analog hardware instruments were seriously considering what added value they were offering to the potential purchaser.

In the case of Moog this was clear enough. The return of the brand name had been much anticipated and it was pretty clear that musicians would pay a premium for a genuine new product under the Moog brand. This was confirmed with the eventual launch of the Voyager, the new name chosen for Moog's Performance Synthesizer. The Voyager superficially resembled a Minimoog, with the same three-and-a-half-octave keyboard and sloping control panel, and has similar basic sound facilities: three genuine analog oscillators plus white noise. But a vast number of updates made the Voyager a much more contemporary machine. It had MIDI, was fully programmable, offered velocity sensitivity, added an X-Y touchpad that sent MIDI controllers to the existing complement of physical controllers, added a high pass filter and a split mode on the low pass filter with a variable range between two independent peaks, had oscillator sync for those screaming sounds from the Moog Prodigy, an independent LFO so modulation was still available when playing three-oscillator sounds, and so on.

What it did not add was polyphony (or built-in effects), and though many, many purchasers were very happy with the classic, heavy Moog sounds it created (along with lighter and more subtle ones now, thanks to

1999 Clavia releases the Nord Lead synthesizer played by Mike Oldfield and Jean-Michel Jarre

the high pass filter and other additions), it has to be wondered whether any other company would have succeeded in marketing such a product at the time, certainly at the Voyager's substantial launch price of over £2000/$3000.

However, the Voyager was a success, had its operating software revised once or twice, and was joined by a keyboardless rack-mount version and an optional VX351 interface, which made many of its internal parameters accessible for patching into a modular system. The Moogerfooger pedal range was also expanded, with the addition of the CVP251 Control Voltage Processor, a small box cosmetically matching the Voyager interface and offering an LFO, voltage mixer, sample-and-hold, and other basic analog facilities. The CVP251 was intended to turn a set of Moogerfooger pedals into an integrated, patchable system, so almost realising the dream (other than offering oscillators) of a new, small Moog modular synthesizer, but that was to come about some years later.

The giant Synthesizers.com Studio 110S closely resembles the classic Moog modular systems.

1999 FutureRetro 777
monophonic synth with
sequencer launched in USA

Bob Moog celebrated his seventieth birthday and the fiftieth anniversary of his business in 2003, but soon after was diagnosed with a brain tumor and sadly passed away in 2005. Before his death he had completed the design of what became known as the Little Phatty, a stripped-down version of the Voyager launched in 2006 at around £800/$1500 (the exchange rate having changed substantially by this time). The Little Phatty offered a three-octave keyboard, two oscillators, MIDI, and 100 programmable memories, omitting the X–Y pad, splittable and high pass filters, white noise source, and most of the panel controllers from the Voyager design, offering only one selectable controller in each area, and successfully replicated most aspects of the Voyager sound. Again, without offering polyphony or effects, it seemed underpowered compared to polyphonic virtual analog keyboards available at half the cost, but once again the Moog brand name was regarded as a highly desirable feature.

Bob Moog in the background at
a Moog Voyager demonstration
during his final UK visit.

In the few years between the release of the Moog Voyager and the Moog Little Phatty, perhaps in part because of the rise of analog software, substantial changes had taken place in the analog hardware scene. In Germany, both Quasimidi and Waldorf had closed (the latter only temporarily), despite more affordable models such as Waldorf's 2U-high Micro Q Omega, a more powerful version effectively of a digital wave shape-equipped, polyphonic Pulse module with effects, which still at the time of writing remains a good second-hand purchase. The powerful Mephisto from Touched By Sound had no impact, and analog designs from Vermona – the controller-free 1U monophonic MARS module with its optional programmer and the four-voice PerFourMer, which had no memories and so in many ways resembled a Korg MonoPoly in rack form – came and went. However, one company that did have an increased impact was Manfred Fricke's Berlin-based MFB. The company had been in the electronic design business for many years, marketing video processing equipment and in the 1980s a very early digital drum machine, vastly undercutting the cost of contemporary models from Linn and others by using commercially available "project box" casings and very limited displays. These were marketed in the UK by Syco Systems, but the MFB models were not widely distributed through the 1990s until their launch of a new range of small musical instrument products. The MIDI-equipped MFB Synth was a wedge-shaped desktop, three-oscillator, fully variable true analog monophonic synth with a micro button keyboard layout and built-in sequencer (again in a commercially available plastic casing) and roughly equivalent in layout to a Minimoog, and the subsequently introduced Synth Lite was even smaller, omitting the keyboard and moving to dual DCOs, but with the same powerful 24 dB/octave fully resonant filter.

Around 2004, both models were updated to Mk 2 versions with fully programmable memories and built-in programmable sequencers. In 2006, the MFB Synth 2 and Synth Lite 2 certainly constituted the smallest analog-style MIDI-equipped synthesizers in production, and had been joined by a matching monophonic sequencer with both MIDI and CV/Gate outputs, a Simmons-style analog drum machine, a sampled drum machine featuring cult sounds from 1980s instruments, a stereo analog filter with step sequencing, and a four-voice polyphonic version of the Synth Lite 2 with no controls but programmable either from the SL2 itself or from any MIDI controller. MFB continued to build on this legacy, more details in Chapter 8.

On the subject of very small synths, Korg, like Vermona, had already looked back to the MonoPoly design for the millennium with the release of the MS2000, and now repackaged the MS2000 circuitry in the compact MicroKorg, which offered three octaves of mini-keys, fewer controls, a simple numeric display, battery power and a microphone for the vocoder on a flexible gooseneck. Again, this was a popular stage and studio instrument, and reflected the interest in "classic" analog sound by labelling its preset banks "techno", "classic", "vocoder", and so on.

The success of hardware instruments using virtual analog made some designers keen to compare its abilities with genuine analog, so back onto the scene came Dave Smith, the designer of the Prophet 5 and head of Sequential Circuits. Smith had moved to the software design arena for some years, but had been called upon by Roger Linn to help specify the AdrenaLinn, a small drum machine with guitar effects facilities allowing it to create rhythmic effects patterns. Smith launched his own new company, Dave Smith Instruments (DSI), and the first new product was a small desktop hardware synth module with similarities to the AdrenaLinn, the DSI Evolver. The Evolver could also act as an effects unit, but was a monophonic analog MIDI synth with a sequencer, two genuine analog oscillators and two digital oscillators offering all the waveforms of the old Prophet VS synth.

2000 Korg launches the MS2000 virtual analog synth and module inspired by their MonoPoly

2000 Dave Smith launches his new company DSI with the Evolver desktop analog synthesizer

The author (right) and classic Prophet T8 with Dave Smith, inventor of the Prophet 5, co-designer of MIDI and now marketing DSI instruments from the Evolver Mono (inset) to new Prophet keyboards, and now Eurorack modules.

The Evolver was a great success. With its strong Prophet- and Moog-like sounds, plus powerful effects, it could sound like it was playing polyphonic patterns. In quick succession between 2002 and 2006, DSI then launched an actual four-voice rack version, the PolyEvolver, a four-voice keyboard version and then a monophonic three-octave keyboard version, the Mono Evolver. By 2006 these represented the most flexible and powerful-sounding instruments with genuine analog oscillators on the market, though at a cost comparable to large samploid workstations, but this was just the start for a re-birth of Dave Smith's designs, of which more in Chapter 8.

Around the time Dave Smith's company relaunch was taking place, a second design called MARS had appeared in the UK: the Monophonic Analog Rack Synth came from a small company called SMS and used an imaginative semi-modular design in a fetching dark blue finish. This comprised a two-oscillator, non-MIDI pre-patched variable analog synth module, with patch points cleverly consigned to the rear panel to save space. These could all be brought to the front with an optional patch bay called BOB (for Break Out Box), and a combined system of two synths with patch bays – or better still the Planet 9 expander, which added patch points, sample-and-hold, lag generator, and more – formed an affordable and powerful way into modular analog synthesis. However, very few systems were built and this was also the case for other UK designs, including the similar Red Square Phobos, again a semi-pre-patched modular rack; the Vostok, a costly tribute to the suitcased EMS Synthi A with MIDI and a pin bay; the small Euro Rack format Black Coffee module, which was almost a complete synth in itself; the Filter Coffee, which was a 1U MIDI-equipped analog filter; and the Macbeth M3X synth, which was very much in the style of a rack-mounted Minimoog with MIDI.

2000 Waldorf releases the Q Rack module, revived in 2006 as Terratec Komplexer software

Much greater market strides were being made in the UK by Nova-tion, which launched various virtual analog keyboard synths in the Nova line, updating the design as the KS4 and KS5 models. The very convincing virtual analog system found in these synths, referred to as "liquid analog", was used in several other models. The K-Station had a very compact two-octave keyboard but was solidly constructed and an excellent bet for playing analog lead and bass lines. Its keyboard was removed to create the rack-mount or desktop K-Station Rack, which resembled Roland's JP8080 module in its mixed use of knobs and sliders, and finally the design was squeezed into a single rack unit for the A-Station, available in a blue or a silver finish. The A-Station repre-sented one of the best virtual analog deals of its time. In 1U of space it squeezed in 25 rotary controls in two rows, and polyphonic playing of strong two-oscillator sounds with effects, vocoder, and arpeggiator. It only had a two-digit display so sounds could not be named, but it had a small keypad for fast sound selection and was used by The Stranglers, Rick Wakeman, Howard Jones, and many others. In late 2018 it's still an excellent second-hand bet.

Very significantly, the software design of the K-Station and A-Station allowed Novation to launch an exactly compatible piece of software, the V-Station, which could load their patches. This was inexpensive and gave an excellent analog sound on a Mac or PC. The author used it extensively on the all-virtual album, *If The World Were Turned On Its Head, We Would Walk Among The Stars*, premiered at the Pittsburgh Planetarium in 2005. There is more in Chapter 7 about the V-Station, which emphasised Nova-tion's very wise decision to get into the software market rather than relying solely on sales of hardware synthesizers. Also appearing in this hybrid market was Korg Legacy, which was actually a software prod-uct offering a patch-compatible software replication of the Wavestation synth along with virtual emulations of Korg's Polysix and MS20 synths, but is mentioned here as well as in the software section in Chapter 7 because initial shipments were accompanied by a hardware bonus, a USB controller keyboard in the form of a five-eighths-sized Korg MS20 with three octaves of mini-keys and mini-jack patch sockets. This was a very attractive little instrument, since connecting cables into its patch sockets actually made corresponding changes on the MS20 software, though sadly it turned out that it was not actually handling voltages and so could not help patch the software directly into a hardware modular synthesizer. The unit quickly started to turn up on the second-hand mar-ket without its accompanying software, as its distinctive shape made it less than ideal as a general-purpose master keyboard, and later it was rather superseded by the similar looking MS20Mini, which did actually create its own sounds.

2000 Touched By Sound previews the Mephisto synth with motorised faders

Novation had also turned largely to the manufacture of MIDI/USB master keyboard controllers intended to run software synthesizers and sequencers, but did not neglect the possibility of re-incorporating synthe-sizer facilities. The short keyboard ReMote 25 was designed to edit Reason and other software packages described in Chapter 7 (hence its name), but was quickly upgraded to run internal synthesis software itself, so as to become a powerful hardware virtual analog synth. This whole package was then revised as the X-Station, which also had a USB audio interface, and Novation successfully started to ring the changes on this design with hybrid controller/synthesizer keyboards in various sizes.

Alesis also took a step back from the pure analog design of the expensive A6 Andromeda with the Ion, a fairly compact virtual analog polysynth with eight-voice polyphony, three oscillators per voice, two multi-mode filters, and effects. The entire Ion spec was then ported into the Micron, a very compact three-octave synth with few controls but potentially very powerful sounds and arpeggio patterns. At the time of writing, this is widely available second-hand and useful as a compact controller keyboard that just happens to be crammed full of strong virtual analog textures. Its design re-appeared little changed on the Akai MINIAK.

Creamware, which had earlier offered the MiniScope as a software emulation of the Minimoog, in an unusual move started to offer desktop hardware instruments that ran software imitations of classic synthesizers including a Minimoog, a Prophet 5, and a Hammond organ. These were known as the ASB series (for Authentic Sound Box) each with a USB connection for computer interfacing, which allows them to run as a remote-controlled hardware instrument.

Access had developed their Virus line with the TI models: the three-octave Polar finished in white, the five-octave TI Keyboard, and a new TI version of the desktop module. TI stands for Total Integration. These models were directly compatible with a VST and Audio Units format software plug-in (of which more in Chapter 7), and could also function as programmable MIDI controllers and as an audio interface. As these had all become rather expensive, a more affordable Virus Classic desktop module without the TI functions and based on the earlier Virus B design was also launched.

In 2006, Novation (by now bought out by the studio equipment manufacturers Focusrite) showed the very compact ReMote LE, which offered two and a half octaves of keys with various controllers, including an X–Y pad and a joystick, but it was only a matter of months before this design too became available in a version with built-in sounds. Novation's tiny so-called XioSynth broke new ground in virtual analog design, offering all these MIDI controller facilities, which were invaluable in working with software synthesizers, plus an audio-to-USB interface for computer multi-track recording applications, built-in sounds using the typical Novation "liquid analog" routines, and battery or USB power. At just £229 ($400) with a four-octave model also available, the XioSynth became a very hard act to beat, the author keeping four for studio and for stage use.

Although virtual analog was making great strides in accessibility and affordability, after a flat period there had been a great development in modular analog synth design. Around the millennium there had been perhaps five substantial manufacturers on the scene, but within a few years 15 or more had appeared. Many of these were in the USA, unsurprising given the great tradition of analog modular design there. Companies included Sound Transform Systems making the Serge lines, Synthesizers.com

2000 A revived Oberheim shows the OB12 virtual analog synth built in Italy by Viscount (also known by their original name MOTM or Module of the Month), Blacet, and in the UK, Oakley, plus several others.

The area of Eurorack and other module manufacture became the most rapidly expanding of recent times, so that the original discussion of around 20 manufacturers in the field from the first edition of this book should now be expanded to encompass around 300, with six and a half thousand Eurorack products alone on the market. Obviously it's no longer possible to discuss all of these in detail (though see Chapter 8 for many updates), but rather than removing the listings and gear photos for these pioneering companies, they have been retained here for historical interest. Prices and mentions of current products should be ignored, the Anyware Instruments products, for example, all being out of production though their website persists, and some of the other companies, such as Bananalogue and Technosaurus, certainly no longer exist.

- **Analog Systems (www.analogsystems.co.uk)** Around 50 modules in Eurorack format using mini-jack patching, and including authentic-sounding EMS filters, voltage-controlled phaser and stereo delay, and MIDI module; £120 for a basic VCO, plus keyboards including an unusual Ondes Martenot-style controller. Through Big City Music in the USA and Fukusan Kigyo in Japan.
- **Analogue Solutions (www.analoguesolutions.com)** Red Square Phobos patchable modular at £586, SEMblance clone of Oberheim SEM with MIDI plus four-voice version, Oberkorn step sequencer, Vostok briefcase synth with MIDI and 484-way matrix pin bay, TBXTB303 Trans-Bass-Express clone of Roland TB303 voice. Concussor modular system with over 30 mini-jack patched percussion and synth modules, roughly compatible with Doepfer and Analog Systems.
- **Anyware Instruments (www.anyware-instruments.de)** German based, Semtex is a rack-mount Oberheim SEM clone with many patch points.
- **Bananalogue (www.bananalogue.com)** Collaboration of Serge Tcherepnin of Serge with Ken Stone and Seth Nemec to build Serge-style modules.
- **Blacet Research (www.blacet.com)** Frac Rac format. VCO2100, $175 kit, $250 assembled. Forty modules plus power and cabinet options, including a bar graph display, the ID2510A "Improbability Drive" noise generator and the Mini Wave digital oscillator, which holds 256 different waveforms.
- **Buchla & Associates (www.buchla.com)** Don Buchla's company, with 200e Electric Music Box modular systems very extensively updated around 2005, based on the 200 series from the 1970s and 1980s, with modules from $700, systems from $10000–$30000 with full MIDI and including a memory module. Also, MIDI controllers, such as the Lightning 2 "air drums" and Marimba Lumina tuned percussion controller, Buchla also designed the Piano Bar for Moog Music to add a MIDI-controlled module to acoustic pianos.

Buchla's large 200e modular
system, as played in 2018 by
Suzanne Ciani.

2000 Alesis shows A6
Andromeda true analog
keyboard synth promoted by
Klaus Schulze

- **Cwejman (www.cwejman.net)** Swedish builder of the S1, a powerful semi-modular rack-mount synth with MIDI roughly resembling a Minimoog but more flexible, and more recently of a small range of individual modules.
- **Cyndustries (www.cyndustries.com)** FracRac size but also MOTM, Blacet, Synthesizers.com, Euro Rack, Modcan versions with adaptor modules for banana jack systems like Serge. Zeroscillator with ten-turn tuning knob can be used as LFO and FM source, $750. Around 30 modules, including Twin Wasp Filter and Nyle Steiner designed Synthacon filter, $260.
- **Doepfer (www.doepfer.com)** Euro Rack format A100 modular system with studio or portable metal racks, over 100 modules available as well as MIDI converters, keyboards and other interfaces.
- **Encore Electronics (www.encoreelectronics.com)** MIDI interfaces and retrofits for Roland Jupiter 8, Moog Source and others, MOTM and Frac Rac style frequency Shifter at $389 and some other modules.
- **Livewire (www.livewire-synthesizers.com)** Short-run filters and other modules in Euro Rack format, including the Dalek Modulator and Dual Cyclotron, a very powerful and flexible LFO.
- **Macbeth (www.macbethstudiosystems.com)** M3X, which resembles a rack-mount Minimoog, and M5, which resembles a rack-mount ARP 2600.
- **Manikin Electronic (www.manikin-electronic.com)** Hardware offshoot of the synthesizer music label, offering the Schrittmacher MIDI-equipped analog step sequencer.
- **Modcan (www.modcan.com)** Over 40 modules in A (black with banana connections) and B (silver with jack connections) series. Dual oscillator $465.

● **Moog Archives (www.moogarchives.com)** History and photos of Moog products, including modular systems and rare prototypes such as the SL8, which was an attempt at a budget Memorymoog from 1983.

The Zeroscillator module from Cyndustries.

Blacet and MOTM/Synthesizers.com module formats compared.

● **Oakley Sound Systems (www.oakleysound.co.uk)** Based in Edinburgh, apparently independent of Oakley Sound Systems from August 2006. Ten different PCBs (or currently just circuit diagrams for them) normally built MOTM compatible, but users have to source panels (usually using the Schaeffer panel design and manufacture service from Germany), and components that are standard and widely available.

Modcan B series modular
system.

A well constructed example of
the Oakley modular system.

- **PAiA (www.paia.com)** 9700 series modular system still in production, with elements combined so that four modules can provide two
  each of VCOs, VCFs, VCAs, EGs, plus noise and MIDI In. Kit $450. Frac
  Rac format. 9730 dual filter kit alone $107. Theremin, vocoder, Fatman
  synth, fuzz and other kits also available.
- **Sound Transform Systems (Serge) (www.serge-fans.com)** No
  official web page for the company at the time of writing, but this site
  does offer a contact phone number, and Serge Tcherepnin himself is
  involved with Bananalogue.
- **Synthesis Technology (Synthtech.com)** MOTM modules, 5U
  high, mostly one-fifth of 19-inch rack wide. Also Frac Rac versions in
  the MOTM-1000 series compatible with Blacet and PAiA (smaller at
  3 × 5.25 in.). MOTM-300 Ultra VCO, $349 kit, $449 assembled. Users
  include Robert Rich. Synthtech also work on DKI Synergy keyboards
  and Mulogix modules, and stock some CEM and SSM chips and data
  sheets.

An MOTM modular system conveniently mounted in an SKB portable rack.

- **Synthesizers.com (www.synthesizers.com)** Over 40 different modules. Rack (in MDF or walnut), portable, and studio systems. Custom systems. Excel spreadsheet for planning. Starter system from $120/month (22 spaces, MDF cabinet, $2658 total, 110 space $14000). Jordan Rudess is a user. Q960 Sequential controller is a clone of the Moog 960 sequencer.
- **Technosaurus (www.technosaurus.ch)** Small and large modular systems and "small monster" series MicroCon, MicroCon 2 synth, Cyclodon sequencer and Effexon, but no developments since around 2000.
- **Wiard (www.wiard.com)** A dozen or so modules in Frac Rac format finished in blue or black, some available only on a long waiting list, including a joystick at $150, a joystick output expander to handle several additional parameters, eight-step sequencer, various filters and power supplies.

Check Chapter 8 for a more detailed discussion of what happened to these pioneering companies in the later explosion of Eurorack and other module manufacture.

Moving back from a consideration of modular synths to more conventional keyboards, in 2004 Roland launched the lightweight five-octave Juno-D, and while this had little in common with the original analog Juno synths, it shared its accessibility for entry-level players and did have controllers for filter cut-off, resonance, and envelope times for the (digital) parameters applied to its (basically General MIDI) sound set. Roland used the success of the Juno-D to revive the brand names of some other products from the analog age. The Juno-G was cosmetically styled to very closely resemble the analog Juno 106, but had almost nothing in common with it, stemming rather from the design of the contemporary FANTOM samploid workstations. All of these synths (as well as the competing Korg Trinity and Triton, and the Yamaha MOTIF keyboards) by this time had convincingly authentic-sounding, smooth analog-style, and usually variable resonance filters all modelled digitally, obviously sported the simple analog wave shapes as well as more complex sampled waves, and offered plenty of controller options, and so could very convincingly produce analog-style pads, basses, lead lines, sequences and much more. Korg updated their MS2000 line to the MS2000B keyboard and BR module, with improved sounds and a new black finish, but later launched the microX, which had almost nothing in common with the MicroKorg, being a miniaturised Trinity-series workstation with the usual piano, guitar, pop/rock and other sounds. But with filter cut-off and resonance controller knobs, the microX was also capable of some analog-like textures, so could seem better value for money than some genuine or virtual analog-only instruments.

For example, the microX was around half the price of the company's RADIAS, launched around the same time. This was a rack unit very much in the style of the MS2000BR, but with an optional keyboard into which it could be slotted. Still with four-voice operation, it added digital synthesis routines to virtual analog and vocoder sounds, and looked terrifically stylish but seemed expensive at launch.

Roland had also continued to market Groovebox products, often with analog-style controls, reviving the brand name of their long discontinued VP330 Plus analog vocoder/strings keyboard for the VP550. This had almost nothing in common with the much earlier analog design, using completely digital format and pitch shifting rather than analog filtering vocoder techniques, but the use of the name did tend to emphasise the extent to which the major manufacturers now wanted to reflect their "glory days" of analog design. The naming of Roland's SH201 also made this very clear. Its claimed heritage was from the SH101 and it slightly resembled the large SH5, offering two-oscillator sounds and a very basic display and effects, but claimed to offer an easy way into learning classic analog techniques. Its complex control layout was baffling at first, though the sounds with long internal delays or reverb added ultimately could be very rewarding.

Despite the vast number of analog and virtual analog products available on the market by this time, and the inroads into the potential market made by virtual analog software, imaginative new products and even new manufacturers continued to appear. The Radikal Technologies Spectralis, for example, designed in Germany by part of the same team that had worked on the huge digital Hartmann Neuron synth, offered analog-style filtering effects applied to drum and polyphonic sequence patterns, vastly expanding on the style of techno-oriented virtual analog products, such as Quasimidi's 309 Rave-O-Lution. True modular systems continued to expand, with new module designs incorporating different types of filter and new ways to handle and modify voltages, while products such as the AdrenaLinn and Evolver demonstrated the attempt to bring guitarists and other instrumentalists into the analog synthesizer fold. Despite the enormously rapid development of analog imitative software outlined in the next chapter, the future for analog and virtual analog hardware instruments continued to look bright.

The period 2007–2012 saw an expansion both in hardware and software products, with many companies marketing both. Waldorf surprisingly re-appeared with a small four-octave analog imitative synth, the Blofeld, in silver or black together with a very compact desktop version, and revised PPG Wave 3 software. Another software synth Largo has features from the Blofeld and the older Q synth keyboard.

In the UK, Novation launched many controller keyboards in their SL and Nocturn lines and in 2010 the Ultranova, with a three octave keyboard and a vocoder. M-Audio's compact analog imitative synth was Venom, a strong-sounding techno dance-oriented four-octave design with four main controller knobs and a very affordable price point.

In the USA, Dave Smith Instruments expanded the Evolver desktop synth into the four-voice Tetra model, re-styled it into an 8-voice Prophet 08 keyboard, which had much in common with the old Prophet 5 and Prophet 600 designs, spun off the single-voice Mopho desktop module and two-and-a-half octave keyboard version, and continued to work with Roger Linn to launch the Tempest analog-style drum machine. Moog expanded their Little Phatty range with module versions and launched the Moog Voyager XL effectively adding the CP251 expansion box and a ribbon controller to the Voyager design. In 2012 the company launched Minitaur, a desktop 2-oscillator non-programmable synth for bass sounds inspired by their Taurus bass pedal products. The Moog 500 Series of rack modules, starting with the Ladder Filter suggested the eventual return in the future to a modular synth product line.

In Japan, Korg launched miniature monophonic desktop synths in the Monotron series, some including a simple ribbon type keyboard (Gakken's SX-150 is a very similar product in kit form), while Roland marketed the flexible three-octave polyphonic Gaia SH01. Akai also returned to keyboard manufacture with the three-octave MINIAK (based on the Micron from Alesis, a company also owned by Numark) which included a vocoder.

Other countries launched analog hardware too. The huge 8-voice Schmidt synthesizer from Switzerland previewed in 2011 came from Axel Hartmann, designer of the MAM lines and the Hartmann Neuron, but was built in very small quantities, perhaps 25 in a batch. Jean-Michel Jarre became a fan. But perhaps the most surprising appearance was the 2012 launch of the ElectronicMusicWorks.com products from Brazil.

These included two desktop monophonic synthesizers with wooden end cheeks, two small modular systems, a host of MIDI accessories and patch storage units for specific MIDI synthesizers, and later on Eurorack modules.

The EMW lines appear solidly designed and indicate a trend towards an involvement in electronic equipment production within up-and-coming economies. However the recent upsurge in software design has been worldwide too, as we will see in the next chapter.

7

Programming and using virtual analog hardware and software

As we have seen in earlier chapters, the strength of the "analog revival" came as a surprise to most of the major instrument manufacturers, whose main and much delayed response was simply to reintroduce analog-style front-panel controls to some of their existing digital synthesizer designs. But for those manufacturers who did wish to offer some analog-style product without the cost and potential unreliability of using genuine analog circuitry, the "virtual analog" synthesizer became the obvious solution.

One of the first implementations of this idea was on Clavia's Nord Lead synthesizer, which was rapidly followed by similarly themed offerings from Access, Roland, Quasimidi, Novation, Red Sound, and others. The attractions of virtual analog rapidly became obvious; tuning and stability problems disappeared, the circuitry was relatively cheap, improvements could be made through software revisions rather than through expensive hardware upgrades, and additional facilities not available on genuine analog synths (such as programmable digital effects and new modulation options) became easy to introduce.

2001 Korg releases the MicroKorg virtual analog synth with built-in vocoder

One question remained: did the virtual analog synthesizers sound as good as the originals? Certainly, the basic approach would seem to be more than sufficient to achieve this. Firstly, the familiar analog waveforms – sine, triangle, square, sawtooth, and perhaps white noise – had to be generated by digital elements rather than by analog circuitry. This is a trivial matter for microprocessor-based designs and digital sound processing chips, which were easily capable of creating such simple waves. Then the analog filter had to be reproduced digitally. This was a much more challenging proposition; the exact tonal response of any given analog filter design (such as that on the Minimoog or the Korg PS3300) is difficult to reproduce, and even if the tonal response is right, the way the filter cut-off point is changed is very different in the digital domain than in the analog domain. A lot of processing power has to be devoted to controlling a filter at any more than (say) 128 different levels, but stepping it in fewer levels than this can result in a quite unauthentic stepping or "zipper" noise while making filter adjustments.

Lastly, the volume envelopes applied to the oscillators and filters have to sound authentic. Analog circuitry can sometimes respond extremely quickly, at speeds that can challenge those generated by the best microprocessors. Insufficient processing speed can result in envelopes that lack sharpness in the attack stage, particularly if the processor has large numbers of voices to handle simultaneously.

Quite apart from the process of sound creation, there is the question of how the sound manipulation facilities are presented. The design of a virtual analog synthesizer gives the possibility of offering facilities not available on the classic analog synth, but economic constraints mean that these new functions are often hidden behind multiple function buttons or menus. In summary, it is quite possible to design a virtual analog synthesizer that has a weak and unconvincing-sounding filter, slow, and flabby envelopes, and inaccessible controls that simply lose much of the spontaneity of genuine analog performance and editing.

Happily, most of the products available in this area to date have pretty much avoided these problems. Clavia's Nord Lead was an early success, with just four-voice polyphony, no effects and a short four-octave keyboard, it looked like a very limited instrument at launch, but of course thanks to the revival in interest specifically in analog sounds it did not have to attempt to compete with contemporary all-purpose instruments. As long as the Nord Lead could handle Minimoog-like basses, Prophet 5-like twangs and Juno-like arpeggiated effects, with good MIDI implementation and stable tuning, the designers were onto a winner.

That's not to say that the virtual analog designs were achieving everything that could be had from a classic synthesizer. All of them could handle a Minimoog-like bass easily enough, but few could compete with the patching options and control possibilities of an EMS VCS3. However, the next generation of virtual analog synthesizers did attempt to go a little further, offering increased polyphony, digital effects, processing of external audio signals and many other new facilities. The Access Virus – first updated in 1999 to a Virus b desktop or rack-mount version, and a Virus kb keyboard – was a typical example.

VIRTUAL ANALOG PROGRAMMING

Let's look at how typical analog effects can be achieved on an instrument, such as the Access Virus, which now is on a sixth generation TI2 Desktop version. This instrument is chosen because of all the recent virtual analog synthesizers it has amongst the widest available range of possibilities, but the most familiar-looking control panel; one that would not give any problems to anyone already familiar with a Minimoog, Memorymoog, or Prophet 5. Of course, the instrument comes programmed with scores of exciting sounds, and most of the obvious categories – basses, lead lines, sequencer notes, pads, sweeps, strings, brasses, and so on – are well represented. In contrast with the days of classic analog synths, the Virus can be programmed with multi-timbral and multi-split set-ups, has effects and offers very complex arpeggio patterns, so playing a Virus preset (like the Multi 1 set-up on the Virus b, simply called "Play") can sound like a whole piece of music in itself.

2001 Novation ships the A-Station rack-mount virtual analog synth and vocoder

The Access Virus b was one of the most powerful early virtual analog synthesizer modules.

VIRTUAL OSCILLATORS

The Virus b sounds on the website, track 74, include some complex multi-layered arpeggios, but let's assume that the task for the moment is to program a single, much simpler sound from scratch. Firstly, go out of Multi mode and into Single mode to make sure that only one sound is being heard during editing; a simple start-up sound is at the end of the A bank of memories, A127. Start with the oscillators to set the basic tone required. On the Virus b, as on some other virtual synths, the oscillator wave shape varies continuously from a sawtooth to a square wave; this is a little more flexible than simply offering one wave or the other. The pulse width of the wave can then be adjusted independently; as on genuine analog synths, shorter pulse settings give thinner, quieter sounds, so you may want to boost the overall volume if using these settings.

Unlike some other virtual instruments, the Virus b also has a more general "Wave" setting at the lower end of the oscillator wave shape controls, which allows access to more than 60 digitally-generated waves that go far beyond the standard sine, sawtooth, and square. These are useful for generating more "realistic" sounds, and for patches more reminiscent of the PPG or Yamaha DX digital synthesizers; the competing Novation Nova as originally released lacked these, though they were intended to be added as part of a software update. Choosing one of these more digital wave shapes may instantly take you out of the area of analog-sounding programming though, so beware.

Genuine analog synths tend to set oscillator pitches in octaves, but on the Virus, as on many other virtual designs, it is set in semitones over a range of 48 semitones. The pitch of Oscillator 2 can then be set independently, and also detuned. Does detuning on a virtual analog synth resemble the same effect on a genuine analog synth? One of the reasons the classic synths sound strong is that the detuning was never very stable, with the oscillators drifting in relation to one another over time, or even from note to note as the keyboard is played. This won't happen on a virtual instrument unless it's programmed to do so, but setting oscillators to slightly different tunings will still effectively thicken up the sound. On

2002 Waldorf releases PPG 2V software based on the PPG Wave 2, reissued in 2006

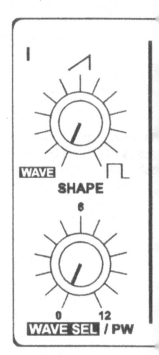

The oscillator controls on the Access Virus b.

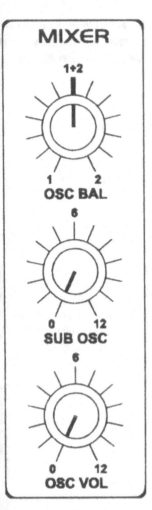

The Mixer section on the Access Virus b, showing the sub-oscillator level control.

the Virus the oscillators run freely, so you can indeed get some of these random phase cancellation effects. As on genuine analog instruments, the volume balance of the oscillators can also be set; on the Virus b there's also a sub-oscillator, and in the Version 3 software, which arrived later for the Virus b or kb, a third audio oscillator (available only with reduced polyphony) and a Twin Mode to use two sets of oscillators per voice with reduced polyphony, in which the oscillators can be panned apart to differing stereo output positions. On some other designs, such as the Novation Nova, a third oscillator comes as standard.

If the oscillators are set to Sync – in other words, the slave oscillator is forced to reset its cycle every time the master oscillator does so – detuning between oscillators doesn't have the same thickening effect. Using Sync is not very interesting if the oscillators are running near the same frequency, but if you change the octave setting of the slave oscillator, new and strong overtones are produced, and these can create some very powerful metallic effects. When you sweep the pitch of the slaved oscillator during the course of a note, a very strong effect is produced, similar to closing down a filter but much more harsh and cutting. On the Virus b, the Filter Envelope Mod control sets the depth of sweep of Oscillator 2 in Sync mode (as well as the FM amount, discussed later).

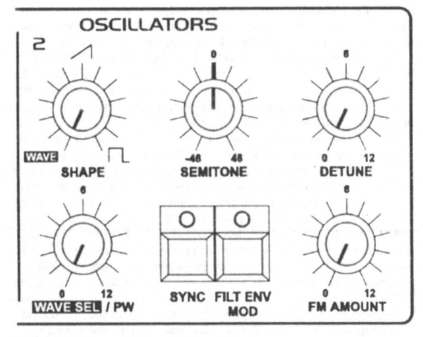

Oscillator 2 and Sync controls on the Access Virus b.

Of course, Sync mode on virtual synthesizers has nothing to do with physically locking the output of one oscillator to another, it is all realised digitally. So are these effects as strong as on genuine analog synths? The answer is generally "yes". Sync on a virtual analog synthesizer can sound every bit as powerful as on a classic analog synth, and on the Novation Nova there are powerful Sync modes which, being digitally controlled, do not even need to tie up the B bank of oscillators. The Virus also implements FM, the pitch modulation of one oscillator by another. This is the principle used by the Yamaha DX synths to create its sounds, but on analog and virtual analog synths it tends to be used just for effects, since it's too difficult to control as the basis of actual sound creation. Adding a little FM can also create some powerful new overtones.

Bending oscillator pitch during performance can be a highly expressive technique, and on the Virus b the Pitch Bend scale can be set to act in either a linear or an exponential manner. Depending on your playing technique, this can give a wider range of expressive possibilities that are not normally an option on a classic analog synth.

Apart from the main oscillators, a major sound source on analog synths is white noise. Of course, this is digitally-generated on virtual analog synths, again, a pretty straightforward task, since microprocessors can easily generate the random (or at least pseudo-random) frequency output necessary to create a white noise effect. But it is said that digital white noise sounds subtly different from analog white noise (which is sometimes generated by a circuit as simple as a single diode), and oddly enough

2002 Peter Bardens of progressive rock band Camel dies in Malibu aged 57

2003 Access releases the Virus, the first of many virtual analog keyboards and desktop modules

David Vorhaus gradually converted the White Noise *Kaleidophon* analog studio of the 1960s to digital synths and finally a laptop-based synthesis system.

the noise generator on the Virus b was extremely simple, it has no controls other than output level. The Novation Nova has no more complex an implementation, but the Red Sound DarkStar did offer pink noise (with reduced high frequencies), white noise (with all frequencies equal), or blue noise (with reduced low frequencies).

VIRTUAL FILTERS

One of the main defining features of the classic analog synths is a subtle distortion caused by overdriving the audio levels of the oscillators into the filters; this supposedly defined the Minimoog sound in particular. On the OSC OSCar, this overdrive setting option was intentionally provided, and the overdrive or "saturation" effect is now common on virtual analog synths. But there's a great difference between a genuine analog overdrive and a digital simulation of the same effect. This is why valve-based amplifiers and effects are still popular with guitarists. On the Virus b, the degree of saturation can be modulated by an LFO to create a tremolo effect.

Filter controls on the Access Virus b, including saturation level.

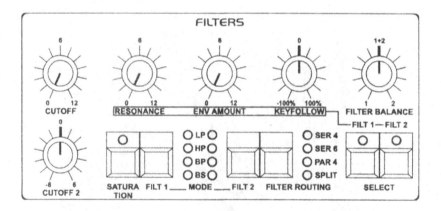

There are also three editing modes: Jump, Snap, and Relative. When you change a setting, such as filter cut-off, you can decide whether the value begins to change upwards or downwards from the existing value as soon as you touch the control, jumps immediately to the current value as soon as you make any change, or adds to or subtracts from the current value regardless of where the control is currently set. This option, denied to most programmable analog synths, helps avoid sudden leaps in parameter levels when making edits and contributes towards a smoother overall performance. A routine called Adaptive Parameter Smoothing also assures that the filter cut-off and other changes aren't affected by digital stepping or "zipper" noise.

On the Virus b there are two filters with independent controls for Cut-off 1 and Cut-off 2; each filter has low pass, high pass, band pass and band reject modes. Push Select for both and you'll edit both simultaneously; set

2003 Cwejman of Sweden releases the S1 rack-mount semi-modular synthesizer

FILTER ROUTING

Filter routing options on the Access Virus b

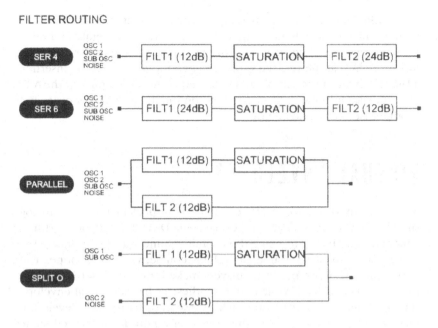

one to low pass and the other to high pass, and centre the filter balance, and you'll hear nothing (!). The filter can also be set to follow the key position; key follow by the filter is available on most small monophonics in order to brighten up the tone as you play higher up the keyboard, but on the Virus b and some other designs it can be used to control many other modulation facilities as well. The filters can also be set to operate either in series or in parallel, and their resonance can be controlled by velocity.

2004 German manufacturers of virtual analog synths Quasimidi and Waldorf both close, though Waldorf is later revived

VIRTUAL LFOs

The Virus b has two main LFOs, each with a flashing rate LED. On many classic analog polyphonic synths, effects such as vibrato are introduced using a single LFO routed to all the currently sounding oscillators. This means that all voices are modulated in exactly the same phase: useful, but not as strong sounding as modulating each with an independent, free-running LFO. On the Virus b all the LFOs are independent, so vibrato and other modulation effects are particularly expressive.

The LFOs can also be set to a one-shot mode so they can be used as additional envelopes, for instance, an LFO set to sawtooth will go through just one cycle when a new note is played, giving a simple rising envelope, the length of which will depend on the LFO speed. This is a very uncommon facility on genuine analog synths, although it is found in the modules of the Serge modular system. There's also a less fully featured LFO 3 (existing in software only) with no dedicated front-panel controls, just intended to create oscillator vibrato in order to free up the other LFOs for more complex applications.

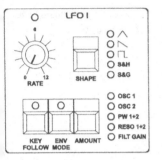

Access Virus b LFO 1 controls, including the sample-and-glide facility.

Another unusual facility is sample-and-glide, a simple alternative to sample-and-hold, which means that the random values generated for modulation purposes change smoothly from one value to the next rather than in steps. This can be realised on an analog modular system by inserting a lag processor in the sample-and-hold output. It's also found on the ARP Odyssey's sample-and-hold section, and on the modern SE Boomstar, but is otherwise pretty uncommon.

VIRTUAL ENVELOPES

As previously discussed, a fast envelope is important for good analog simulation. The Virus b's envelopes have an ADSTR format; the "T" (time) parameter can be set to make the overall volume fall or rise during the sustain stage over a variable period. Of course, virtual envelopes can have many elements, but most players make little use of settings other than the basic ADSR. If your virtual synthesizer has additional envelope stages, it's worth experimenting with these in order to make levels and relative mixes change during the course of a note for increased sonic variety.

On the Virus b there's also a Hold mode, which makes notes continue to sound until all keys are released and more notes are played. This has an important application in simulating, say, the abstract effects sounds from an EMS VCS3, which can keep playing regardless of any activity on the keyboard. The Definable 1 and Definable 2 rotary controls on the Virus could also be programmed to simulate the facilities found on the VCS3's joystick, so the Virus can be a good abstract sound generator as well as a more keyboard-oriented synthesizer.

Amplifier envelope controls on the Access Virus b.

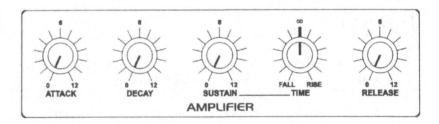

VIRTUAL CONTROLLERS

On virtual analog synthesizers, which are invariably MIDI equipped, it's generally possible to control many parameters through MIDI commands. This has advantages and disadvantages as compared to the classic analog synthesizers or modular systems. Obviously, on a medium-sized analog synthesizer, such as the ARP 2600, there are plenty of knobs, switches, and sliders – for envelope stages, oscillator pitches, filter settings, and so on – that can be grabbed and varied while the synthesizer is being played, but the classic problem of analog synthesizers is that none of these changes can be recorded as part of a performance or exactly reproduced when required. Modern analog and virtual analog synthesizers can put all these parameter changes under MIDI control, which means they can be recorded, edited, and reproduced using a sequencer. Scores of new parameters can be added, since controlling each new parameter may only require the addition of a few lines in the controlling software. But the temptation on the part of the designers is then to make many of these parameters accessible by MIDI control only, leaving off the physical components that would have put up the cost of the instrument. For instance, there is no way that a modern 16-part multi-timbral virtual analog synthesizer will have 16 sets of physical envelope controls.

2004 Michael Garrison, American pioneer of space music, dies aged 47

So, on many virtual analog synthesizers, there can be enormous possibilities for sound control, but many of them will require learning new techniques in order to access them. On many designs (at least on the Access Virus, Novation Nova, Red Sound Systems DarkStar, and other polyphonic virtual analog synthesizers, and on the Waldorf Pulse, Novation Super Bass Station, and some other monophonic MIDI synthesizers), the movement of any front-panel controller is transmitted as MIDI information, which can be recorded and played back as part of a sequence. Accessing other parameters that do not have their own front-panel controls can be more

The author's 2005 hybrid analog, digital, and MIDI modular system, "El Monstruo". Built around a Delta Music Research Studio Series 2 analog modular system, also seen are a Synare PS1 drum synth brain, step sequencer and mono synth, Korg MS2000 module, Quasimidi Groove-X 309 Klaus Schulze Limited Edition drum machine, Kawai MM16 MIDI programmer, Moog Taurus 2 brain, two Ult Sound DS2 Pro Muzer dual drum synthesizers, test equipment, Quasimidi Polymorph, a second Korg MS2000, TASCAM MM1 mixer, and Roland MKB300 MIDI keyboard.

complex; increasingly, the major software packages, such as Cubase, Logic, or Ableton Live, running either on a Macintosh or a PC, have special facilities (called "mixer maps" in Cubase) dedicated to providing sets of MIDI controllers that can alter synthesizer parameters gradually during record- and playback, or in a "snapshot" mode that sends several parameter analog synthesizer changes almost simultaneously in order to completely change the current sound (of course, this could also be done simply by sending a patch change command to select a different memory).

2004 Alesis releases the Ion, a compact four-octave virtual analog synthesizer

Whether using a controller map on a computer will ever match the degree of spontaneity and enjoyment associated with physically adjusting knobs and sliders is another question. As mentioned elsewhere, there are a few products available that give back physical control over the parameters of virtual analog synthesizers (whether in the form of hardware, software, or computer sound cards). These include the Peavey 1600X MIDI controller, the Control Freak from Kenton Electronics, available with eight or 16 totally programmable slider controls, and the Phat Boy from Keyfax Hardware, which is more specifically dedicated to editing General MIDI parameters as well as the sound parameters of the Steinberg Re-Birth software package discussed at the end of the previous chapter. Roland made specialised control panels for some of their keyboards and modules such as the D110, D50, and MKS50 in the past, but with these modern, more general controller products there is less chance of them becoming out of date because the manufacturers often make set-ups for new models of synthesizer available through their websites. MIDI controller functions are now much more commonly being built into USB-equipped controller keyboards from Edirol, M-Audio, Novation, Behringer, and many other companies, and some of these, in addition to rotary or slider controls, have joysticks and touch pads, which are excellent controllers for the more abstract analog-style sounds.

The Control Freak from Kenton offered more control over MIDI-equipped analog synthesizers, modules, and software.

MIDI hardware controllers have been around for a while: 35 years ago J.L. Cooper launched the Fadermaster, intended as a mixer to send MIDI volume changes, but (almost incidentally) also programmed with sound editing facilities for modules such as the E-Mu Proteus and Korg M1.

The idea did not take off at the time, but now that there are so many more virtual synthesizers available and since they have so many additional parameters, getting back some means of physical control is becoming vital if any of the spontaneity of analog sound creation is to be retained. Korg's KAOSS Pad (currently up to the KP3+ model) may also come in useful here, though primarily intended as an effects unit with its parameters controlled by an illuminated X–Y touch pad, it also sends MIDI controller information and is seeing a lot of use as a MIDI version of Moog's old X–Y touch voltage controller pad.

2004 Roland releases the Juno-D, reviving a name from their classic analog product lines

ALTERNATIVE APPLICATIONS

With all these voices and processing power available, it would be tempting to think that modern virtual analog synthesizers would be great as generators of more complex and evolving, abstract sound effects. But one problem with simulating these kind of effects produced on analog modular systems using virtual instruments is that it is often difficult to persuade them to make sounds without constantly sending them Note On instructions. In other words, they remain rather organ-like, in the sense that they are dependent upon the input from a keyboard. On modular synthesizers and even on the smaller EMS Synthi VCS3, rather the opposite was the case. Their oscillators played all the time, unless silenced by the application of a filter or volume envelope. Even the Sequential Pro One had a Drone mode, though the original Minimoog disappointingly did not.

At least on the Access Virus (from V2 software) there is the Keyboard Hold mode mentioned earlier, which permanently holds all notes played, until they are all released and a new note is played. This is a handy way of making it easier to set up a complex abstract sound, then play with the filtering and effects without having to hold down a key with one hand, or send a constant stream of notes from a sequencer.

However, many of the virtual analog synthesizers are able to offer facilities completely denied to the classic analog synths. This is simply because their facilities are based on general-purpose digital signal processing chips driven by software, and imaginative reprogramming can realign these to carry out other audio tasks.

One of the most common new facilities is vocoding, found on the Access Virus, the Novation Nova and the MicroKorg amongst others. On the Virus b, all the front-panel filter controls are redefined in vocoder mode to create a 32-band vocoder that uses up the processing power of four voices (the remaining voices and effects carry on as normal). The vocoder's modulators, envelope followers and carriers are each controlled by the filter front-panel controls, and the signal to be processed is fed into the external audio input. On the Red Sound Systems DarkStar, a new EPROM chip gave optional vocoder facilities, while a more flexible and powerful monophonic synth option on an EPROM chip was also promised.

So far, none of the other virtual analog synths have attempted to reproduce the facilities of the Nord Modular, which simulated a very

2005 Bob Moog dies having launched the Voyager and designed the Little Phatty

flexibly patchable modular analog synth using PC- or Mac-based editing software. However, you could try to do this yourself by configuring your multi-timbral virtual analog synth as a compact modular system. Rather than treating each part as an instrument in itself, treat it as a single element in a modular system, limiting each part to a single note and controlling each from a different MIDI channel. Set each part to a different mix of oscillator waveforms and some parts to low pass, some to band pass and some to high pass filtering. Try routing LFOs to unconventional destinations, such as filter resonance or oscillator balance, or (available on the Virus and recent Studiologic Sledge) to Wave Select, so the oscillator waveform repeatedly changes, creating a sort of wave sequencing effect. On the Virus b, the two LFO Rate controls are near the Definable controls, making it possible to put all the most important parameters for editing close together. On the Novation Nova, all the modulation sources and destinations are controlled from a single Modulation section on the control panel, giving quick and easy access to the most frequently used parameters.

In these ways the modern virtual analog synthesizers can give just as many possibilities for experimentation as the classic analog designs, and in some ways will be less problematic, giving far fewer tuning, stability, and servicing problems. But they still require the application of a great deal of imagination to come up with original sounds, as much in the area of mastering the use of MIDI controller information as in terms of patching techniques comparable to those of the old days. Happily, even in the world of virtual analog, it will remain down to the user to apply his or her imagination to the creation of new and exciting analog-style sounds.

ANALOG EMULATION SOFTWARE

We mentioned in Chapter 6 that analog simulations purely in software began to become a significant force around 1999. Unless very attached to the idea of playing hardware analog instruments, the software approach had a lot to recommend it. For those already owning a computer, the addition of analog sound emulation software seemed relatively inexpensive, and since recording techniques were already going in the direction of hard-disk multi-tracking on computer, the opportunity to completely integrate analog sound creation within this format, rather than investing in large and potentially unreliable external instruments, appeared very attractive.

Just as the analog hardware revival was driven by an interest in the sound of the Roland TB303 Bassline and similar instruments, the trend for software emulation of analog sounds really took off with a TB303 simulator, the Re-Birth from PropellorHeads, later licensed by Steinberg. The Re-Birth software package set out to simulate a Roland TR808 drum machine driving a Roland Bassline, an extremely expensive second-hand purchase at the time. The drum machine sounds were sampled in the software, while the synthesizer sounds were made highly variable, and the screen display simulated the front panels of both instruments.

2005 Arturia releases the Prophet V software package, including Prophet 5 and VS sounds

Re-Birth was a logical design for the relatively limited computers of the time. Producing the drum sounds was simply a matter of replaying some sampled waveform memory, while synthesizing a monophonic bass sound without effects was not too demanding for contemporary processor speeds. The computer's other major task was to support the screen display, which was straightforward enough, and so Re-Birth operated simply and reliably. The success of the package gave many other software engineers food for thought, and subsequent developments have simply followed in their power and complexity the very rapid increase in processor speeds, memory, and storage capacity that have been clearly seen in the computer field over the last decades.

For example, the second generation Re-Birth version RB338 was able to simulate two TB303s and a TR808 and TR909 drum machine, running either on a Macintosh or a PC. Steinberg's Cubase, already an internationally popular MIDI sequencing and hard-disk audio recording package, was revised with the addition of the VST (Virtual Studio Technology) standard, and now VST software "plug-ins" from Steinberg and from third-party designers were released to add many extra facilities and effects to the package. The Cubase package itself eventually arrived equipped with A1, a flexible monophonic analog synth emulator, and JX16, a polyphonic analog synth emulator with chorus effects, which made it reminiscent of the Roland Juno models, but one of Steinberg's first stand-alone VST plug-ins was Model E, a very close imitation of a Minimoog (which, of course, was originally Model D). This was quickly joined by other imitative packages from Native Instruments. NI had launched several interesting software instruments, including: "Generator", comprising as many as 150 different modules to simulate popular monophonic and polyphonic synthesizers and vocoders; "Dynamo", which worked as plug-in or stand-alone effects software; and "Transformator", which was a modular sampler with unusual effects such as the ability to slow down a sample until it completely freezes. "Reaktor" combined elements of both packages, offering analog or FM synthesis and granular sample processing. After establishing these packages, NI began to offer VST plug-in instruments that specifically emulated individual classic synth designs, starting with Pro5, simulating a Sequential Circuits Prophet 5 both in appearance and facilities; they then did the same for the Hammond B3 organ on "B4", which originally ran on PC Windows 95/98/NT/2000 and on Mac OS8 or higher.

2005 Alesis releases the Micron virtual analog synth, a compact version of the Ion

From the time of the release of these Moog, Prophet, Hammond, and other emulations, buyers would have a serious decision to make: did they really want to own the original hardware instrument, taking on tuning, servicing, storage, and other problems, or were they happy with the sounds offered by a software simulation? The flow of new software instruments became extremely rapid.

Steinberg's B-Box analog-style sampling drum machine software included TR707, 808 and 909 sounds, and in 2000 Steinberg launched a superb PPG Wave 2 simulator, Waldorf PPG Wave2V, which provided all the facilities of the long discontinued upmarket digital wavetable/analog filter synthesizer in the simple form of a Cubase VST plug-in. PPG Wave 2V even fixed various programming oddities of the original and simplified

JX16 was the analog polysynth emulation in early versions of Steinberg Cubase.

the use of its arpeggiator, reviving a whole set of hybrid analog/digital sounds that had been hard to obtain for many years.

Re-Birth was also updated to Reason, now a combined analog synth, sampler, pattern sequencer, drum machine, and loop player with mixer, effects, and the facilities of Steinberg's ReCycle! Loop editing software. Waldorf's D-Pole, based on the company's hardware synthesizer filters, was a software analog filter emulation for Cubase, while NeoSynth worked with Cubase 3.7 and upwards on the PC and had an attractive front-panel emulation resembling Moog or Prophet synthesizers. Meanwhile, the Cycling 74 company from the USA offered "Pluggo", a set of 74 Cubase VST plug-ins initially at $74 or equivalent including delays, filters, pitch shifters, distortion, granulation, spectral effects, synthesizers, visual displays, and a vocoder.

2006 Arturia releases Analog Factory software with sounds from all their analog emulations

Many other computer-based systems also introduced analog-like effects processors. E-Magic (later Apple) Logic from version 4.0 offered analog-style resonant filters with LFOs, while Creamware's Pulsar audio card and software system for hard-disk recording, sound synthesis and processing included virtual modular synths such as "Inferno", which emphasised the use of effects including filter and chorus, and "MiniScope Mk II", which had a front-panel simulation closely resembling a Minimoog, as well as an Akai-compatible sample player and effects software. Creamware also marketed: "UKNOW 007", based on the Roland Juno 106; FM One, based on the Yamaha DX7 digital synthesizer; Modular V2, a powerful modular analog synthesizer imitator; SB404, a 16-step sequencer and synthesizer; EDS8i, an eight-channel analog drum machine imitator; and many filter, vocoder, effects, and other routines.

Bitheadz marketed the Retro AS1, a virtual analog synthesizer in software with up to 32 simultaneous voices, three oscillators and four LFOs per voice, seven filter types, effects, an arpeggiator, and 100 editable parameters available in Macintosh or PC versions. Digidesign, together with Access Music, announced an Access Virus plug-in (in TDM format only for Pro Tools-based systems), which was virtually identical in sound and functions to the Access Virus b and kb hardware synthesizers, a system progressively updated to the current TI "Total Integration" version.

Mil Productions, a French software house, was developing Modularing 2.0, a set of software emulations running on the Macintosh of the typical facilities from a large analog modular system. Modules include analog- and matrix-style sequencers, a virtual keyboard, a General MIDI editing bay, a scratchpad polyphonic sequencer and a MIDI digital delay. Koblo offered Vibra 9000, a monophonic synthesizer emulator; Gamma 9000, a multi-timbral drum synth/sequencer software package; and Stella 9000, a polyphonic software sampler modifying any sound file with filters, envelopes and LFOs, all on the Macintosh only.

Some software even became available free of charge on the Internet. Yamaha's website (www.yamaha.xg.com) featured "SoftSynthesizer", which turned a suitably equipped PC into a multi-timbral synthesizer. There were also many editing packages available for General MIDI and XG compatible synthesizers, modules, and sound cards, which made parameters, such as filter cut-off frequency and resonance, envelope times, and effects levels editable in a very "analog" manner.

2006 Cwejman releases Eurorack format professional analog synthesizer modules

One modern software approach to analog sound synthesis: Mark Jenkins in concert at the Ocean County College Planetarium, New Jersey in 2006 with Apple iMac running Korg Legacy software, and guest, the space shuttle pilot Dr. Franklin Story Musgrave.

The introduction of Windows XP for the PC and OSX for the Apple Mac proved something of a watershed in the development of sound synthesis software. Some early packages were never converted, but those which were, plus new products introduced to join them, became incredibly powerful, and the trickle of sound synthesis software coming onto the market turned into a flood, so the history of synthesis software became even more difficult to summarise comprehensively than that of hardware instruments. Software companies, which have very limited start-up and virtually no tooling costs, can spring up apparently out of nowhere, market one or two good products, then disappear again. Mergers and buyouts between larger businesses are common, but tiny one-man businesses can also come up with wonderful products and software which, unlike hardware, can be sold on a number of different principles: commercially, as shareware (for which a contribution is expected), as freeware (which, as its name suggests, is free, though more powerful payable versions are often made available) and by download, which means the software company does not even have to replicate disks or print a cardboard box in order to make a sale. Several pieces of software, such as Tassman and Native Instruments Reaktor, also became available, which allowed even the inexperienced user and non-programmer to define new "instruments", and these were often made available through extensive websites on a freeware basis.

For these reasons and others, the market for analog and other types of sound synthesis software has continued to move incredibly rapidly, and in a situation where new commercial, shareware or freeware products appear every day, it is impossible to describe all the available products in a book of this length. All the major software sequencer packages, such as Steinberg Cubase, Apple Logic, Cakewalk SONAR, Ableton Live, and MOTU Digital Performer (which, for example, had a "Polysynth" plug-in cosmetically resembling a Roland Juno 106), included their own battery of soft synthesizers: SONAR, on the PC only, featured the particularly powerful Pentagon 1 synth, which offers filter morphing. It's also important to note that there are now many, many varieties of sound synthesis software, including FM, harmonic, granular, sampled, wavetable, percussive, modular, and other techniques, sometimes combined with analog-style synthesis, with many very useful soft synths coming and going from the market, including Steinberg D'Cota, Plex, and XPhraze, the excellent harmonic synthesizer VirSyn Tera, the modular TASSMAN, the spectral synthesis and analog hybrid Spectra from Kjaerhus in Denmark, Rob Papen's analog style Albino, Cakewalk Rapture, and a whole army of "ROMpler" software, which plays back sounds – whether they may appear analog, digital or hybrid – from hard disk under a wide variety of themes, including atmospheric, spacey, looped or drum sounds, world music packages, basses and dance music elements, and much more. That's not even to mention the enormous range of iPad music apps, which includes imitations of specific analog synths, modular systems, ROMplers with analog style filtering, Theremin-like touchscreen instruments, and much more.

A good example of a musician moving inexorably towards the use of software through this period was the multi-instrumentalist Mike Oldfield, whose *Tubular Bells* album launched the Virgin Records label in 1973, broke new ground in the area of instrumental rock music, and became

2006 Korg launches RADIAS rackmount virtual analog synth with optional keyboard

identified with the movie *The Exorcist*. In a long career featuring some very imaginative works and several revisits to the *Tubular Bells* theme, Oldfield, who is primarily a guitarist, incorporated many emerging technologies, including quadraphonic sound, digital sampling on the Fairlight Computer Musical Instrument, vocoders, and MIDI synthesizers, automated mixing, guitar-to-MIDI control, and more recently software synthesis, though it's interesting to note that the original *Tubular Bells* featured no synthesizers at all but plenty of electromechanical instruments, including Hammond, Farfisa, and Lowrey organs and the Mellotron.

The author with Mike Oldfield at the Roughwood Croft studio in 2002.

In 2002, Mike was re-recording *Tubular Bells* for the album's 30th anniversary and had just released the album *Tr3s Lunas* (sic), accompanied by a PC-based computer game, Music VR:

> "I use PCs for game development but I prefer the Macintosh for music. I use E-Magic Logic, and this time, as well as using all my usual guitars and synths I have added a lot of virtual instruments. Logic has its own instruments and can also run Cubase VST-compatible instruments now; they're a bit of a pain because they all use different commands, but I have a little Mark Of The Unicorn box which is dedicated to the virtual synths, sorts out the MIDI destinations and keeps them separate from everything else."
> (Mike Oldfield, interview with the author, 2002)

Although Mike does play keyboards, he added MIDI control to his guitar set-up and often controlled the virtual synths this way (one earlier album, *Guitars*, was played, including drum parts, entirely from the guitar controller):

2006 Novation launches the XioSynth virtual analog synth in two- and four-octave versions

> "There's a very good alto sax virtual instrument which is on most of the tracks of this album. The genre which they call chillout music can't have

really complicated melodies for some reason, it just doesn't sound right – the melodies have to be very simple, with very few chords as well, so a sax part often seems to fit. This time it's all about the atmosphere of the music rather than the complexity of it. My albums have always been about changing dynamics, but this one has to keep a more consistent atmosphere."

Constantly developing new guitar as well as synth sounds, Mike by this time had started to move away from hardware altogether:

"I haven't really even looked at any of the new amp and cab simulators like the Line 6 models. At the beginning of every album I normally ask around about what are the latest boxes you can get, but the latest things as far as I am concerned are the virtual synths and effects in the Macintosh."

Mike's studio at this time had an early giant plasma screen as a central feature, which showed the Mac display for Logic and the various virtual synth plug-ins, and doubled (at the click of a simple switching box) as a larger display for the Fairlight Merlin, a 48-track hard-disk system used to dump off audio material for mass storage. A Korg Trinity was the main controlling keyboard, along with a Nord Lead synth that featured strongly in stage performances, and already consigned to the floor, a Korg M1, and Roland XP30. In the racks were an Akai S6000 sampler and a Roland JV880 and JV1080 module, along with several JD990s, the module version of the JD800 synth, which featured strongly on the *Songs of Distant Earth* album and appeared at that time to be Mike's favorite analog-style synth. But after this period Mike moved even further towards software synthesis; his *Light and Shade* double album recorded in 2005 featured all virtual synths and the Fruity Loops sequencer software, and in 2006 he was planning his first stage performances in seven years and selling his huge Neve automated mixing desk and many of his original guitars and effects in favour of a much simplified, virtual synth set-up offering sampled and digital as well as analog-style sounds. Ultimately this resulted in a move to a much simpler studio setup in Barbados from where he planned his live performance at the 2012 Olympics, staged by a longtime fan, the movie director Danny Boyle.

Mike Oldfield is an excellent example of a musician who moved away from a very large and complex studio set-up towards a software system that can offer him all sorts of new textures and effects, as well as the analog sounds that characterised his early career. However, simply to describe all the analog-style packages now available to musicians has become an impossible task. In 2018, kvraudio.com lists 324 freeware synth packages for Mac alone. Here we will mention just some of the earlier successful pieces of analog simulation software, with a few updates in Chapter 8.

APPLE

Apple's simple Garageband sequencer initially given away as part of the iLife software suite with all new Macs offered many analog-style sounds and loops, and later became compatible with Logic Pro X, which included EVB3, a Hammond organ with drawbars, EVD6, a Hohner Clavinet, EVP88, a Fender Rhodes electric piano, EVOC which is a vocoder, and any one of hundreds of soft synths in AU plug-in format.

2006 Mark Jenkins' *This Island Earth* CD released using analog, virtual analog and hybrid sounds

ARTURIA

The French company Arturia launched with a sequencer package called Storm, but quickly came to specialise in analog imitative software, each suffixed "V" for Virtual. Their Minimoog V was endorsed by Bob Moog and was quickly followed by the Moog Modular V which originally replicated a Series III Moog Modular, though this suffix was quietly dropped when a few bonus features were added. Modular V actually has to be patched with virtual cables before playing, though there are plenty of factory presets included. A slight update allowed some modules to be swapped and different options as regards the display of the keyboard, the main modules and the sequencer. Modular V creates a superb sound and was used on stage by Tangerine Dream displayed on plasma screens almost the size of the original hardware instrument. Although Bob Moog endorsed these

The Arturia Moog Modular software was personally approved by Bob Moog.

products, he remained committed at the time to analog technology for his own products:

"Of course you can take a laptop these days, switch it on and have a whole modular synthesizer appear, complete with patch cords to connect everything together, and get a great sound. In some ways it doesn't matter how you get the waveform. On my products it's created by an actual electronic circuit, but you can also get it by counting tables of numbers. But if you ask the musicians today who play the Minimoog Voyager, they'll tell you all about the feel you get from playing a real instrument. It's an all-analog sound source, but with digital control for the areas where it's needed, like adding memory capabilities. You just don't get that kind of attachment to a laptop."

(Bob Moog, interview with the author, 2004)

Arturia's Yamaha CS80V was also highly successful, though rapidly receiving an update to improve its sounds further. The implementation of the long pitch bender strip and the addition of an arpeggiator were particularly impressive. Arturia's Prophet V emulated not only the Prophet 5 but the Prophet VS, with a complete set of digital waveforms, offering a Hybrid mode processing the waveforms of one synth through the filters of the other. ARP 2600V effectively emulated the semi-modular synth and added a wonderful emulation of the ARP Sequencer, so complex rhythmic patterns could be created and manipulated. Analog Factory launched in 2006 offered 2000 preset sounds from all these virtual synths, controlled by a "vanilla" keyboard with just a few variable parameters (and with 1000 new sounds added by the Analog Factory Reloaded package launched at the San Francisco AES in October 2006).

At the time of writing in 2018, Arturia still leads the way in analog imitations in software. It can be argued that sometimes these are less than totally authentic. On the other hand, hardware analog instruments often went through several revisions with slight changes of sound and facilities, so the question arises of exactly which model is being emulated. Arturia also adds facilities, such as effects on all models, polyphony on some that were originally monophonic, and arpeggiators on others, though oddly omitting the arpeggiator that did exist on the Prophet VS. Arturia initially had very little competition for most designs, though there was another slightly less powerful ARP 2600 emulator, the TimewARP 2600 from WayOutWare in the USA. More on Arturia in Chapter 8.

2007 Dave Smith Instruments launches the Prophet 08 offering a truly all-analog signal path

Big Tick Audio

Cheeze Machine is an excellent, freely downloadable and quite variable emulator of a 1970s string synth such as the ARP Solina or Hohner Logan LE. Try downloading it from www.kvraudio.com. It's undemanding on the computer's processor and sounds even better if two instances are run layered an octave apart. In 2018 the company was also offering a powerful synth called Rhino for $99.

G-Media

This British design team was never prolific but had three wonderful and distinctive products: M-Tron has the layout of an M400 Mellotron complete with coffee mug stain on the top panel, and offers wonderfully authentic choir, string, brass, and other Mellotron sounds (including the rare rhythm, guitar and other tape sets), while Oddity (now Oddity 2) is an excellent replication of the ARP Odyssey with totally authentic filter, sample-and-hold, and other parameters, though these are a bit demanding on the computer's processor, and ImpOSCar is an equally rigorous (and patch download compatible) recreation of the OSC OSCar, though now polyphonic and ideal for great digital/harmonic PPG Wave-like chordal textures, not something you could have played on the original. The company also markets MiniMonsta, a complex Minimoog emulation that faces stiffer competition from Arturia and others.

IK Multimedia

Sonik Synth 2 offered a huge range of sounds from classic keyboards, including the ARP 2500, Moog, E-Mu, Steiner-Parker and Serge modulars, ARP Quadra, Roland Jupiter 8, Yamaha CS80, Polymoog, PPG, Prophet 5, Gleeman, Minimoog, OSCar, EMS VCS3, and Oberheim SEM, among others, though without specifically identifying which is which, except through cryptic file name abbreviations. The company also offered Elektric Piano, which handled Fender Rhodes and Wurlitzer sounds. In 2018 they had moved on to Syntronik, which emulated 17 different instruments and also had several hardware lines on sale. See Chapter 8 for more.

Best Service Cult Sampler and Zero-G Nostalgia had similar selections of sounds from 1970s, 1980s, and 1990s synths and samplers.

ARP Odyssey enthusiasts welcomed the option of using G-Force Oddity instead.

2008 Moog Music launches the Moog Guitar with analog filters and MIDI

KORG

As mentioned in Chapter 6, Korg Legacy launched with a hardware element in the form of a miniaturised MS20-style USB controller keyboard, later becoming available simply as two software-only packages, *Legacy Analog*, with the MS20 and Polysix set-ups, and *Legacy Digital*, with the Wavestation synth and M1 workstation emulations. While the digital package was superb (and added an attempt at a resonant filter to both synths, which originally lacked them), the analog package is of more relevance here; the MS20 becomes polyphonic, which is a huge bonus, while the Polysix remains a more than adequate replication of the original. These proved a strong start to Korg's involvement in software and continued hardware innovations. See Chapter 8 for more.

Korg's Legacy software series included a wonderful MS20 emulation, originally with this MS20iC hardware controller.

The Polysix synth emulation from Korg's Legacy software series.

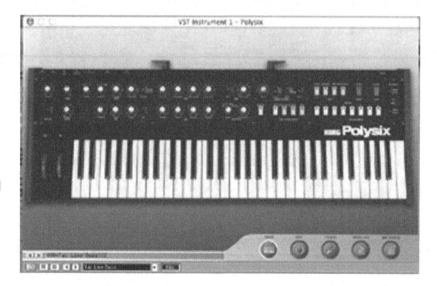

2009 Moog Music launches Taurus 3 MIDI bass pedal, Akai returns to keyboard manufacture with MINIAK based on the Alesis Micron

NATIVE INSTRUMENTS

NI was amongst the first in the analog simulation software business. Their Prophet emulation was updated to Pro 53, their Yamaha DX7 package FM7, and Hammond organ imitator B4 (later updated) remain available, and all three were repackaged in simplified form with only a handful of variable parameters as XPress Keyboards. NI's Absynth 3 (now on V5) was also a powerful soft synth that combined sampling techniques. By 2018 NI was offering Komplete 11, a package with 45 different software instruments.

The Native Instruments Pro 53 was an excellent early emulation of the Sequential Prophet 5.

NOVATION

As mentioned in Chapter 6, Novation made a patch compatible software version of their A-Station keyboard, the V-Station, an excellent, not too processor intensive, arpeggiator and effects-equipped monotimbral polysynth package useful for almost any style of music. A simple monophonic Super Bass Station software package was also available and often given away with the company's controller keyboards, or as part of other software instrument packages. Novation still offered these in 2018 but has concentrated mainly on hardware instruments, such as Peak. More in Chapter 8.

NUSOFTING

A VST or AU format plug-in for Mac or PC, DaHornet was an excellent EDP Wasp imitator, which started as an affordable downloadable synth but was later free and offered full polyphony. By 2018 the company offered a huge range of synth software at www.nusofting.com. See Chapter 8 for more.

2010 Apple launches the iPad, Beep Street releases the iSequence and Sunrizer (later Horizon) apps, Mark Jenkins' *iPad Album* is the world's first all-iPad CD release

ROLAND

Roland blocked the marketing of some very close emulations of their early synth designs, preferring to market their own authentic Jupiter 8 and TB303 packages only within the VariOS hardware package, which wasn't in production for long. There was a D50 emulation card for the Roland V-Synth and the company had close ties with Cakewalk, so Roland GM-style software appeared within SONAR and other sequencer packages. In 2018 the company remains more interested in hardware, though five different plug-out synths for the System 1 and System 8 keyboards will also run independently on Mac and PC. More in Chapter 8.

SCARBEE

RSP73 was an electromechanical piano imitator, and the company had several other imitative and sample-based packages. By 2018 they were marketing Classic EP88M and were involved in creating the Alicia's Keys package for Native Instruments.

EMS, OBERHEIM AND OTHERS

At the time of writing in 2018, no software emulations of the EMS synths the VCS3 or Synthi A have reached a wide market, though there have been incomplete attempts from France and Germany. The difficulties here probably include how to replicate the pin bay with sufficient clarity and how to reproduce the control functions of the invaluable X–Y joystick. No doubt this is something that will be solved eventually. Other analog synths that would appear ripe for software emulation include the older Oberheim designs and maybe Korg's big polyphonics, the PS3200 and PS3300.

But the nature of the software field is such that established companies have to be very careful about making a commitment to any distinctive new products. There are a lot of clever programmers out there, many of them willing to give away designs very cheaply or free of charge. Big Tick Audio's free of charge Cheeze Machine, for instance, made it almost pointless to put any marketing effort into the detailed emulation of an ARP Solina string ensemble, and no company wants to launch a detailed emulation at the full market price if an inexpensive or free one is likely to appear at any moment.

The KVR Audio site (www.kvraudio.com) keeps an updated list of new software releases categorised by type and operating system. This includes freeware, shareware, small company products, and limited time or limited function demos of commercial products otherwise available in stores, by mail order, or by download. Since it's impossible to include an up-to-the-minute list of every software product of relevance to the analog synthesis field, it's advisable to check out this site and others frequently to see what's new.

For example, at the NAMM and Frankfurt music fairs in early 2007 there were many significant new software launches. These didn't completely eclipse new hardware products. Korg's R3 was an update of their MicroKorg keyboard, again with vocoder but with three octaves of full size keys and technology from the larger R.A.D.I.A.S. synth; Cwejman

2011 Apple iPad 2 and Garageband launched, Moog enters the iPad apps field with Animoog

launched a range of Eurorack size modules while still showing their semi-modular rack-mount synth from which they derive, the S1; and FutureRetro launched a closely comparable semi-modular two-oscillator rack-mount MIDI/analog synth, the XS, an ideal accompaniment for their Mobius sequencer, or a more sonically flexible alternative to their Revolution, a Roland TB303 Bassline synth/sequencer imitator with a distinctive circular control layout.

However, the software scene was, if anything, becoming more lively. Korg added to the Legacy software series with an excellent recreation of their powerful four-oscillator MonoPoly synthesizer, while Arturia ended the seeming ban on replications of Roland product designs with Jupiter 8V, an excellent reproduction of the Jupiter 8 polyphonic. Rob Papen showed Predator, an update of his soft synth Blue, while M-Audio showed the Virtual String Machine by the creators of the Oddity and ImpOSCar packages G-Force, which replays samples from the Freeman String Symphonizer, Eminent 310, Elka Rhapsody, Polymoog, Korg PE2000, and many similar instruments, as well as WayOutWare's KikAxxe, which joins their TimewARP 2600 imitation of the ARP2600 to offer the sounds of the smaller ARP Axxe combined with the ARP Sequencer and a tape delay. Soft synth designs for the Pro Tools system also proliferated, with the Hybrid synth, Velvet electric piano and Strike virtual drummer appearing.

The relative ease of new software production led to a vast number of new products appearing in the period 2007–2012. Many of these appeared as freeware, which perhaps limited interest from large companies in producing commercial software. The TAL Elek7ro monophonic and Uno62 polyphonic VST/AU freeware plug-ins released around 2010 are excellent, for example. More complex was the Sylenth VST plug-in from Lennar Digital (2008), or Audio Simulation's freeware DreamStation DXi2 for DirectX host software on Windows PCs. ImageLine launched Morphine for Windows, while Ichiro Toda's Synth1 ran on Mac or PC as freeware.

2012 Moog Minitaur bass module and Arturia's two-octave MiniBrute launched, first analog synthesizer hardware products from Brazil launched by EMW

Some new software offered programmable modular facilities, such as Chris Wolfe's Jasuto, a programmable modular freeware synth on the Mac, or VAZ Modular by Software Technology (Windows, $130 in 2011). Formant Classic Advanced by Ftec-Audio (Windows, 2011) specifically simulated the Formant analog modular system with an impressive screen display, and other imitations of specific analog synths included JEM SX1000 by Pianovintage emulating the Jen SX1000 mono synth (Windows freeware, 2011) and GForce ImpOSCar2 from 2011, while Steinberg's Model E Minimoog emulating VST plug-in became freeware. Another company launched a whole line of software inspired by specific analog keyboards Easytoolz.de offered EasyMuug, a MiniMoog emulator, Protege a Moog Prodigy simulator, MiniTron a Korg Monotron emulator, and emulations of the Jomox SunSyn, Roland Jupiter 8, and many others, all free or very inexpensive. A superb resource was www.hitsquad.com, listing many new software releases with links to freeware and demo versions, now archived but still useful.

The most significant development in software music was undoubtedly the 2010 launch of Apple's iPad (and in short order the 2nd to 5th, Mini, AIR and Pro generations), which after a short period running mainly existing iPhone apps began to see very significant specially-developed music software appearing. In 2011 the field was initially dominated by Korg (with a version of the MS20 synth and SQ10 sequencer, and Electribe drum machine) and by the Polish company Beep Street with the powerful 8-channel sequencer iSequence, which has many excellent built-in analog-style synth sounds and filters, and Sunrizer (later Horizon), a powerful polyphonic synth with effects based on Roland's JP8000 keyboard.

Since the iPad processors resemble those of modern virtual analog synthesizers, the sound quality to be obtained from these extremely affordable apps is fully professional. MIDI interfacing to the iPad quickly appeared in the form of hardware from Line 6, IK Multimedia and others, while the Alesis I/O Dock offered the iPad MIDI In, balanced microphone inputs, quarter inch audio outputs, composite video out and many other facilities, turning the iPad (running Horizon, or the analog synths and electronic keyboard emulations in Apple's own Garageband software) into a powerful stage or studio instrument. Behringer and others marketed iPad interfaces later on.

The Android mobile phone and tablet platform began to see music packages appearing too from 2011, such as the very simple Synthoid from Fleetway 76. Nick Copeland's Bristol Audio Software also has a stunning looking line in synth imitations for Android including one for the Moog Sonic 6, costing literally pennies through the Android Market, though the limitations of the Android operating system sometimes made playing response for all these designs unacceptably slow.

The story of analog synthesis came full circle in late 2011 when Moog followed up their Filtatron for iPhone and iPad, based on the Moogerfooger filter module, by entering the iPad/iPhone synthesizer field with Animoog, a powerful polyphonic synth, which expanded on the basic analog sound set with many types of cyclical modulation and effects.

The iPad's appeal for analog (or FM, or additive, or sample-and-synthesis) sound production lies partly in its responsive touch control, and performing with Animoog and similar apps should have much of the appeal of the early days of sound synthesis. Certainly the iPad and other tablet instruments will continue to be a popular resource for analog-style sound creation in the immediate future.

But however much flexibility these software packages offer, it's unlikely that in the near future they will either completely replace the modern generation of hardware-based keyboards and modules or eliminate interest in the classic analog instruments. In the next few years there will undoubtedly be many more releases in the analog emulation software field, as well as in the area of analog emulation hardware instruments. But the experience of many of the world's leading musicians, who continue to show their enthusiasm for the classic analog instruments despite the increased reliability of digital imitations and the decreasing cost of software emulations, seems to prove that there will always be a place for "real" analog synthesizers.

8

2013-2019 updates: new instruments, Eurorack and movies/TV

The period since this book was last updated has been a highly significant one for the whole area of analog sound synthesis. Not only have new artists arisen with a deep interest in analog sound, but new instruments have appeared at such a rate that the "analog revival" can now be said to be pretty much complete, with a healthy balance between new analog, virtual, software, and digital hardware products now established.

Calls for the major manufacturers – Korg, Roland, Yamaha, and others – to re-issue some of their classic instruments, which were never likely to be met 20 years ago thanks to the constant change in manufacturing methods and component availability, have finally been addressed more recently, with many instruments appearing which pay explicit tribute to classic designs. This hasn't had as much of an impact on prices for second-user, classic instruments as might have been imagined, as for every musician satisfied by the modern re-issue instruments there is at least one still interested in owning a classic, the prices of which remain buoyant as they become more and more difficult to find in full working order.

New fully analog instruments are also appearing at a great rate, predominantly in the area of Eurorack and other modules, as it's far easier and cheaper to create a small specialised module than a full scale polyphonic instrument. Meanwhile, software instruments become more and more versatile, though increasingly escaping from purely analog-style sound into much more exotic areas.

In this chapter we'll look at many of these new instruments and the musicians using them, stopping first, in the style of the Academy Awards ceremony, to mark the memories of those who have passed since we were last here.

OBITUARIES

With the modern history of electronic music now spanning some 60 years, inevitably some of the leading figures in the field have in recent times left us, each leaving a creative legacy either in instrument design or performance, which will be difficult to match.

Robert Moog had passed away in 2005, leaving Moog Inc. in healthy shape ready for a whole series of new product releases which continues to this day, notably the faithful re-manufacture in limited numbers of the original large Moog Modular systems. Fellow instrument designer, Don Buchla, passed away in 2016, leaving a son, Ezra, who performs around Los Angeles and elsewhere using viola, voice, and software. More on Buchla's instruments in the interview with Suzanne Ciani later in this chapter.

Josef "Joe" Zawinul of experimental jazz group, Weather Report, passed away in 2007 just after the conclusion of a five-week European tour, and fellow jazz musician, George Duke, died of leukaemia in 2013, the same year as his last album, *Dreamweaver*, appeared. Isao Tomita, interviewed in Chapter 5, passed away in 2016 aged 84.

Bebe Barron, who composed the "electronic tonalities" for the movie, *Forbidden Planet*, with her husband, Louis, died in 2008. Experimentalist, computer musician and accordionist, Pauline Oliveros, who invented the term "deep listening" died at the age of 84 in 2016, as did Else Marie Pade, Denmark's first electro-acoustic composer who worked with Pierre Schaeffer, Karlheinz Stockhausen, and Pierre Boulez.

Pauline Oliveros, who died in 2016, was an early pioneer of electro-acoustic music.

Jon Lord of Deep Purple died in 2012 after suffering with cancer. His distorted organ sounds became a staple of prog rock players and his fluid lines on ARP Odyssey, as well as on piano, were evident on his solo albums, such as the excellent *Sarabande*. In a very different area of music, Canadian performer, Nash The Slash (Jeff Plewman), died of a suspected heart attack in 2014 aged 66. Plewman generally appeared as his alter ego wrapped in bandages, wearing a top hat and tails, and playing electric violin and mandolin as well as keyboards and electronics. His music spanned atmospheric movie soundtracks to electro-pop, with early appearances on the Dindisc label, support slots for Gary Numan, and work with producers Bill Nelson and Daniel Lanois.

The recent past has also been particularly unkind to the German music scene, with three members of Can passing away, Michael Karoli in 2001, Jaki Liebezeit in 2017, and Holger Czukay in 2017, as well as Kluster's Conrad Schnitzler in 2011, who was also a member of Tangerine Dream, and

Dieter Moebius who continued along with Hans-Joachim Roedelius in the renamed Cluster (2015). Klaus Dinger of Kraftwerk, Neu! and La Dusseldorf died in 2008, his everlasting achievement being the invention of the "motorik" drum beat which influenced Stereolab and many other bands, though it remained less well known that he had also played the keyboard and synthesizer parts on many recordings under the pseudonym Nicolas Van Rhein. Pete Namlook, who founded the Fax label and released the *Dark Side Of The Moog* CD series with Klaus Schulze, died of a heart attack in 2012 aged 51. Rainer Bloss who collaborated and performed on keyboards alongside Schulze and also worked on his orchestral arrangements died in 2015. Ingo Bischof (Karthago, Kraan, Guru Guru) sadly died in early 2019 aged 68.

Else Marie Pade was Denmark's pioneer of electro-acoustic music.

Most notably, 2015 also marked the passing of Edgar Froese, founder of Tangerine Dream, not long after the completion of an Australian tour. It's difficult to over-estimate Froese's contribution to the popular electronic music field. Named after the 1967 album release by a British psychedelic band Kaleidoscope, which sounded similar to early Pink Floyd (and not to be confused with an American band of the same name), the early Tangerine Dream used Froese's guitar alongside drums, sax, the cello of Conrad Schnitzler, and other instruments to create free-form rock improvisations, before starting to incorporate effects units and electronic keyboards, then early synthesizers and sequencers to forge a lush, romantic but driving sound, which later found a place in many European and Hollywood movie soundtracks. More about this period on the website, in the interview with Steve Jolliffe who played along Froese in the early days of the band and again in 1978 on the *Cyclone* tour. While TD's signature sequencer sound with its distinctive "ratcheting" repeated notes was very much the work of Christoph Franke, it was Froese who always defined the overall direction of the band – sometimes into extremely over-commercial areas, though by the time of his last Australian tour Froese had brought about the re-introduction of a slightly more experimental sound – and his early solo albums, including *Aqua*, *Epsilon In Malaysian Pale* and *Ages*, remain object lessons in how to combine the analog synthesizer, sequencer, guitar, Mellotron, and other elements into an atmospheric whole.

The following year an even greater shock arrived with the evident suicide of Keith Emerson. In 2010 Emerson had played an Emerson Lake & Palmer fortieth reunion show in London and had been active both as a duo with Greg Lake and with his own Keith Emerson Band. He was forced to take time off later the same year for health reasons, which included the re-occurence of nerve problems in his hands and arms which had affected him for some time. Emerson continued to work on orchestral music and with Moog Inc. as they worked towards re-issuing the classic Emerson version of their modular synthesizer, but despite several operations, the nerve problems never cleared up and, depressed about possible criticism of his performance on an upcoming Japanese tour, Emerson shot himself at his Santa Monica home in March 2016.

Emerson, widely acclaimed as the most technically accomplished rock keyboardist of all time, demonstrated most clearly how to integrate the synthesizer with piano, organ, and other keyboard instruments into a rock format so strong that the three-man ELP created an onstage sound more than matching other bands with five or six members. Carl Palmer's drumming and Greg Lake's guitar, bass, and vocals can never be underestimated in that regard, though sadly, Greg Lake himself survived Keith Emerson by only nine months, passing away in December 2016.

Despite the loss of these and other important designers and musicians from the electronic music field, the area as a whole remains buoyant with many new names and manufacturers making an impact. Happily, many of them seem to be well aware of the lessons which can be learned from these most prominent figures of the recent past.

NEW INSTRUMENTS, 2013 ONWARDS

In the early days of the "analog revival" there were many calls for manufacturers, such as Korg, Roland, and Yamaha, to re-issue classic instruments. This was unlikely to happen because manufacturing techniques change constantly, specific components become unavailable, and economies of scale dictate whether or not a product is worth launching or re-launching.

After a very long delay though, perhaps partly driven by uncertainty as the "analog revival" built very slowly, from 2013 or so there was a gradual and ultimately a very substantial response from manufacturers to the continuing interest in classic instruments, mainly taking the form of "lookalike" re-issues using modern technology to closely reproduce the design layout and sound of some well known keyboards and modules. To do this in the quantities demanded by modern manufacturing techniques required some degree of certainty about the level of interest in analog sound. This seemed to be confirmed by a huge burst of activity in the areas of TV and movie soundtracks. For the first time since the 1980s major productions were being soundtracked using obviously synthesized sounds rather than orchestral sounds, in perhaps an inevitable re-discovery of the styles of John Carpenter and other composers from the 1980s.

Simultaneously, enough impetus had built up for a few companies to create very up-market new products, so that Moog Inc. was able to build some new modular systems from scratch following the original designs, and the Prophet and Oberheim lines from Dave Smith and Tom Oberheim once more came into prominence.

In this section we will look at the very exciting new product issues from the period 2013–2018, as well as speaking to many musicians using these new instruments alongside classic designs.

Moog

Moog Inc. continued with production of Minimoog Voyager lines alongside the Little Phatty – a white Stage II version was the last to be produced – and from 2013 the Sub Phatty, a compact 2-octave instrument.

The Taurus 3 bass pedal synth (2011) was discontinued but the Minitaur bass synth module received a software update enabling it to save and recall presets from the panel, as opposed to having to use a software editor.

In 2014 Moog introduced the Sub 37. This was a powerful duophonic three-octave keyboard with a generous complement of controls and a versatile sequencer/arpeggiator. Lying somewhere between the Phatty and Voyager lines in complexity, the Sub 37 offered two LFO's so that very complex, modular system-style modulation could be created. The Tribute Edition retailed at $1500 at the time of writing, and the Subsequent 37 (black) and Subsequent 37CV (white) limited edition versions (around 2000 built) offered improved mixer headroom, greater overdrive options, and CV patching, and many more improvements. A video demo of the Sub 37 is on the book's website: www.routledge.com/cw/jenkins.

In contrast, 2014's Werkstatt-01, launched at the Moogfest event that year, was extremely simple: a kit-built single oscillator monophonic desktop module with pushbutton keys and pinbay patching. Soon a minijack patch expander became available, and later on, pre-built Werkstatts. Priced around $200, the Werkstatt offered not only an entry into analog sound but into the area of understanding and modifying the construction of simple analog synths. A busy forum developed in modifying and rackmounting Werkstatts, and the design was expanded the following year into the Mother 32, a rackmount module with a sequencer, and more extensive patching possibilities. The Mother 32 gives the appearance of a tiny modular system, with various official and third-party mounting options available, and is often seen in groups of two or three instruments. It was joined in 2018 by DFAM – Drummer From Another Mother – a matching desktop module dedicated to analog percussion sounds, again with the help of an extensive minijack patchbay.

As part of a long-term project, Moog Inc. had been working alongside Keith Emerson on a re-creation of his classic modular system, finally released in 2014 at around $150,000. This was precursor to the re-issue,

Building System 55 modulars at the Moog factory.

in very limited numbers, of several original Moog Modular designs, including the Model 15 ($10,000, 150 made), System 35 ($22,000), System 55 ($35,000), IIIC and IIIP ($35,000) and Sequencer Complement. These were all hand-built to the original specifications with a minimum of modern modifications, spanning the studio and portable design of the 1960s and 1970s. These giant modular systems still have enormous appeal, but since they're out of reach of most musicians, Moog Inc. also looked at the software field, offering an app version of the Model 15 modular, which gives some terrific sonic possibilities for around the $30 mark. With 4-voice polyphony and an arpeggiator, it can be said to be more flexible than its hardware equivalent.

Moog also introduced an app version of the Model D Minimoog for 64-bit iOS devices, this time at around $15 and again offering an arpeggiator. But hardware development wasn't being neglected. While a limited number of original Minimoog Model D instruments were released over a period of about a year, 2018 saw the launch of the Grandmother, a striking fusion of the Werkstatt, modular, keyboard synth and Moogerfooger effects unit lines. With its multi-coloured panels, the two-and-a-half octave Grandmother differs somewhat in appearance from the traditional Moog, going back to the appearance of the old Rogue or Realistic MG-1. It has something of a homebrew look to it and is very much oriented towards sonic experimentation as well as processing external sound sources.

Moog's Grandmother, a new genuine analog semi-modular synth.

Unusually it incorporates a genuine spring reverb unit, alongside dual oscillators, low pass and high pass filter, arpeggiator and sequencer, envelope and LFO, MIDI and USB MIDI, and 41 patch points arranged at the top of each "module" rather than in an integrated patch bay. It can be used without any patching required, and has no memories, but it's particularly adept at patching into Eurorack or larger modular systems. At around $1000 it provides a flexible approach to stage or studio sound creation. See Appendix B for Moog's 2019 release of the new Moog One polyphonic instrument.

Moog, in recent years, has shown a flexible approach to the development and updating of new products both in hardware and software, while maintaining very high construction standards and audio quality. However newer players in the field with a from-scratch approach to electronic design are responding with massively competitive products, as we'll see.

www.moogmusic.com

KORG

Korg had a busy period from 2013 onwards, starting with the launch of the RK-100S – a keytar version of the MicroKorg virtual analog synth – and the MS20 Kit, a partially built version of the already popular MS20Mini reproduction of the original MS20 synth, quickly followed by a keyboard-less desktop version, both frequently seen accompanied by a small analog-style sequencer, the SQ-1. Perhaps even more exciting was the news that Korg had been collaborating with the original designers of the ARP instruments to recreate the Odyssey in a similar, slightly scaled-down version. Eventually this became available in three models to represent the original white face, and later black and black/orange versions of the Odyssey, each with slightly different filter characteristics.

The Odyssey had been a highly significant instrument in the 1970s, seeing use by jazz musicians Chick Corea, Herbie Hancock, and Neil Ardley, experimentalists Tangerine Dream and Klaus Schulze (who used it

Korg's re-imagined ARP Odysseys with various filter types and keyboard sizes have been a hit – seen here with a limited edition ARP graphics version of the SQ-1 sequencer.

extensively for a high-pitched windchime-like sound, which characterised many of his pieces, and for an echoing repeated rhythm on the track *Totem*) as well as by many progressive rock players including Caravan and Jon Lord. Surviving instruments were very much in demand, one constant problem being that sliders and keys would tend to snap off, making a reproduction Odyssey at affordable prices a winning proposition. Later, Korg introduced a keyboardless desktop version, and full size key Odyssey FS versions, following these with Minilogue in 2016. Minilogue is a compact three-octave, dual oscillator four voice polyphonic (or mono Unison) fully variable and programmable analog synth with a polyphonic sequencer, tape echo style effects unit, and a tiny oscilloscope-style display, selling at around $400. Korg had always enjoyed success with four-note polyphonic analog designs, from the MonoPoly to the MS2000, MS2000R, and MS2000BR, so it made sense to have a new offering in the area and unusual possibilities, such as controlling delay length with the expression slider, gives the Minilogue, quickly followed by the simpler two-octave Monologue styled in black or white with a row of 16 flashing sequencer buttons like those on the MS2000, a lot of appeal.

Also launched in 2015 was Korg Gadget – a modular studio software app for the iPad integrating many different soundmaking devices – and this later became available for the Mac. An iMono/Poly was also introduced for iPad, but Korg hadn't neglected larger keyboards and workstation designs, with the Kronos and Krome lines and the M3 workstation and RADIAS synth obviously offering many variations on analog sound as well as typical sample-based textures. The King Korg was a lightweight 24-voice, five-octave virtual modelling synth with vocoder, still in the catalogue in 2018, but more specifically analog was the Prologue, launched in early 2018 in 8-voice or 16-voice versions at around $1000 or $1500 respectively. The Prologue's styling is very much like that of the old Sequential Prophet 5 and Prophet 600, with the use of wooden end cheeks, a black on black panel and many similar facilities, such as poly glide and a versatile arpeggiator. The Minilogue's tiny oscilloscope style display also appears, the designs combine two analog oscillators with a third digital oscillator, plus on the 16-voice model an analog compressor.

At the time of writing, Korg also offers many entry level instruments, such as the Volca series of desktop modules including a monophonic synth, drum machine, sampler, and mixer, and the Little Bits set of a dozen or so uncased modules, including a tiny keyboard and four-step sequencer, which can be freely patched together, costing around £130/$150, and so clearly intends to remain equally active in the fields of affordability, re-issue style instruments, software, and upmarket hardware designs. It's tempting to think that other re-issue style instruments may appear – the Mono/Poly seems an obvious possibility, the Poly 6 and similar models perhaps less so because their capabilities are well covered by the Minilogue and Prologue designs – the last of which most obviously aims to compete with Dave Smith's continuing Prophet series, of which more later.

ROLAND

As mentioned earlier, calls for re-issues of classic instruments had gone largely unanswered until manufacturers came up with ways to combine well known classic facilities with modern design and manufacturing techniques. Roland had tried re-imagining the Jupiter 8 and TB-303 Bassline, for example, in the VariOS hardware/software system, but that had been far ahead of its time. The large Jupiter-50 and Jupiter-80 keyboards available from 2012 had included sound sets emulating the Jupiter-8, SH-101, TB-303 Bassline, Juno-60, Jupiter-6, Juno-106, and the digital D-50, as had the excellent Fantom workstation, and the Juno-Gi offered a combination of virtual analog and digital sounds. The compact 3-octave GAIA virtual analog synth, which could be edited using Mac/PC software, remained in the catalogue for some years, but it was the System-1 launched in 2014 which definitively looked back towards its analog forebears the System-100, System-100M and System-700 modular.

The System-1 "Plug-Out Synthesizer" is a compact 2-octave monophonic with two oscillators, effects, and illuminated controls, rather similar in appearance to the much earlier all-analog SH-101. The all-digital System-1 can be completely reconfigured with a software plug-in, including specific SH-101, SH2, ProMars or System 100 Modular characteristics. All of these, as well as the System 1 software, can also run stand-alone on Macs and PCs. The "plug-out" changes the routing assignment and filter characteristics, so users can make the extra investment in the plug-out when wanting a more specific emulation of a classic synth.

The System-1 overlapped with the company's AIRA product line, which included new versions of the TR-808 drum machine (the TR-8) and the TB-303 Bassline (the TB-3) as well as a vocoder, a mixer, and other products. Combined with a keyboardless module version of the synth, the System-1M, the Aira system offered easy, reliable access to a "cleaned-up" version of the classic analog sounds used in techno, acid, and house music. Illuminated controls made these instruments particularly visually appealing in stage conditions, this also applied to the compact JD-Xi synth. With three octaves of mini keys, this occupied the same market area as Korg's

MicroKorg and Akai's MINIAK, offering a fusion of virtual analog and digital sounds, sequencing and percussion effects. A larger four-octave JD-XA version was also launched.

Roland's JD-XA is a compact analog/digital hybrid with sequencing and illuminated controls.

In 2016 the larger and very impressive-looking System-8 Plug-Out synthesizer was launched. This was a polyphonic design which could host up to three different plug-outs, including emulations of the Jupiter-8, Juno-106, SH-2, SH-101, ProMars, and others. A polyphonic step sequencer, vocoder and CV/Gate outputs made the synth particularly adept at interfacing to modular systems and other instruments, and a choice of wood or aluminium end cheeks allowed its cosmetic appearance to be altered to suit the user's studio.

Roland's Boutique line finally saw the long called-for re-issue of some classic synth designs.

By 2018 Roland also had a wide range of offerings in the "Boutique" series. These were tiny desktop module recreations of classic synths – in some cases with knobs and controllers so small that they seemed difficult to use in stage conditions – though the sounds produced are excellent. Joining the VP03 vocoder, which was a reproduction in module form of the old VP330 keyboard vocoder/strings (a small keyboard, the K25m, is an option and works with all the Boutique modules), is the SH01a, which emulates the SH101 monophonic synth (also playing polyphonically and appearing in all the originally available colours), the JX03 which emulates a JX3P, the JP08 for the Jupiter 8, the JU06 for the Juno 106, the TB-03 for the TB303 Bassline, the TR-09 for the TR909 drum machine, the SE02, designed in collaboration with Studio Electronics in a clear tribute to the Minimoog, the TR08 for the TR808 drum machine, the D05 for the D50 digital synth, and the A01, which is a basic MIDI and analog controller with some built-in sounds. Prices ranged from $300 to $500 so the modules were immediately attractive. Going into the future they may have a major impact on the perceived second-hand value of the originals, particularly since many of them offer new features, such as additional LFO waveshapes, though several of them quickly found a lot of competition from new upcoming manufacturers.

Roland also entered the Eurorack arena with the System-500, firstly in the form of re-programmable effects modules and later with synth modules. See the upcoming section on Eurorack developments.

JAPAN, 2013-2018

Yamaha had exited the analog field more definitively than Roland though their popular MOTIF, MOX, MX, and S-Series workstations all retained some ability to create analog-style sound. A single exception to this appeared in 2015 with the ReFace series, minikey portable instruments emulating past Yamaha products specifically the DX7 FM synth, the YC series of organs, the digital piano ranges, and the CS80 synth. All of these gave a good impression of the original instruments, though perhaps falling short in some significant areas: the original CS80, for example, depending very much on its individual note aftertouch, voltage controlled ring modulator and long pitch bend strip for its most distinctive sounds. The ReFace CS also lacked memories, though these can be added using an iPhone or iPad.

https://usa.yamaha.com

Akai made a partial return to analog style sound with the Timbre Wolf, a compact 2-octave multitimbral 4-voice instrument with a rotating voice assignment mode and a step sequencer offering fill patterns. Discontinued at the time of writing, along with the Tom Cat drum machine and Rhythm Wolf drum machine and bass synth, it remains in some stores in direct competition to Korg's Minilogue.

www.akaipro.com

Perhaps the most long-awaited Japanese synth of the period has been Deckard's Dream from the Black Corporation. This is a rackmount emulation of the Yamaha CS80 (named after the central character of *Blade Runner*, which heavily features CS80 on the Vangelis soundtrack) with 16 VCO's and the ability to support MPE (MIDI Polyphonic Expression), which makes it well suited to playing from the Roli Seaboard. No sign of the voltage controllable ring modulator though, which would appear to be vital for some archetypal CS80 sounds, such as those played by Klaus

Deckard's Dream is a powerful rackmount recreation of the Yamaha CS80.

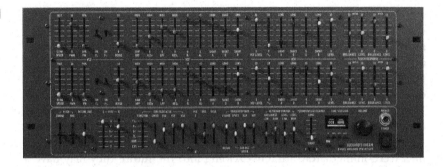

Schulze. At the time of writing the instrument is just coming onto the market at a price of around $2500 (or slightly less in kit form) and a more generalised semi-modular instrument, the Kijimi, with SSM chip-based filters, was also being previewed.

www.deckardsdream.com

As in other parts of the world there is also a busy market in Japan for very short-run experimental analog synths aimed at creating abstract sounds. Perfect Circuit Audio in the USA carries the lines, for example, from JMT Synth by Tetsuji Masuda, drone, and noise box products, some of which don't have any function labelling so are really for experimentalists, though many are more sophisticated and interface readily enough, for example, to a Korg SQ-1 sequencer.

www.jmtsynth.com

USA, 2013–2019

The period 2013–2019 saw the resurgence in the USA of two early synth designers, Dave Smith and Tom Oberheim: Smith through his company DSI, which started to issue upmarket keyboards under the familiar Prophet brand alongside his existing Evolver lines, Oberheim in the form of updated versions of his Two-Voice synth, as well as a powerful polyphonic design the OB6 and its desktop version in collaboration with Dave Smith.

DSI

DSI's Evolver lines, starting with the Evolver desktop and adding two-voice and polyphonic keyboard versions and the compact Mopho module and keyboard lines, had become very powerful, but there was always an impetus to issue a design specifically recreating the original Prophets from Sequential Circuits. This would demand genuine analog sound, which was always going to be an upmarket proposition, and so the Prophet 6, when it appeared in 2015 after DSI regained use of the name from Yamaha sold at something over $2000, with a module version a couple of hundred dollars lower.

The design explicitly resembles the old Prophet 5 but with many modern additions. With 6 voices, built in digital effects and stereo analog distortion, arpeggiator, polyphonic step sequencer, a sub-oscillator and two independent filters per voice, a versatile PolyMod section with added destinations, Unison mode, 500 factory and 500 user memories, and a four-octave semi-weighted velocity and channel aftetouch sensitive keyboard, it's a powerful proposition.

The Prophet 6 was quickly joined by a Prophet 08 model and then a Prophet Rev 2 16-voice model with desktop version. With a five-octave keyboard this was less compact, though the three and a half octave two-voice Pro 2 and similar format Mopho x4 also remained in the catalogue.

Dave Smith also worked for the DJ equipment company Pioneer on the AS-1 Toraiz, a small desktop monophonic synthesizer with a built-in keypad, and the matching SP-16 desktop sampler.

At the time of writing, DSI is just introducing the Sequential Prophet X, which looks like becoming a very significant instrument. This is a bi-timbral, stereo 8-voice (mono 16-voice) five-octave keyboard with digital oscillators and 16-bit sampled sounds processed though analog filters. There is 150GB of internal samples created by 8Dio, covering all sorts of electronic, acoustic, ambient, and cinematic effects, and 50GB of user sample memory. The digital waveshapes can be modulated and there's four syncable LFO's, four loopable five-stage envelopes and a 16-source modulation matrix.

Dave Smith's Prophet X combines digital oscillators and samples with analog filters.

The Prophet X has no fewer than three OLED displays so there is plenty of access both to the analog and digital aspects of its sound. At around £3500 it's an upmarket purchase but no doubt one which will become very significant, both on stage and in recording and movie soundtracking studios over the next few years.

Dave Smith's DSI also started to offer Eurorack format modules, more about which in the following section.

www.davesmithinstruments.com

OBERHEIM

After some years operating as Marion systems, Tom Oberheim re-formed TomOberheim.com and set about re-issuing his classic SEM module and Two Voice synth, the most compact models from the early years of the company. The new Two Voice Pro in black or white finishes has a three octave keyboard, memories for the built-in 16-step sequencer, which can now play two patterns at once and chain them into songs, velocity, and pressure sensitivity on the keyboard, a separate vibrato LFO, and mini-jack patch points for CV control of various parameters. The US price was $3495. The SEM module exists in three versions with MIDI, minijack patch points, or both, with prices from $899 to $1199. All of these retain the classic Oberheim 12dB/octave low pass filter, which has a slightly different tone from the Moog 24dB filter.

Tom also called in Dave Smith to collaborate on the OB6, cosmetically something of a hybrid between the Prophet lines with their distinctive wooden end cheeks and the black-with-blue-stripes styling of the earlier Oberheim OBX. The OB6 is a compact four-octave 6-voice dual analog VCO synth with effects, step sequencer, arpeggiator, 500 factory and 500 user memories and retailed at around $3000 with a desktop module version around $2300.

Tom Oberheim's reborn TVS Two-Voice synth.

The current Prophet and Oberheim lines are favourites with the *Stranger Things* composers, Michael Stein and Kyle Dixon, who lead their live performances with them. More in the section on movie music.

www.tomoberheim.com

STUDIO ELECTRONICS

We looked at the current Boomstar Mk2 SEM version from Studio Electronics in Chapter 2. The company started by modifying and rackmounting Minimoogs, then developed their own programmable monophonic synth rackmounts, such as the SE1x, Red Eye, ATC (Analog Tone Chameleon, introducing the idea of swappable filter types), and ATC-Xi. All of these are now out of production but the company continues with the large rackmount Omega 8 and CODE polyphonic synths, and an Orion

Galaxy version of the Omega 8, which really does look like something out of *2001: A Space Odyssey*. The company also co-produced the SE02 alongside Roland, basically a small module version of the Minimoog for the Boutique series.

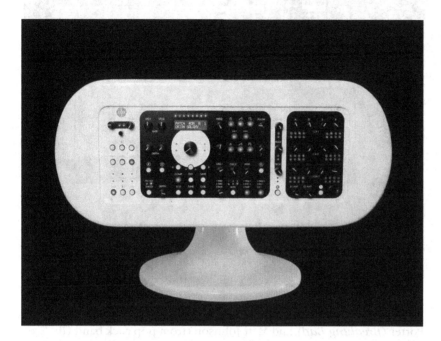

The Orion Galaxy version of the Omega 8 from Studio Electronics.

More recently the company returned to its roots with the Boomstar Mk2 line, which comprises desktop monophonic non-programmable synths with a choice of filter types; the models are 3003 (Roland Bassline TB303 filter), 4072 (ARP 2600 filter), 5089 (Moog 24dB/octave filter), 8106 (Roland Jupiter and Juno style filter), SEM (Oberheim style 2-pole 12dB/octave filter), and SE80 (Yamaha CS80 style filter)

All of the Boomstar Mk2 versions help to interface MIDI to analog technology with a modest number of patch points, giving accessibility without the complexity of a patchcord modular system. Boomstar is a two oscillator monophonic synth with a good choice of routing options to create either powerful bass and lead sounds, or more abstract experimental synth effects. A diode-based feedback circuit in the mixer section allows the sound to be dirtied up as required. The company also has several related products in the Tonestar and Boomstar Modular lines, more about those in the section on Eurorack, and there are some great sounds and much more on the Boomstar SEM on the book's website: www.routledge.com/cw/jenkins.

www.studioelectronics.com
www.mslpro.co.uk (UK distribution)

John Bowen has a long pedigree programming and designing for Sequential, Korg, Yamaha, and Creamware, and around 2014 got his own keyboard product, the Solaris, into production. In its seventh production run at the time of writing, Solaris sells for $4199 and has been used

One of several Boomstar Mk2 variants and modules in production from Studio Electronics.

by JJ Abrams (*Star Trek*), Harry Gregson Williams (*The Martian*), Dave Porter (*Breaking Bad*), and Matt Johnson (from pop rock band The The) among others.

The Solaris from John Bowen.

Solaris is a five octave synth, finished in black or white, with six LCD displays to access ten voices of polyphony with four oscillators per voice, intended to act as a modular synth, with all of the fast access of a pre-patched keyboard synth. It uses CEM VCO chips as in the original Prophet 5 and models the SSM filter chip as one of six filter types assignable to four different filters, including multimode settings. There is an eight-stage loop-ing generator for Wavestation-like textures, built-in effects, audio inputs, two mod wheels, a joystick controller, and a pitch strip.

www.johnbowen.com

Kurzweil, Alesis, and other US manufacturers have been well out of the synthesizer business in the recent period, though having plenty to do with domestic and stage pianos, workstations, controllers, and home keyboards. Most of the other US activity in the field is within the area of modular synths, with Synthesizers.com, for example, still offering a huge range of Moog format modules which can be built up into compact or giant systems intended for the studio or travel, in portable formats, or to expand existing Moog systems. The compact Thoughtbox TB22++, for example, offers 13 different modules plus one black space for a module of the user's choice and looks particularly enticing with its folding cabinet allowing patch cords to be left in place during transportation. The price is around $3700.

The Thoughtbox from Synthesizers.com is a compact Moog-style modular.

A starter system can be purchased at $155 per month with one module arriving per month (the MOTM "Module Of The Month" scheme from Synthesis Technology having disappeared as they moved into Eurorack). A Synthesizers.com Portable 22 system with three oscillators comes in at a little under $3000, less than a third of the price of the comparable Moog Model 15. The company's website also has the best selection of *synthies* (owners pictured with their modular systems) that you could ever ask to see.

www.synthesizers.com

Some US companies are expanding upwards from the modules field into stand-alone instruments. In late 2017, Malekko, a company which already had an excellent line in Eurorack modules and pedals, launched the Manther at $649, a desktop monophonic synth with a CEM3340 oscillator, SSM2044 filter, and built-in 64-step sequencer reminiscent of a modernised Roland MC202.

www.malekkoheavyindustry.com

Malekko's Manther is a desktop monophonic synth using archetypal analog chips.

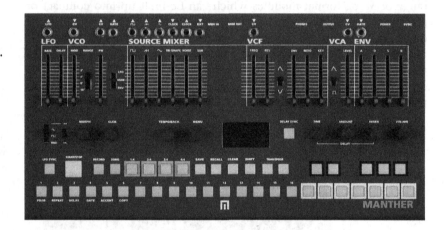

Much more about other US-built modules in the section on Eurorack, but first a look at recent developments in the UK and Europe.

UK AND EUROPE, 2013-2019

DUBREQ

Typically of the UK market, eccentricity takes pride of place alongside technological innovation and one of the most quirky of the early analog instruments has re-appeared more recently in a versatile new format. The Stylophone from Dubreq was a paperback-sized monophonic instrument launched in 1967 and played using a conductive stylus on a miniature metallic touch keyboard. The lack of moving parts made it a sturdy instrument and it's clearly heard making a rising glissando on David Bowie's *Space Oddity*, later very prominently on Kraftwerk's *Pocket Calculator*. Later on a two-voice 350S model was released along with a bass version and many music books and tuition courses. After a long period out of production, the classic Stylophone was re-issued in 2007, followed by a much more synthesizer-like S2 version in 2012.

But it's the new Stylophone GenX-1 version released in 2017 which looks set to have a greater impact. A cross between the original Stylophone and a compact analog synth like the Korg Monotron, it features a touch strip as well as the stylus keyboard, an analog filter with resonance, variable envelope, and modulation, and a delay line. Still able to work from battery power, the GenX-1 at around £60 is certainly capable of playing melody parts but can now create a wide variety of abstract synthy sounds too. It's a great purchase as a portable soundmaker or first instrument for

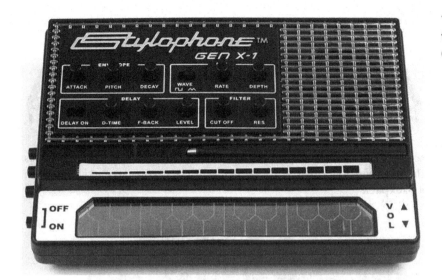

The GenX-1 from Dubreq sees the Stylophone reborn with great new capabilities.

youngsters wanting to get into sound creation. A full demo on the book's website: www.routledge.com/cw/jenkins.

www.dubreq.com

As in the USA, several British stand-alone synth designs were expansions of Eurorack and other modules. Analogue Solutions, for example, launched Treadstone (it's a *Bourne Identity* reference) a flexible single oscillator synth with an SSM2044 chip filter as a Eurorack module at around £400, but also offered it stand-alone in a desktop version at around £500. Larger instruments, such as the 2-oscillator Telemark, the Leipzig S with its built-in 8-step sequencer, the Nyborg, which resembles the old Oberheim SEM desktop module with a similar 12dB/octave filter (and a 24dB version available), the 3-oscillator Fusebox and the 4-oscillator Polymath remain available alongside the company's sequencers, pedals, and Eurorack products, some under the Medic Modules brand.

www.analoguesolutions.com

Modal Electronics

In massive contrast to two of the most compact and portable instruments around, Modal Electronics from Bristol took some years to bring to market the very large and impressive 002, a hybrid analog/digital 8-voice synth. The 002 is an eight-voice instrument with built-in effects retailing at something over £3000.

In 2018 the company was also bringing to market the 008 – a slightly simplified version styled in black – alongside rack versions of both, and a compact three-octave two-voice version called 001.

In contrast was Modal's successful Kickstarter campaign to launch Skulpt, a very compact battery-powered four-voice instrument using some of the same voice architecture in a design costing just $300 and

Modal's 002 is a very upmarket UK design but the company does have mass market analog products coming too.

reminiscent of a futuristic EDP Wasp. The company is describing this as the only battery-powered, four-voice synth that you can carry around in a backpack.

Modal's Skulpt may become a powerful entry level instrument in 2019.

This looks like being an interesting design; the company is also marketing the tiny Craft Synth and Craft Rhythm, affordable microprocessor processor-based kit format designs for a monophonic synth and sampling drum machine with a minimum of hardware elements leaving most of the circuitry exposed, with an optional iOS/Mac/Windows/Android editor which also allows several synths to be played as a single polyphonic instrument. Both kits are easily assembled in about ten minutes with a UK price of around £80 each. More on the book's website: www.routledge.com/cw/jenkins.

www.modalelectronics.com

Synth Restore under Adrian Dolente and based in Bedford, UK, specialises in repairs and modifications for a wide range of instruments and under the Harpur Instruments brand offers two all-analog monophonic keyboards, the three octave Aurora AR57, and the compact ZV-10.

The Aurora AR57 is based on earlier AR1 and AR51 versions and has three transistor-based oscillators and a 24dB/octave low pass/high pass filter. The LFO includes a random mode and a "pattern" mode, which creates riffs and arpeggios, and there are CV, Gate, and Filter CV inputs and an audio input to the filter. The price is £995.

The ZV-10 at £599 is a slim 3-oscillator monophonic synth without performance controls but with many modulation routing options between the oscillators. Equipped with CV and Gate inputs for interfacing to modular systems, there's an audio demo available on cassette – very retro.

Harpur ZV-10 from Synth Restore, a compact keyboard with modular-style sounds.

As with the AR57, there's a lead time of a few weeks on each unit. The company also supplies a range of ARP2600 and ARP Odyssey equivalent circuits for filters, sample-and-hold, VCA and envelope shaper, a Juno 106 filter circuit, and various Minimoog replacement circuits.

https://synthrestore.co.uk

M-Audio remains a leading manufacturer of MIDI and USB controller keyboards and the author's studio uses two of their Keystation Pro 88 piano-weighted controllers as well as the pad-equipped Air 25 from their very popular Axiom range. M-Audio's recent entry into the virtual analog synth stakes has been and gone though at the time of writing. The four octave Venom, launched in 2011, was discontinued after a reasonably short production run, having been mildly criticised for a gritty sound quality, limited editing without relying on computer software, basic effects, and over-complex multi-layered presets very much aimed only at the dance music sector.

However, many players love this grittier element of analog, the Venom presets form a great basis for composing the more hard-edged type of dance music. So at current second user prices, which are not much over £100, a Venom could be a very pleasing purchase.

www.m-audio.com

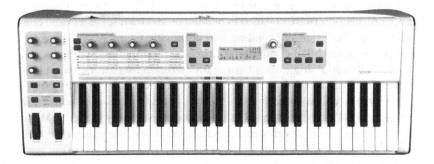

M-Audio's Venom has been and gone but may now be one to look out for second-hand.

Novation discontinued their Nova series modules but continued with the MiniNova and UltraNova virtual analog keyboard synths, as well as launching the grid-based Circuit, which has built-in polyphonic virtual synths, and Circuit Mono Station, which has a more readily accessible set of synth controllers The Bass Station keyboard was upgraded to Bass Station 2 but the company's lead product became the Peak desktop module.

Novation's Peak is a powerful
and flexible virtual analog
desktop module.

Launched in early 2017 at around £1200, Peak is an 8-voice instrument with three "numerically controlled oscillators" and two LFOs per voice, and the ability to add distortion at various points in the audio chain. Like all Novation synths it's highly versatile, making a good all-round studio instrument capable of anything from lush pads and fat basses to fast techno patterns.

www.novationmusic.com

S-Cat is a small company originally associated with "circuit-bent" modified instruments, but called in well-known electronics engineer John W. Oram to help design Dub Siren, a small desktop noise generator featuring a LARGE red button to trigger off fairly abstract sounds. It's certainly an analog synthesizer, though not a specifically musical one: it creates deep drones, falling or rising and modulated tones like sirens and syndrums. At around £175 it's an affordable way to create analog sound effects on demand and at the time of writing a more versatile Dub Synth model was also available at around twice the price.

www.spacecataudiotechnologies.com

Retrowave R-1 has an unusual pedigree coming from a British company Trax Controls, otherwise engaged in making model train set electronic control systems, but is an outstanding and affordable (at £340 with free UK delivery) shot at a MIDI controlled true analog monophonic synth.

The R-1 has a single oscillator with two sub-oscillators, two LFOs, sample-and-hold and white noise, and comes with a universal power supply which will work anywhere in the world. The sound is clear and smooth, great for electro-pop styles along the lines of early Depeche Mode as well as for more abstract and acid effects. There's a short queueing period as each batch is built but that's hardly surprising given the very affordable price point.

http://traxcontrols.com

Over to Italy where Studiologic continued the grand tradition of Italian synth design with the Sledge, released in 2015 and revised as V2 in 2017. The company was well established making MIDI controller keyboards, some with built-in sounds, such as the slimline Numa Compact 2x

The Retrowave R-1 from Trax.

and turned to Waldorf in Germany to design an internal sound system for their new synthesizer, and this collaboration proves a great success. The mechanical design is very compelling, with a superb controller layout and great construction, which nevertheless imposes only a very modest selling price, while the sonic architecture is incredibly flexible. Two virtual analog oscillators offering all the usual waveforms are matched with a third capable of playing a whole wavetable of sounds, offering the wave-shapes of organs, pianos, acoustic instruments, drums, human voices, and much more. Dual LFO's can be routed very quickly and flexibly, including routing both to the same destination, which is an incredibly rare find much needed for creating abstract sounds. Basic digital effects come in three layers – for modulation including Chorus and Phasing, for Delay, and for Reverb – and there's a simple and flexible arpeggiator.

Studiologic's Sledge V2 offers virtual analog at an incredible price point.

The Sledge (particularly in its V2 version) stretches virtual analog to its ultimate, for example, in offering polyphonic glide, which is often omitted from competing designs. The keyboard can play in mono or poly, single or multiple note triggering modes, making lead line performances (for example, if you want an archetypal monophonic Minimoog style sound) very flexible, and there's both velocity and aftertouch sensitivity on the keyboard, making it possible to create either huge analog filter sweeps, or dramatic wavetable movements with changes in pressure. The filter has low pass, high pass and band pass modes for a great deal of textural diversity.

The Sledge factory sounds are arranged in a delightfully haphazard manner but just go to show how versatile the synth can be. Sitting side by side are great PPG-style digital sweeps, cutting analog-style lead lines, convincing electric pianos and drum kits, fast techno arpeggiated patterns, crazy random FX patches, powerful monophonic basses, stabbing analog-style brass, splits with drum rhythms in the lower octaves, and much more. Some genuinely analog techniques are denied to the Sledge, for example, you can make the filter resonate to create a sine wave, but you can't modulate it with the oscillators turned down because it doesn't really exist in analog terms. However, the oscillators have a perfectly good sine wave mode themselves and this can indeed be double modulated, making echoed abstract effects of the Tim Blake/Jean-Michel Jarre/EMS Synthi variety readily accessible.

The most recent synth design with this sort of versatility and power may have been as far back as the Yamaha AN1X, but the Sledge has by far a more accessible panel layout resembling nothing more than a large chunky polyphonic Minimoog with some extra modulation options. The most unbelievable aspect of the Sledge is its retail point – just £900 from some retailers but under £700 elsewhere – which in the area of creative sound synthesis, as opposed to workstation or songwriting sorts of applications, makes it probably the best value all-round keyboard synthesizer currently in production anywhere in the world. Extensive video demos on the book's website: www.routledge.com/cw/jenkins.

www.studiologic-music.com
www.mslpro.co.uk (UK distribution)

Also based in Italy, IK Multimedia continued with a diverse range of software including SampleTank 3SE and various audio/MIDI interfaces and controllers, but also in July 2018 launched their first hardware synth, the monophonic Uno at just over £200. With an arrangement of keyboard buttons on the front panel it has two analog oscillators, resonant multimode filter, noise, overdrive, random sample-and-hold, step sequencer, arpeggiator, USB MIDI, audio input, and 100 memories (80 reprogrammable), and looks like being good value for money. More on the book's website: www.routledge.com/cw/jenkins.

www.ikmultimedia.com

In 2016 in Sweden, Mellotron previewed the Micro model which has a 2-octave keyboard and digital samples of all the usual Mellotron tape-based sounds built in, but a rather steep price point at around €900.

www.mellotron.com

IK Multimedia's Uno is the company's first hardware synthesizer.

The most compact format of the digital Mellotron is the two-octave Micro version.

Elektron, also based in Gothenburg in Sweden, started life in 1998 with SIDStation, a desktop synth based on the sound chip of the Commodore 64 computer. This was followed up with the Monomachine keyboard in 2003, of which 500 were built, and various drum machine and sampler designs. The company's first analog synth was the Analog 4 module from 2012, updated to a Mk2 version in 2017 and later joined by Analog Keys.

This is a three octave, four-voice design with two oscillators, two sub-oscillators and two LFOs per voice plus white noise, three effects sends, step sequencer, and arpeggiator. The Analog Keys has extensive interfacing possibilities thanks to MIDI and CV outputs and a flexible joystick controller, and has been selling at £1249 and downwards. Although it's a flexible

Elektron's Analog Keys became a favourite for creating ambient music.

enough design it seems to have found favour mostly among musicians working in ambient styles, though a busy forum group, www.elektronauts. com, had plenty of information on other types of user. The company also manufactures analog overdrive units Analog Drive and Analog Heat, and a small polyphonic desktop digital synth module the Digitone.

www.elektron.se

In Chapter 7 we looked at the French company, Arturia, founded in Grenoble, and their analog emulation software. More recently the company launched into hardware first in the form of the compact Keylab and other keyboard controllers for their software packages, with full size 88 note and other intermediate size version now available alongside the compact sequencers, Beatstep and Beatstep Pro, and drum machines Drumbrute, Drumbrute Impact, Drumbrute Creation, and Spark/Spark LE.

Arturia's launch into hardware synthesis came in 2012 with Minibrute, a compact two-octave true analog monophonic synth with multiple waveforms available simultaneously from the single oscillator, sub-oscillators and a multi-mode filter. At £429 it competed with second-hand classic monophonic synths and was later replaced by two new models: the Minibrute 2, with a very extensive patchbay, and the Microbrute, with slightly simplified features and a mini patchbay at a lower price point.

All of these instruments have proven very popular and are among the very few all-analog compact monophonic keyboards now on the market. Arturia also launched an alternative version in the Minibrute 2S, without a keyboard but with step sequencer pads, and the Rackbrute powered cabinets, which mount sets of Eurorack modules in 3U or 6U format above a Minibrute or similarly sized instrument. No doubt Arturia will have Eurorack module products of their own eventually.

The *piece de resistance* for Arturia arrived in early 2017 in the form of the Matrixbrute, a 20kg, four-octave monophonic synth with a tilting control panel, a 16x16 programmable matrix of routing buttons, which doubles as a step sequencer display, a four octave full size velocity and aftertouch sensitive keyboard, three analog oscillators, two programmable LFOs, one with delay to program, for example, a fading in vibrato, plus built-in effects, MIDI and USB, initially at £1779 (€2299). This put it at around half the price of the cosmetically similar Moog Voyager XL and with a similarly extensive range of programming options. With 24 patch points on the rear panel, it's well suited to integration into other analog systems and was being described by the manufacturers as "a modular system with presets".

Arturia's Matrixbrute is probably the world's most impressive relatively mass market synth.

Matrixbrute has since found favour with many of the major studios and composers around the world. Sonically, it's flexible, offering both Moog-style 24dB/octave and Steiner-Parker style multimode filters, and the arpeggiator and sequencer modes running across the programming matrix are very useful for creating repeated and varying atmospheric patterns. Adam Lastiwka has used one on the soundtrack for the Canadian TV show *Travelers* (see interview later) and other users include Vince Clarke and Jean-Michel Jarre.

The success of Arturia's hardware lines doesn't mean the company neglected software development over the last few years. Their iPad app line now includes four products emulating the Minimoog, Oberheim 2-Voice, Prophet VS, and their own Spark drum machine, and there are three software filters emulating Moog, Oberheim Matrix 12, and Oberheim SEM designs, as well as three software pre-amp emulations, but the Mac/PC software lines which run stand-alone or as plug-ins in the formats VST2.4, VST3, Audio Units, AAX, or NKS (on 64-bit audio workstations only) have expanded so widely there is hardly space to list them all.

We looked at some of the Arturia software lines in Chapter 7 and while most of these are still on the market, some have been revised, while many new software instrument products have been introduced. These are gathered together in the affordable Analog Lab, which runs some of the best presets from 21 different software synth designs with some basic editing capabilities, and the truly awesome V Collection 6, €499 at the time of writing, which offers some 6000 presets in fully editable versions of Arturia's entire soft synth range.

As mentioned earlier, Arturia's approach is to emulate the original instruments as accurately as possible but with no qualms about adding useful facilities such as polyphony, effects, built-in sequencing, added modulation options, alternative filter types, and so on. So, it's quite possible to turn to the Arturia lines for a convincing simulation of a Minimoog, but the package won't be limited only to what an original Minimoog could do. It's a sensible and compelling approach to software synthesis, and affordable, with most instruments at around €150.

DX7V

Yamaha's FM synthesizer but with a welcome simplified approach to programming.

CMIV

Fairlight's Computer Music Instrument sampler complete with 600 original sounds and the Page R sequencer.

Buchla Easel V

Exotic analog/AM/FM synth with sequencer.

Clavinet V

Hohner's archetypal funk machine modeled in detail.

ARP2600V

The semi-modular three-oscillator synth here with the ARP Sequencer built in.

CS80V

Yamaha's polyphonic giant for those Vangelis and Klaus Schulze sounds.

JUP8V

Roland's Jupiter 8 plus step sequencer and added modulation possibilities.

MiniV

The Minimoog with added modulation routings and vocal formant filter.

Modular V

Moog's modular system including step sequencer.

SEMV

Oberheim's Two-Voice keyboard with sequencer included.

PROPHET V

Sequential's Prophet V also with the digital Prophet VS emulated.

VOX CONTINENTAL V

Archetypal combo organ.

FARFISA V

Convincing emulation of the combo organ design as used by early Pink Floyd.

WURLI V

The Wurlitzer electric piano with added guitar effects pedals and Leslie rotary speaker simulation.

SOLINA V

The lush sounding string synth with vastly expanded modulation possibilities.

Arturia's Matrix12V is a stunning recreation in software of a powerful original.

MATRIX 12 V

Oberheim's Matrix 12 keyboard complete with 15-mode filter system.

Stage73V

The Fender Rhodes stage electric piano with added effects.

Synclavier V

NED's Synclavier sampler complete with many of the original FM synthesis sounds.

B3V

Hammond's B3 organ in stunning details.

PianoV

Various acoustic pianos with added effects.

www.arturia.com
www.sourcedistribution.co.uk (UK distributors)

In mid 2018, also in France, Baloran started to take advance orders for The River, a very large 8-voice multi-timbral analog synth due to sell at around €5000 with a proposed production of around 20 units.

www.baloran.com

In Greece, Dreadbox started marketing in 2011, first with effects units and soon with semi-modular synths. FreaqBox Murmux was shown in 2014 and also had an unusual bass pedal synth version. Erebus debuted in

Baloran's The River, a giant polyphonic from France.

2015: this is a desktop paraphonic (two note) synth which so far has gone through three incarnations, the latest with an expanded patchbay and third oscillator. A slightly simplified kit version was also available. Nyx is a slightly smaller desktop instrument with onboard reverb, while Medusa is a collaboration with Polyend adding a 64-position digital matrix of programmable pads and up to 6-note paraphonic playing.

The Medusa from Dreadbox in Greece is an imaginative collaboration with Polyend.

Dreadbox also entered the Eurorack market with their White Line and other modules, see the Eurorack section for more details.

www.dreadbox-fx.com

GERMANY 2013-2019

The Schmidt synthesizer appeared in 2013, designed by Stefan Schimdt, who was behind the MAM MB33 clone of the Roland Bassline. It was quickly taken up by Hans Zimmer and Jean-Michel Jarre and a few other top end players at a retail price of €20,000 and with only 25 in each production run it is not a mass-market product.

The Schmidt offers 8 voices of fully analog sound, four oscillators, and three filters per voice, eight part multi-mode, 61 velocity and aftertouch sensitive keys, MIDI, and USB. In 2018 a white panel version was announced.

www.schmidt-synthesizer.com

The prototype of the Schmidt synthesizer from Germany went to Hans Zimmer.

Waldorf continued to innovate throughout the period, offering Streichfett and Rocket, a string synthesizer emulator and monophonic synth, both in compact desktop formats and a matching analog style filter called 2Pole, followed by STVC, a compact string synth, and vocoder keyboard roughly based on Roland's old VP330.

Waldorf's Quantum is a powerful digital synth with many analog-style abilities.

In 2018 the company's lead product was Quantum, a powerful five-octave eight-voice synth with splits and layers at around £3500, while the more affordable Blofeld 8-voice virtual analog synth and module remain in the catalogue. Waldorf also entered the Eurorack field with a handful of modules, plus, unusually, a keyboard controller incorporating Eurorack spaces. More on Waldorf's Eurorack lines later.

www.waldorfmusic.com

RADIKAL TECHNOLOGIES

Radikal launched in 2004 with the all-digital Spectralis synth and module, which is now up to a Version 2 featuring four oscillators per voice, a fixed filter bank, as well as variable state filters, 32 voice stereo polyphony, up to 4GB of sample memory and a flexible realtime sequencer. Imminent at the time of writing is Accelerator, aimed more at live performances and with 3-band parametric EQ rather than the fixed filter bank, step sequencer, and arpeggiator, and a motion sensor which creates modulation from movements of the whole instrument. While these are fully digital designs, the company is also going into Eurorack with a combination of analog and digital technology.

Radikal's Delta CEP A is a small rackmount paraphonic synth with analog-like controls fronting a digital interior. Due for release in 2019 at around €900 it is in a pre-patched desktop format and has MIDI, but can be demounted and used in Eurorack format too. Radikal also announced three Eurorack modules in 2018: a dual multimode filter, a multifx processor, and a "swarm oscillator", which has memories and can recall several oscillator settings simultaneously to create chords and tone clusters.

www.radikaltechnologies.com

The success story of German synthesis in very recent times has certainly been Behringer, a company famed for cutting price points in the

The Radikal Delta CEP A, an analog-style front end for digital technology.

areas of amplification, mixing, sound processing, guitar effects, and similar lines. Their Deepmind 12 keyboard synth and its module version were partly developed in the UK and designed to offer the accessibility of the old Roland Juno lines, with a compact three-octave 6-voice version quickly following.

Behringer's DeepMind 12D offers Juno-type sounds plus great graphic display facilities.

Nothing controversial there, but it did spark controversy when Behringer announced their next synth design would be monophonic and called Model D, closely resembling in style the original Minimoog. Since Moog Music was in the process of re-issuing the original Minimoog around the same time this led to some intense speculation. The Model D, when it finally appeared, turned out to very closely emulate the Minimoog's sound, though in a much more compact format – that of a desktop or Eurorack module about half the size of a genuine Minimoog panel – and at an astonishing price point just under £300.

At the time of writing the Model D has only just hit music stores and is proving very popular. It prompts questions as to how Behringer can achieve such affordable price points in competition with companies launching instruments at $2000 or $3000, though mixer and amplifier manufacturers have had to live with a similar question for some years now.

Behringer's Model D with its straightfoward no-extra-cost Euroracking option and low price point gives pause for thought to many competing companies.

Behringer rattled the cages of the competition even further with Neutron, a semi-modular desktop synth in a similar compact format with CEM3340 oscillator chips and BBD delay line, and with a similar budget price point under £300. Achieving a similar system to a Neutron with individual Eurorack modules could cost over twice as much. The instrument, which features a huge patching matrix, is highly flexible.

Behringer's Neutron simplifies the Eurorack concept while retaining great patching options.

Behringer's next planned product was a line of Eurorack modules very closely patterned along Roland System 100M lines, which with Roland launching System 500 modules of their own, must have seemed highly challenging. The company also previewed VC340, a three-octave clone of the Roland VP330 Vocoder/strings, which offers a stunningly accurate recreation of its sounds. This goes directly up against similar module and keyboard products from Roland and Waldorf respectively. Rumoured to be among Behringer's next releases are an Oberheim polysynth clone, which again will present a great challenge to the longer established synth companies, and maybe an EMS VCS3 clone, which would be very welcome and probably not too much of a challenge for the Behringer design team, given the existing constituent elements of the Neutron.

Behringer's aggressive entry into the analog synth market and their slashing of established price points promises much for the affordability of these areas in coming years, though at the same time, up-market and very much more costly instruments still seem to be more than holding their own. A great deal more detail on the Behringer instruments, including in-depth video demos, on the book's website: www.routledge.com/cw/jenkins

www.musictribe.com/behringer

Roli

Finally in closing this section on new instruments, a look at a device which may point the way forward for the expressive control of both analog and digital sounds. Roli's Seaboard was designed in the UK by Roland Lamb, a graduate of the Royal College of Art in London, and introduced in the form of the upmarket Seaboard Grand in 2015. It features prominently in the Academy Award-winning movie, *La La Land*, played by Ryan Gosling, who was introduced to it by the movie's music director, Marius DeVries. Seaboard is a multidimensional controller, a touch keyboard responding not only to velocity and aftertouch but to movements upwards and along the keyboard.

Although Bob Moog at Big Briar had been interested in alternative types of keyboard control, and the Korg Z1, for example, introduced a KAOSS Pad type controller for alternative styles of modulation, the field of keyboard control in the purely analog area has been pretty much stagnant, as we saw in Chapter 2 on keyboard modules. The Seaboard's innovation is mostly in the digital realm and even in that area, a modification had to be made to the MIDI protocol in order to handle more advanced expression options. MPE is the MIDI Polyphonic Expression protocol, which assigns performances to multiple MIDI channels simultaneously so that more controller information can be sent to the receiving module. Remaining backwards compatible with earlier MIDI instruments, MPE also appears on the Linnstrument controller from Roger Linn and the FH-1 polyphonic MIDI interface from Expert Sleepers, plus an increasing number of other devices, while the Seaboard comes accompanied by a powerful software synth, *Equator*, which takes full advantage of the new protocol.

Roli laid on a run-through of the new instrument at their London HQ. Expert demonstrator, Heen-Wah Wei, gives a great overview of the expressive capabilities in a video on the book's website: www.routledge.com/cw/jenkins. Like the original Seaboard Grand, the compact Rise 25 and Rise 49 models have four programmable faders which can mix different parts of the sound or the depth of different modulation types, while the smaller two-octave keyboard Block accompanies other compact modules called Live, Lightpad, Loop, and Touch, in creating a readily configurable system to control sounds from an iPad or computer.

While the controllers are well set up to create huge digital sweeps from the Equator software, which can be bought independently though it certainly benefits significantly from their multi-dimensional control techniques, they also have a lot to say about straightforward analog-style sounds. Something like a TB303 Bassline sound becomes highly velocity sensitive, bending, and modulating, according to aftertouch and finger movements up and down the keys. Resonance can be programmed to

The Roli Seaboard Rise 49 and the Block system in action at the company's London HQ.

increase according to velocity and with both hands free to play and modulate rather than the left hand being devoted to performance controls, an expressive analog style sound, which might otherwise have demanded several layers of programming, can be generated in a single pass, making the workflow much faster.

When similar techniques are applied to layered polyphonic sounds, the effect can be tremendously impressive. Techniques denied to players since the days of the Yamaha CS80 and Ensoniq SQ80, calling for individual note aftertouch for expressive top line melodies, are instantly restored. Seaboard/Equator performances often instantly resemble instant movie music, with atmospheric ambient pads swelling dramatically, or brass smoothly replacing strings according to the performer's playing technique.

Many of the interviewees for this new edition of the book, particularly Dixon and Stein on their live performances of the *Stranger Things* music, and Adam Lastiwka in his work for the Canadian TV series, *Travelers*, are raving about the Seaboard and its potential for bringing a whole new degree of expression to sounds both digital and analog. More on this later in the section on movie music, more extensive Roli demos on the book's website: www.routledge.com/cw/jenkins.

www.roli.com

2019's NAMM show saw the emergence of several new trends, including breadboard style pin cable patching on Korg's compact West Coast (Buchla) style Volca modular with low pass gates and wave folding, making it compatible with the Moog Werkstatt, the sub-Eurorack sized AE Modular from Robert Langer's Tangible Waves in Germany, and various compact digital synths, such as the Bastl Kastl, NS1, Tinsizer and Olegtron, and EDP Wasp-style touch keyboards on Arturia's four-voice paraphonic MicroFreak, which has digital oscillators but a single analog filter, on Modal's V2.0 of the compact Craftsynth, and on Dubreq's Gen

R-8, an exciting dual oscillator, modular compatible expansion of the Gen X-1 featured elsewhere in the book.

Expanded versions of existing synths included Moog's Sirin, an extension of the desktop Minitaur without its limited octave range, making it better suited to melody parts, and Korg's XD version of the Minilogue keyboard synth, with a much expanded effects and sequencer section.

Behringer launched the long expected MS-101-RD and VC340 Vocoder along the lines of Roland's classic SH101 and VP330 Vocoder/Strings, and previewed Crave, a semi-modular analog synth reflecting the Neutron design with a sequencer and a sub-$200 price point, and previewed around 80 Eurorack modules along the lines of Roland's System 100M, to be released over a period of two years, alongside clones of the Sequential Pro 1, Roland TR808 drum machine, and ARP Odyssey. Sadly, ARP designer, Alan R. Pearlman, passed away just before the event at the age of 93.

MODULES 2013-2019: THE EURORACK EXPLOSION

Eurorack exploded in the period 2013–2019, with more than 300 manufacturers now in the field at the time of writing and more than 6000 modules available. There are perhaps 60 more manufacturers also releasing Moog, Frac Rac, Modcan, Buchla, MOTM, Serge, and other formats. With many companies already offering the basic oscillators, filters, and amplifiers, the field has proven fertile for more imaginative products particularly those interfacing the traditionally fully variable analog module field into MIDI, USB, multi-channel digital transfer and programmability.

There are excellent resources for helping keep up with new products in the field of Eurorack and other modules. Muffwiggler.com is the very busy discussion group for all matters modular. In the USA, modularaddict. com lists and supplies DIY kits and accessories for building Eurorack modules. Modulargrid.net remains prominent and offers great search facilities to narrow down what modules may be suited to a particular job by format – including Moog, Frac, Modcan, Buchla, MOTM, and Serge, as well as Eurorack – or by function, cost, and rack size, and includes a versatile modular system planner (Eurorack module widths are measured in HP, horizontal pitch, of 5mm or 0.2 inches per unit, so modules are generally 4, 8 or 12 HP wide, though many 2HP modules are now appearing, and cabinets will usually offer 84 HP of width which is equivalent to 19-inch systems, but can easily be larger or smaller).

Eurorack users starting to work with custom size rackmounts regularly began to come up with unfilled space, which offered an opportunity to turn some small modules sideways and redesign them into the 1U format normally associated with 19-inch modules (1.75 inches or 1U in height inconveniently being fractionally over 8HP). So, there's now a busy area in developing 1U modules (known as "tiles") which will typically fill just a single row above or below a Eurorack system. Intellijel, Synthrotek, Makenoise, Syinsi, and many others now have offerings in this area often intended as multiples, pedal inputs, or other types of link, but increasingly as additional LFO's, ribbon controllers, such as the RIB from Synthrotek, and so on.

The Eurorack and desktop semi-modular field is now mature enough that some manufacturers have already come and gone, or at least no longer remain in production in the field of modules, for example, BugBrand in the UK, or Bananalogue in the USA. But many products from these companies are still in the second-hand market and seen on eBay, Craigslist, on Gumtree in the UK, and so on. For example, there was a second-hand BugBrand Modular 4 on eBay UK for £1300 at the time of writing.

London-based Mara is typical of a new generation of analog users making experimental and ambient music with Eurorack and desktop synthesizers, releasing through Bandcamp and Soundcloud. Eurorack cabinet by Noise Inn, with Behringer Model D, Korg SQ-1 sequencer, Korg Nanocontrol 2, Digitone, and in the background a Roland TR909 drum machine.

Given this huge expansion in the Eurorack field it's now impractical to mention every current Eurorack product, see the book's website, www.routledge.com/cw/jenkins, for links to around 300 companies. Here are more detailed profiles of just a few companies making particularly interesting innovations in the field. It is pleasing to be able to include representatives from England, Scotland, France, Belgium, the Netherlands, Italy, Portugal, Germany, Spain, Romania, the Czech Republic, the USA, Canada, Thailand, and elsewhere. Very detailed video demonstrations of these outstanding products are also on the book's website, www.routledge.com/cw/jenkins, along with some ideas about commercial rack options and custom conversions.

4MS

Founded in 1996 and currently based in Portland, Oregon, 4ms is one of the most firmly established of the US module manufacturers with around 20 modules and almost as many module kits in production at the time of writing, as well as pedals and rack accessories. Module designs include WAV Recorder, which uses Micro SD cards to record stereo audio, and three larger (18–20HP) modules to handle audio, echoes, and loops: Tapographic Delay, Dual Looping Delay, and Stereo Triggered Sampler. There are detailed video demos showing the tremendous versatility of these three modules on the book's website: www.routledge.com/cw/jenkins.

"Rather than replicate concepts from the past, we strive to turn unique ideas into playable musical tools that expand the possibilities of music", writes founder, Dan Green. Crucially, the company is one of the few also offering power modules. The Row Power 30 or Row Power 40 can handle systems of most sizes making it possible to create modular systems without too much concern about cabinet depth and capacity, again, there's a demo of just how easy these are to set up on the book's website: www.routledge.com/cw/jenkins.

https://4mscompany.com

ADDAC

ADDAC has been based in Lisbon, Portugal, since 2009 and offers some tremendous modules as well as racks and frames, including an attractive foldable, portable 15U frame system the 901PF.

At present there are more than 50 modules in production, mostly with an eye-catching bright red finish, though there are 700 series modules in black too, including an oscillator, filter, mixer, and the 709C Organ Machine, which creates polyphonic organ sounds from almost any input.

The 800 Series includes a Quad Spatializer for quadrophonic placement and panning, in the 200 series there's an Intuitive Quantiser with a micro keyboard to help create various scales, while in the 500 series Marble Physics, which creates voltages simulating the movement of a bouncing object. Detailed demos of these last two on the book's website: www.routledge.com/cw/jenkins.

Construction quality on all of these modules is very high and ADDAC seems a friendly company, encouraging musicians visiting Portugal to get in touch.

www.addacsystem.com

BARD SYNTHESIZERS

A British one-man operation with a background in creating guitar pedals, Bard has come up with a winner in the Ivan Phosphor Drive, a valve overdrive module with a glowing front panel display using a Vacuum Flourescent Triode. In only 3HP it's also useful as a level indicator. The Austin Ring Modulator/VCA is also available, with some sales going through Patch Audio (https://patch.audio). A demo of Ivan is on the book's website: www.routledge.com/cw/jenkins.

www.facebook.com/bardsynth

BEFACO

Based in Spain, Befaco has more than 20 modules in production, all with a gorgeous glossy black panel finish. There are some unusual options here: the BF22 VCF, based on the Sallen-Key filter of the Korg MS20, an Instrument Interface module featuring an envelope follower to process incoming signals for use in a modular system, again reminiscent of the MS20 design, and standards, such as Slew Limiters and Mixers, alongside a nice Joystick and Spring Reverb module.

Muxlicer is a bit different though, superficially an 8-step sequencer but actually capable of routing different inputs to a wide variety of outputs, while Rampage is a Serge/Buchla style ramp generator, which can act as an oscillator or as an AR/ASR/ADSR envelope generator. Detailed demos of these last two are on the book's website: www.routledge.com/cw/jenkins.

www.befaco.org

DIVISION 6

Operating from Jamestown, NY, since around 2009 by Scott Rise, Division 6 must take the award for some of the most compact designs available in the synthesizer field. The Business Card Sequencer is just that – a 16-step sequencer about the size of a business card which can also be used as a micro keyboard – and it was apparently no problem to mount two of them in Eurorack format as the Dual Business Card Sequencer. This is a great looking design reminiscent of two tiny Roland TB303 Basslines and the two sequencer channels can be interfaced together to play in series or run a pattern of transpositions.

The company also offers VM, simply an LED ladder volume meter in a mere 2HP, which is a very eye-catching addition to any modular system, and has many more designs in ready built or kit form. Worldwide distribution is very extensive but the company will also supply by mail. A detailed demo on the book's website, www.routledge.com/cw/jenkins, of the Dual Business Card Sequencer, which at around $200(built) must be one of the most entertaining Eurorack modules around.

https://division-6.com

DOEPFER

Doepfer introduced Eurorack in the modular synthesizer field and continues to lead the way with more than 140 modules available at the time of writing, alongside various discontinued models. Vintage versions in black are becoming more widely available in contrast to the usual silver panel finish. Various cabinet options are also available, including one of the few compact power systems intended for just four modules, so systems of any size from the very compact to the enormous can readily be constructed.

New modules introduced for 2018 include a quad poly VCF, quad poly ADSR, and octal poly VCA, all intended to help build up polyphonic modular systems, alongside a new Mk3 version of the semi-modular desktop Dark Energy, which in effect combines half a dozen modules into a semi-patchable desktop format.

www.doepfer.de

Cosmetics cases in the process of being converted to Eurorack: in silver as a remote modular system controller with Synthwerks FSR 4B/C quad sensor panel and programmer, 10-way minijack snake, Doepfer A-174 Joystick and A100MNT mini power supply concealed under a blanking panel, or in black as a modular drum machine with Vermona Random Rhythm and TwinCussion, and 4MS Row Power 40 about to be installed.

DREADBOX

Dreadbox, based in Greece, is becoming better known for its desktop synths, including Nyx and Erebus, but does have a line in Eurorack modules with over a dozen in the White Line series – oscillator, filter, slider-controlled ADSR, and so on – and two versions of Drips, a dual VCO analog drum synthesizer, using slightly different circuitry. 94HP and dual row 168HP powered cases with MIDI to CV conversion are also available.

www.dreadbox-fx.com

DSI

As the manufacturers of the Evolver, Prophet, Sequential, and Oberheim/ DSI synths, Dave Smith's company needs no introduction, and the company's entry into the Eurorack world with a Prophet style filter module was widely welcomed. Two subsequent introductions have been a bit more unusual: the DSM02 Character's controls are labelled Girth, Air, Decimate, Hack, and Drive, and aim to add low or high frequency harmonics, sample rate reduce, distort, and add saturated warmth to a sound, while DSM03 Feedback includes a resonant low pass filter and a triggerable noise source for Karplus-Strong style string sound synthesis.

Both modules are unashamedly aimed at "sound mangling" and derive from the effects sections of the Pro 2 and Prophet 12, so adding them to a Eurorack system can bring some of that big-keyboard sound to your rig. Distribution is worldwide. There is a detailed video demo of both modules on the book's website: www.routledge.com/cw/jenkins.

www.davesmithinstruments.com

EXPERT SLEEPERS

UK-based Expert Sleepers is a company which may be overlooked by a lot of analog modular enthusiasts, because their dozen or so products are on the whole digital and far from knob-heavy, though anything beyond a superficial glance reveals how devastatingly powerful and original many of their Eurorack products can be. ES is all about digital interfacing and that spans MIDI, USB, ADAT Lightpipe. S/PDIF, and more, as well as Din Sync, Audio to CV, and CV to Audio.

Want to stream multichannel audio from a digital audio workstation to a modular system, process it through a filter module, and send it back again? ES3, which uses an ADAT-style Lightpipe connector will do the job. Do the same with the S/PDIF stereo digital audio output of a Mac computer? ES-40 at around £95 is the answer, while ES8 handles control signals generated by software such as Silent Way, Max/MSP, Reaktor, CV Toolkit, and Bitwig Studio, converting them into control voltages for your modular system.

Users coming up against requirements such as these will probably come across ES anyway because in many cases they're the only manufacturers of a solution. Slightly more hidden away is Disting Mk4 at around £135, a very unassuming 4HP module with just two knobs, five sockets marked A, B, C, X, and Z, a Micro SD card slot and a tiny visual display. It has nothing to do either with distortion or with distribution.

Disting is simply a microprocessor which runs any one of 84 different algorithms and applies them to incoming and outgoing voltage signals. Just listing all these is a task: it can act as a Quantiser, Dual Waveshaper, sample-and-hold, Slew Limiter, Pitch and Envelope Tracker, LFO, Vocoder, Phaser, Tape Delay, Tuner, or CV/MIDI convertor. It can emulate Logic's ES-1 and ES-2 software synth, act as a stereo reverb, or play back audio or MIDI files from a MicroSD card, and much more. A fantastic toolkit for any modular system.

Even more striking is General CV, launched in 2017. In 12HP it also looks very unassuming with just five knobs, eleven sockets, an SD card slot and a tiny display, but conceals a vast amount of processing power allowing it to create and modify sounds from the General MIDI sound set, play audio files from its SD card slot, act as a polyphonic VCO, or act as a drum sound bank with nine trigger inputs. Just as on most GM instruments there are internal EQ, reverb, and chorus.

General CV makes it possible to introduce polyphonic playing and more recognisable instrument and drum sounds into the modular system arena, and Expert Sleepers is well worth checking out for all those modular system problems you thought it impossible to solve. There are detailed demos of General CV and Disting Mk4 on the book's website: www.routledge.com/cw/jenkins.

www.expert-sleepers.co.uk

A giant CD cabinet in the process of conversion into a dual 194HP Eurorack.

Top Row: GR11 by REBACH, S090, D112 and F711 by LADIK, Pedz, Stompz and Hertz by MONDE SYNTHESIZER, Dual Looping Delay, Stereo Triggered Sampler and Tapographic Delay by 4MS, Pattern Generator, Philter, Panning VCA and Dual Envelope by KILPATRICK AUDIO, Marble Physics and Intuitive Quantiser by ADDAC.

Bottom Row: General CV and Disting 4 by EXPERT SLEEPERS, Cassa CV by LEPLOOP, Echo by SYNTHROTEK, Logic, PMult, WF, Mult, AR2 and Clock 2 by YORK MODULAR, Ivan Phosphor Drive by BARD.

FEEDBACK MODULES

It's a pleasure to be able to include a mention of a line from Romania. Feedback, under Andrei Stratone, offers almost a dozen designs ready-built as well as many in kit form. These all look eye-catching with busy yellow legending and white knobs on black panels, covering a range of applications from gain control with the distortion characteristics associated with the Roland TB303, or well known mixer designs, to chorus and multitap delay, an attenuator/invertor, and a 10HP, multi waveform VCO at €130.

Multi Dimension though at €150 is one of the most interesting designs, a chorus/flanger/short delay in 14HP based on the MN3007 BBD chips often used in guitar pedals. More about Feedback Modules on the book's website: www.routledge.com/cw/jenkins.

https://feedback-modules.myshopify.com

The MultiDimension from
Feedback Modules.

INSTRUO

Instruo is based in Scotland, has terrifically sumptuous panel design with gold legending and Scots Gaelic derived names for each module, and offers some of the most innovative modules coming from the UK.

At the moment there are around a dozen modules in the range, most of them quite large in HP terms though there is a 4HP module Tanh, which is a simple wave shaper. There is some very advanced design work here though. Aither creates an internet network which can send voltages under the control of a connected device running a web browser, and Scion is a biofeedback sensor incorporating a quad voltage generator, which can allow plants, animals, or humans to generate music spontaneously.

Io-47 (£260, built to order) is a multimode filter and Cs-L (£510) is a dual oscillator with cross modulation available without the need for any patching. Harmonaig is a voltage quantiser with many built-in scales, which can turn incoming voltages into meaningful musical passages. Traigh (£190) is a standard 24dB low pass filter but incorporates the mixing of three audio inputs to save a lot of patching, while Tona (£190) is an oscillator, which incorporates a wavefolder to add unusual harmonics.

Instruo's modules have the look and feel of a high class product and an all-Instruo system should be a very impressive item indeed. See the end of this section for a photo of Instruo modules in a cabinet by Noise Inn. A very detailed look at Instruo's Cs-L, Tona, Harmonaig, and Traig is on the book's website: www.routledge.com/cw/jenkins.

www.instruo.media

KILPATRICK AUDIO

Andrew Kilpatrick launched Kilpatrick Audio, based in Canada, in 2008 and the company is probably best known for Phenol, a desktop dual VCO semi-modular with digital delay, patched with banana plugs and popular amongst the more experimental and improvisational type of musician, particularly since it was paired with an interface to portable USB power banks. The company has a busy line in Eurorack modules, as well as in a new Kilpatrick Format comprising almost a dozen modules each 4U high, which looks very impressive. Both product lines have great solid looking designs with spacious layouts on unfussy silver panels.

Many of the Eurorack designs are spin-offs from the Phenol, including the K6501 Philter and K3021 Master VCO. Kilpatrick's very first Eurorack design, the K4815 Pattern Generator, was immediately attention-grabbing though. In 20HP it features an 8x8 LED dot matrix display, which helps create scales, arpeggios, sequences, and transpositions under a number of different types of voltage control. A handy Panning VCA is another agreeably no-fuss module, but the 16HP wide K6101

Dual Envelope is much more sophisticated, with selectable modes to act as an audio oscillator or LFO.

Kilpatrick sells direct from its own web store but also has good distribution worldwide. A very detailed look at the Kilpatrick Pattern Generator, Philter, Panning VCA, and Dual Envelope is on the book's website: www.routledge.com/cw/jenkins.

www.kilpatrickaudio.com

KLAVIS

Klavis is based in Belgium and at the time of writing has five attractive looking Eurorack modules in production as well as a small adaptor to generate +5 volts from the -12V supply. The modules, all with intricate black legending on silver panels, include: Quadigy, a quad programmable envelope generator with attractive sliders rather than rotary controls; Caltrans, launching in late 2018 and described as a "Calibrator and Transposer"; Logica AT, which is a programmable three-input voltage controlled logic processor, adding, gating, or randomising inputs as required; and Mixwitch, a dual 4-into-1/2-into-1 mixer, which conveniently produces variable offset voltages if there's no input.

Twin Waves at €229 is probably the most impressive Klavis module, a dual VCO/LFO with a tiny waveform display, ten octaves of tracking, suboctave output, additive synthesis mode with 7 stacked oscillator sounds, ring modulation using an independent internal oscillator, and much more. Klavis modules unusually complete with a neat colour manual. There is a detailed look at Twin Waves on the book's website: www.routledge.com/cw/jenkins.

www.klavis.com

LADIK

Based in the Czech Republic, Ladik started sales, including through eBay UK, a few years ago and has made some impressive inroads into pricing and module specifications using relatively simple, plain metal panel designs and straightforward hardware.

Astonishingly there are more than 250 Eurorack module designs currently available and they are organised into various series covering VCAs, switches, logic modules, CV and envelope generators, drum sound generators, LFOs, meters, mixers, sequencers, and so on.

All of the basics are here and there are some very innovative designs too. There's an 8-step trigger sequencer, the S-180, in just 4HP of width; a VCO with LFO mode in 4HP called O-111; a MIDI Thru module in 6HP; mixers with either rotaries or sliders; a handy speaker in 12HP; LED ladder meters; a micro keyboard the K-010, and much more.

Prices are very reasonable: from £55 for an ADSR with Hold and most of the 4HP modules, or £75 for a Recording Pattern Generator and many of the more complex modules, plus a few pounds for shipping.

Klavis Twin Waves, part of a new generation of Eurorack modules using tiny graphic displays.

Ladik seems an approachable company and also offers power cables, accessories, and blanking panels. Panels are all silver though various knob designs are used, and a whole Ladik system of Ladik modules would be solid and not too hard on the wallet. On the book's website, www.rout ledge.com/cw/jenkins, there is a detailed video demo of the O-112 oscillator, F-711 filter, and the handy S-90 Skipper, which randomly passes or blocks voltages, but huge numbers of the Ladik module designs are well worth checking out.

www.ladik.eu

LEPLOOP

Leploop is best known for the desktop synthesizer/sequencer/drum machine of the same name and has started spinning off elements of its design into Eurorack modules under the name Lumanoise. Cassa CV in an 8HP format is unusual, it's a self-oscillating filter but mainly designed for creating drum sounds and offering CV control over frequency and resonance. There's a simpler version, Cassa, without the voltage control, and a 3-voice version, the Multicassa, in desktop or Eurorack format. This has dual clock dividers to create complex rhythms played by the three fully variable percussion voices; it's like having a Roland TR808 with much more variable analog drum sounds, the company also offers a TR808 style cymbal sound generator module.

All of these analog percussion sounds are of the dirty and grimy variety, they're really intended for those who prefer the more industrial side of sound synthesis and do the job well. All of the Leploop modules are well distributed worldwide and worth taking a look at; there is a detailed video demo of Cassa on the book's website: www.routledge.com/cw/jenkins.

www.nazzilla.com/lep-loop

MFB

MFB, under Manfred Fricke, has long been at the cutting edge of (often very affordable) analog and hybrid digital/analog instruments in the areas of drum machines, samplers, desktop synthesizers, and more recently in Eurorack modules.

Some of the earlier designs, such as the triple-oscillator Kraftzwerg in Eurorack format, and the very useful Power module, which doubled as a MIDI input, have sadly become discontinued and some of these earlier MFB designs (alongside their little desktop Synth Lite and Synth Lite 2) have become quite sought after. By late 2018 a new line in Eurorack modules was available and this usefully included a digital delay alongside an oscillator, ADSR, VCF, and various percussion sound modules, a VCA with a tiny visual display, and the more substantial Nanozwerg Pro module, which is effectively a complete monophonic synthesizer voice.

www.mfberlin.de

MONDE SYNTHESIZER

Monde is our Canadian representative among the Eurorack community, launched in Ottawa in 2009 by John Emond, who has more than 35 years of experience in high tech design. Currently there are more than half a dozen modules in the catalogue and all pay a lot of attention to skiff compatibility, elegant panel design, and carefully thought out normalisation of the available features. Monde modules have white panels with pale blue legending and white/silver rotaries, they're very well constructed and a little reminiscent of the very early E-Mu Modular System.

Hertz, at just $219, is a frequency counter with LCD display equally able to recognise sub-audio modulation speeds or pitches of audio signals, Voltz is a similarly formatted voltage counter. Ribbonz, at $329, is a ribbon controller with two modulation LFOs in the external control box, a terrific recreation of the unobtainable Moog ribbon controller, and more affordable than Doepfer's equivalent. Inz is an input module which can handle any audio input including guitars, create an envelope shape from the output, or send signals to Hertz for tuning purposes.

Pedz is an input module too, this time for foot controllers, such as keyboard or guitar volume pedals. There's also a panning VCA called Panz, and a send/return module called Stompz, intended to loop signals through guitar pedals and other external effects processors.

The Monde products with their gleaming white panels look hardy and attractive in your rack and are well worth checking out. There are detailed video demos of Hertz, Stompz, and Pedz on the book's website: www.routledge.com/cw/jenkins.

www.mondesynthesizer.com

REBACH

Rebach has been established just a couple of years, is based in the Netherlands, and offers some sales through modularsynthesizers.nl.

They offer an S-Trigger to V-trigger convertor called Catch, have a filter design in the new Wolfram series coming up, and market a matching set of modules in the GR series: the GR11 VCO, the GR21 VCF, which is a two-pole low pass 12dB/octave type, in other words Oberheim style and able to self-resonate to make deep bass drum sounds, and the GR61 3-into-1 mixer. They're all skiff-friendly and range from 8HP to 18HP in width.

The VCO is particularly interesting, it's also tuned primarily for bass applications and goes down to semi-LFO speeds. Construction on all of these is very solid with large, easily adjusted control knobs and there's a full demo of the VCO on the book's website: www.routledge.com/cw/jenkins.

https://rebach.eu/

ROLAND

Roland's initial entry into Eurorack was with a series of re-purposable effects units associated with the Aira line, the Demora delay, Torcido distortion, Scooper loop recorder and Bitrazer bit crusher. By 2018 all of these were discontinued and the company had launched the 500 series of more recognisable synthesizer modules. Roland has a long pedigree in modular synths from the massive early System 700 to the 100 and 100M systems, so plenty of design experience went into the 500 series, designed and built in the USA in collaboration with Malekko Heavy Industry.

There are five modules in the series: 512 VCOs, 521 filters, 530 VCAs, 540 ADSRs, and 572 Delay/Phaser with LFO, and an optional rackmount with cover and e, xternal power supply for the whole set, at a total cost of around £1920. Generally the construction and sound of the modules have been acclaimed, with a distinctively Roland sound quality slightly different from Moog and other modules easy to perceive, and for purchasers without a more restricted budget the 500 modules have proven popular.

www.roland.com

STUDIO ELECTRONICS

We looked at SE's MIDI modules in earlier chapters, as well as at the Boomstar line of desktops, which took the company back to its roots in nonprogrammable monophonic synthesis. The company is now also offering Eurorack modules and there are three variants of Tonestar, a complete synthesizer voice with filter elements from ARP2600 or Roland synths, plus an expansion unit Toolbar, which adds an LFO and delay.

Individual modules in the Boomstar Modular line now number more than twenty and unsurprisingly very prominent among these are SE's various filter designs reproducing ARP, Moog, Roland, and Oberheim types.

The Studio Electronics Boomstar Modular system is impressive looking to say the least.

There is a lot more to the Boomstar Modular line with MIDI, LFOs, a quad digital oscillator, a ring modulator called Sci-Fi, a slimmed down oscillator called SLIMO, a rotary sequencer Charcot Circles, and something called Grainy Clampit (a *Beverly Hillbillies* reference probably), described as a "granular and phase distortion additive oscillator". With a variety of studio and portable racks available the Boomstar Modular system looks very impressive.

www.studioelectronics.com

SYNTHROTEK

Synthrotek is one of the better established of the US Eurorack manufacturers with more than 25 module designs on sale, a dozen or so more in collaboration with George Mattson, various lo-fi synths and pedals, many kit-build modules and other products, and a good range or connectors, cabinets, and accessories. The company ships worldwide but also has a strong network of dealers.

Synthrotek's entire system can be bought in kit or ready-made form.

Most Synthrotek modules are finished in black and the Mattson modules in silver. One early highlight was Eko (later Echo), which was a rare delay unit in Eurorack form taking up just 4HP, an invaluable module which would overdrive and create unpredictable results at extreme settings. Currently there's a low pass gate in 4HP also, and an oscillator after Buchla designs, which features illuminated sliders rather than rotary controls. There are multiples and an interface to a ribbon controller, MIDI to CV convertors, and filters, both clean and "dirty".

Synthrotek also offers a very flexible range of power units and powered racks ("skiff boats") – it's possible, for example, to buy a whole 84HP of module kits to build into a supplied powered rack at $1350. Complete Synthrotek modular systems look busy and highly flexible. More detail on Synthrotek modules on the book's website: www.routledge.com/cw/jenkins.

www.synthrotek.com

TAKAAB

Takaab hails from Thailand and along with a couple of other manufacturers has very successfully gone along the route of seeking extreme economy and compactness in module construction. The panels are in fibreboard rather than metal and there's a glossy black finish. Controls tend to be widely spaced but some of the modules are very compact, only 3HP in width. VC-EG is a voltage controlled envelope generator based on a digital chip the ENVGEN8 from Electric Druid, and so needing a +5V supply rail, but at 12HP that's one of the larger modules. There's a dual Vactrol-based low pass gate in just 3HP, various multiples and attenuators, but probably the best value is 3LFO, offering square and triangle outputs from three LFOs running at different speed ranges in a module just 6HP in width. At just over £40 including shipping to the UK, that's tremendous value.

The company is well worth chasing up for modular builders on a budget, an all-TAKAAB system would be very lightweight as well as extremely affordable. A detailed look at 3LFO and several other modules is on the book's website: www.routledge.com/cw/jenkins.

www.siammodular.com

VERMONA

Vermona has been in the analog design game for longer than most manufacturers, starting life in what was East Germany, aiming to design budget competitors for Western instruments – and coming up with some very interesting electric pianos, string synths, and other instruments along the way – moving into analog effects units, and more recently into synth modules, including the polyphonic PERfourMER series, and into Eurorack with a range of just over a dozen black-finished modules which look and feel professional and solidly constructed.

An oscillator, dual multimode filter, dual VCA, overdrive, and quad modulation generator take care of the basics, while some of the other modules are aimed at interfacing Vermona's desktop instruments, such as the Lancet mono synth. There's also a quad MIDI interface.

More unusual are twinCussion, a dual drum synth module which gives many of the effects of the now much sought-after Pearl SY1 Syncussion or various Synare models, and the randomRHYTHM, launched at NAMM 2018, a matching pattern generator module under the slogan "Rhythm is no accident but planned randomness". Four sliders for each of two rhythm sections set the likelihood for quavers, eighths, sixteenths, and triplets to appear within a basic rhythm, with an internal clock or the ability to sync to external clocks. This allows anything from 4/4 patterns to extreme variations to be generated without any need for advance programming. A detailed look at this pair of modules is on the book's website: www.routledge.com/cw/jenkins.

At the time of writing, Vermona is also introducing the '14, a three and a half octave monophonic synth expanded from the Mono Lancet and PERfourMER Mk. 2 designs.

www.vermona.com

WALDORF

We discussed Waldorf's current keyboard lines elsewhere. The company launched tentatively into the Eurorack field with a wavetable oscillator module NW1 but caught up quickly later on with more releases and a KB37 three octave keyboard controller at around £649 with mod wheels and interfacing for mounted Eurorack modules, which has proven very popular.

Now it's possible to build an entire Waldorf synth voice in Eurorack. VCF1 is an analog filter with distortion based on the Rocket desktop synth and 2Pole desktop filter, while DVCA1 includes a "colour" control for subtle filtering, as well as an inverted linked mode for stereo panning, and CMP1 is a hard or soft knee compressor with side chaining. MOD1 at as little as £169 provides both the envelope shapers and LFO.

www.waldorfmusic.de

Jean-Michel Jarre with Eurorack modular system and Waldorf's KB37, a rare example of a keyboard controller aimed at mounting Eurorack format modules.

YORK MODULAR

Simon Ward's York Modular (YoMo), based in the UK, has addressed questions of cost and size of Eurorack modules by making available some very affordable and very compact products. Largely, this is achieved by using black-backed 3mm perspex rather than aluminium for the panels. "It's much easier and cheaper to get perspex cut to the tolerance I need than it is to get aluminium cut", Simon explains, "and for anyone who wants metal panels I make my designs available in DXF AutoCAD or FPD Front Panel Designer formats so they can make their own".

The York Modular range includes many units fitting into only 2HP, and many unpowered modules, so some great price points are achieved, for example, under £30 for a 2HP sample-and-hold module and under £20 for a 2HP resonant low pass filter. Sales are through eBay UK (yorkmodular) and Etsy (YoMoCircuits). There is a demo video on the book's website: www.routledge.com/cw/jenkins.

Yomo.interzen.co.uk
yomo@interzen.co.uk

Cabinetry for Eurorack and other modular systems has become a field of its own, benefitting from a long tradition stretching back to the period of the early do-it-yourself module designs by Digisound, and others in the UK. The whole point of modern modules is total flexibility along with interchangeability (which is why the owners of synthesizers.com, for example, rail against the inclusion of non-standard half-size modules in some of the Moog systems) and so some users are inevitably tempted to go out on their own in the area of cabinet design rather than sticking with factory products.

Custom Eurorack setup posted by subultresk on the muffwiggler.com forum.

For a few years this was a source of problems because Eurorack demands a relatively complex power system providing +12V, -12V and ground, and usually +5V too in order to run some modules with digital components, and commercial power supplies along with their power distribution bus circuits tended to be on the bulky side, limiting the possibilities particularly for making very shallow "skiff" type cabinets. Eventually, this was solved by companies including TipTop with the microZeus and 4MS with the Row Power simply providing the power distribution circuit as another module, mounted alongside the rest of the sound generating modules, and using a "flying bus" comprising nothing more than a multistrand cable punctuated by module connection sockets. Mains power is then handled by an external "brick" usually requiring a 15–20V DC model typical of many laptops, so it has become possible to design very slim and shallow cabinets, or deeper ones, which can still accommodate modules with larger circuit boards along with the more substantial traditional mains-connected power supplies if preferred. A compromise is the Doepfer A100MNT mini power supply, which handles up to four modules and also relies on an external power "brick".

Mark Jenkins on stage in London with custom compact briefcase style Eurorack system.

All sorts of existing cabinets can provide inspiration for DIY designs. The author's portable Eurorack system is built into a very shallow EMS Synthi A-style black combination briefcase from the German industrial equipment company Friedr. Dick, thanks to the use of a TipTop microZeus power system, and plastic or metal hard shell briefcases of this type, particularly from companies such as Samsonite, can often be a good prospect for such conversion. Aluminium photographic equipment briefcases can often handle a dozen or more modules, perhaps accommodating an additional 19-inch format strip for "tile" type modules too, and even much smaller metal cosmetics cases can handle two or three modules, maybe incorporating the use of a Doepfer MNT power distribution circuit with an external power brick.

In the case of cabinets and power supplies for Moog, MOTM, and other larger module systems, sticking to commercial suppliers, such as synthesizers.com, is probably more straightforward, but users of Eurorack might like to note that a Eurorack module is very nearly the same height as a CD case, and many types of CD cabinet exactly accommodate Eurorack modules, simply requiring the installation either of metal rack bars with inserted loose nuts or threaded strips, from companies, such as Schroff (https://schroff.nvent.com) or Gie-Tec (https://gie-tec.de), which makes the aluminium cabinets used by Doepfer, or even more easily, strips of 3/8ths-inch wooden dowel, permitting the use of self-tapping wood screws to mount the modules rather than the usual 3mm miniature bolts. Since many people are now dumping their entire CD collections in favour of downloads, some spectacular fibreboard fake wood finish or even genuine solid wooden CD cabinets are turning up in charity shops for just a few pounds, many are ripe for Eurorack conversion.

Many industrial companies offer module mounting hardware, and local carpenters, some with experience in making custom studio furniture or amplifier cabinets, are now often marketing Eurorack and other module mounting designs through eBay or through websites of their own.

TWISTED HEAD

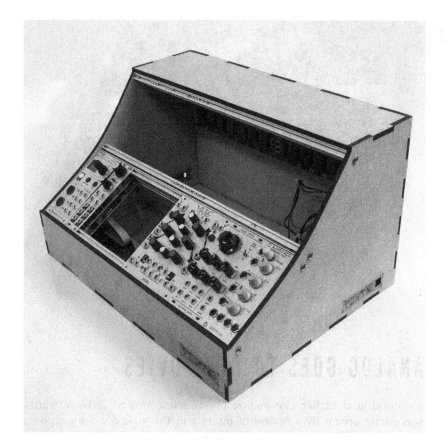

A dual 3U mounting cabinet from Twisted Head.

Based in Manchester (UK), Warren Edmond's company offers a range of cabinets which arrive flat-packed and are constructed from neatly laser-cut panels. Prices for Eurorack cabinets vary from £90 to £190. More on the book's website: www.routledge.com/cw/jenkins.

www.twisted-head.com

NOISE INN

Based in London, Noise Inn uses wood sourced from Poland to make very lightweight cabinets in cherry, maple, or oak. A 6U (two row) 84HP (roughly 19-inch) cabinet is £149. The design can be used vertically or horizontally and readily accommodates TipTop or 4MS power modules. This makes for a great lightweight system ideal for either studio or stage use.

www.ebay.co.uk – search for Noise Inn

On the book's website, www.routledge.com/cw/jenkins, you can see an example of the rack being populated along with detailed video demos of all the modules shown installed.

The attractive and lightweight 6U Eurorack cabinet from NOISE INN. Clockwise from top left: Dual Business Card Sequencer by DIVISION 6, Muxlicer and Rampage by BEFACO, Feedback and Character by DAVE SMITH MODULAR, ADSR by YORK MODULAR, 2LPG, Mix and 3LFO by TAKAAB, Traigh, Harmonaig, Tona and Cs-L by INSTRUO.

ANALOG GOES TO THE MOVIES

As mentioned earlier, the analog boom, from around 2010 onwards, was partly driven by a revival of interest in the style for the purposes of movie and TV soundtracks. Traditionally, a movie or TV production with any budget would go for an orchestral sound and hire a composer and arrangers trained in working with the classical orchestra. Because of the huge budgets and time restrictions involved, studios were extremely wary about taking on composers lacking experience in movie work, and the field was dominated by established names, such as John Williams, Thomas Newman, Trevor Jones, Jerry Goldsmith, later his son Joel Goldsmith, and Hans Zimmer, alongside his many assistants and collaborators (note that many of these, for example, Zimmer, had extensive experience with synthesizers, but often used them just to mock up scores prior to recording with an orchestra). This tendency overlooked the great success of many composers in the 1980s who took a more guerilla approach to soundtracking and created their own music almost solo, generally with whatever synthesizers were available at the time, notably John Carpenter for the low budget *Dark Star*, then *Assault on Precinct 13*, *Halloween*, *Escape from New York*, and many others (right up to *Ghosts of Mars* in 2001), the British director, Harry Bromley Davenport, for *Xtro (1982)* and many others.

Around 2005 Hollywood movie producers, perhaps initially with a view to making budget savings but later clearly out of artistic choice, had started to commission electronic music composers for their scores. *Stealth* (2005), with a $135 million–budget, used rock songs largely by

A very youthful Hans Zimmer in the studio with Roland System 700, Sequential Prophet 5, Yamaha electric grand piano, Korg VC10 Vocoder, Roland MC Micro-Composer, and more.

Incubus alongside a score by Brian Transeau (B.T.) who although having a classical background had become best known for creating the "stutter edit" in the trance music field. This involved slicing music and vocals almost into individual "grains" and re-combining them into new textures. Tom Holkenborg, *aka* Junkie XL, came to prominence with a remix of an Elvis Presley song in 2002, moved to Los Angeles the following year, and started to use his vast array of analog and other synthesizers (sometime in collaboration with Hans Zimmer) on movies such as *Domino* (2005), *Inception* (2010), *Divergent* (2014), *Justice League* (2017) and *Alita; Battle Angel* (2018).

Trent Reznor, who founded industrial rock band, Nine Inch Nails, in 1988 had built up a vast array of modular and other synthesizers and started a parallel career, first composing for video games and later for

Deadpool's Tom Holkenborg, *aka* Junkie XL, with just part of his giant modular system.

movies with *Natural Born Killers* (1994), *Lost Highway* (1997), *The Social Network* (2010), *The Girl With The Dragon Tattoo* (2011), and *Gone Girl* (2014).

Trent Reznor continues to innovate both with Nine Inch Nails and in movie soundtracks.

Given this increased acceptance of pure electronics for use in TV and movie music (though obviously keyboard and sampled sounds had been in use all along, though in a much less prominent way), and the fact that fashions go in cycles, it was hardly surprising to experience a huge burst of renewed interest in the visual and analog-led musical styles of the 1980s, initially through two hugely influential TV shows, *Halt & Catch Fire* (AMC, from 2014) and *Stranger Things* (Netflix, from 2016).

PAUL HASLINGER ON *HALT & CATCH FIRE*

Paul Haslinger's history as one of the most successful new young composers in Hollywood is well known. Born in Austria and classically trained, he auditioned for Tangerine Dream in 1985 when Johannes Schmoelling left the band, offering pieces for the *Underwater Sunlight* album which relied largely upon his use of the Yamaha QX1 sequencer and TX816 bank of FM synth modules. After scoring many movies with Tangerine Dream, including *Miracle Mile*, *Near Dark*, and *Shy People*, he went on to a very successful solo career, first through the team of Graeme Revell (who had his own background in electronic music through the band SPK in Australia) working on *Pitch Black* and *Lara Croft; Tomb Raider*, then solo on the *Underworld* movies, *Resident Evil: the Final Chapter*, and for TV, the primetime Emmy nominated *Sleeper Cell*.

Your first well known tracks came from a digital period using a lot of FM synths, but had you used analog electronics earlier?
My first synth proper was a Korg MS-20, which I used in combination with an SQ-10 hardware sequencer. Then came a Moog Prodigy, a Moog Liberation and a KORG PS-3200. The Prophet 600 I believe was the last VCO-based synth I bought before digital instruments came into play, starting with the Yamaha DX7.

Paul Haslinger in his studio in
Los Angeles.

The work with TD helped give you access to a lot of different instruments, what preferences did you develop?
It definitely was a playground, especially for a studio rat like myself. The TD studios at the time were set up around a Soundcraft mixing console, an Otari 24-track analog tape machine, and scores of keyboards and modules. My personal favorites were the Oberheim Xpander, Matrix-12 and OB-8, Minimoog, Prophet-5, Roland Jupiter-8 and MKS-80, and the PPG Wave 2.2.

When you moved into solo and movie music, how did your instrumentation and workflow change?
Moving to Los Angeles, and meeting some key people like Jon Hassel and Brian Williams, definitely broadened my musical horizons and gave me a taste of music outside the Tangerine Dream bubble. It had a major impact on the way I understood music, and in consequence, how I went about making music.

By the time I arrived in LA, I had switched from hardware sequencers to sequencing software, all my writing work was done with an early incarnation of Steinberg Cubase. In every setup I've ever built, I always had hardware instruments as part of the constellation. As we moved into the 21st century, the digital side of those hardware units got absorbed into software, whereas the "analog portion" has remained, and over time has acquired more prominence as a necessary antidote to life with computers and screens.

Did you notice a specific revival of interest in analog sounds later on? And even more so just a few years ago when it seemed that movie and TV soundtracks became more open to electronic rather than only orchestral sounds and textures?
It certainly seems cyclical. When I arrived in Hollywood in the 90s, "electronic music" in film was used as a derogatory term. You better not put it on your resume! A couple of decades later, *Stranger Things* happens and John Carpenter goes on tour. And people all of a sudden remember that there was some good electronic music. It's the process of history I guess,

to move past a certain phase only to come back to it later and remember that there was something good in it.

Halt & Catch Fire tells the story of young hardware and software engineers in the early days of the personal computer revolution, and your music for each season seemed to track the historical development of music technology almost to the exact year as the plot developed – did you research and use specific instruments?
I was definitely excited about the project and the opportunity. After getting hired, I went through a lot of my old 80s recordings (with and without TD) and analysed sounds/instruments used at the time. I then proceeded to prepare an array of instruments and patches (both actual and virtual) and did a lot of sampling with the less stable units I have, like my Memorymoog and Oberheim SEM. My trusty old Minimoog came in handy, as did the Roland MKS-20 piano and JD800.

One of my creative ideas for this project was to start with a sound that is very particular to the period depicted, and as the story and the seasons of the show evolve, to stretch and expand that sound, both into the past and into the future. This is why, in the fourth and final season, you hear a lot of early (pre-80s) sounds, but paired with looping and ambient techniques more typical of the 90s and beyond.

What's your feeling now about the balance between analog, digital/software and orchestral/sampled sounds in compositions?
Well, I am working on both sides of the equation frequently and I find both domains rewarding. Ultimately, they are means in the service of an idea. If the idea is good, the means will follow.

Of course, if I look at it purely from an inspirational angle: sitting down at an instrument and starting to play, for me, beats loading up some plug-in and mousing around to find a sound. The initial effect is simply more immediate and more gratifying, a more direct path to making music.

What's your current studio instrument lineup and favourite analog instruments?
My studio is based around several PCs running Cubase and Vienna Ensemble Pro software. I use BURL Audio converters, SPL Summers and a Yamaha TF-1 digital mixer for live audio management.

Live keyboards and modules currently in use are: Oberheim TVS-Pro and SEM 4-voice, Arp Omni 1 and Soloist, Minimoog, Memorymoog, Studio Electronics SE-1X and Omega, Yamaha CP70, Fender Rhodes 88. Recent favorites are the new ARP re-issues from Korg.

Each one of Paul's studio and soundtrack projects has something exciting to say about the imaginative blending of orchestral, digital/sampled, and analog sound. At the time of writing he is also working on new projects with fellow ex-Tangerine Dream member Peter Baumann.

www.haslinger.com

And for the _Halt & Catch Fire_ soundtrack:

www.firerecords.com

While _Halt & Catch Fire_ was still on the air, Netflix debuted _Stranger Things_, heavily influenced by 1980s movies such as _E.T._ and _The Goonies_.

The pilot episode used a library music piece by Kyle Dixon and Michael Stein of the Austin, Texas-based electronic band S U R V I V E as its theme (Emmy Award 2017, Outstanding Original Main Title), and when the show was picked up, the duo was approached to soundtrack the whole series which was an immediate success, with actress Millie Bobby Brown becoming the youngest ever inclusion on the Time list of the world's 100 most influential people.

DIXON AND STEIN ON STRANGER THINGS

A performance at London's Barbican Theatre saw the stage festooned with glowing red neon lighting in a tribute to the show's surrealistic setting, referred to as the *Upside-Down*.

In concert we play *Stranger Things* stuff but we also do a lot of improvising, it's something for us to play in the linking passages, with this show we're just doing what we want to do. We came from an experimental background so the show is very much that, it includes the TV music and we never planned on playing *Stranger Things* music live, but there was a demand for it. But it felt a little bit cheesy to bring the logo out, or bring on cast members or dress someone up as the *demogorgon*, we felt that was a bit like selling out, though we wanted to keep the weird textures because there's so much of that in the show in the bedding, particularly in the scenes where you're being terrified. So we have this really cool stage setting, just an abstract reference to the red title sequence of the TV show.

Are your listeners now a very varied group?
Before the TV show most of the people who would come to see us were also making electronic music with similar gear, we had a few fans but it was mostly people who were into the genre and wanted to see us come out with a whole bunch of synths. But after the TV show lots of people turn out now and they're just getting into the music. As S U R V I V E we played the Moogfest while it was still in Asheville, and later on we did a different version of this Season 2 *Stranger Things* show, and a Q&A with Dave Smith about his DSI and Prophet synths. So we still have this approach of bringing out a lot of equipment in performance – the first Moogfest year we ran into Malcolm Cecil (T.O.N.T.O.) and invited him to come along, he was remembering some moments from the 1970s with all these top line melodies going on and that was great, it was really such an honour. That was the first trip, even before *Stranger Things*, where we got to meet some of the people we'd been really looking up to, on the way back we played with (Italian progressive rock soundtrack specialists) Goblin and they were really enthusiastic too.

So was electronic soundtrack music your main early influence?
Kyle – In our early influences there's a bit of John Carpenter, but for me the main reason I wanted to start the band was a song called *MASS* by Yellow Magic Orchestra. I'd been listening to Aphex Twin and all that stuff growing up, but when I heard YMO they were doing some of the same things – or at least some of the textures – that we wanted to do, certainly

in the percussion sounds and the level of complexity, and I thought "this has really been going on for a long time", that song has a toughness and a seriousness to it. So not all of YMO but some songs, if you listen to YMO then one of our albums you'll hear a lot of similar FM sounds, they're atonal but they're so ring modulated that they sound tonal. And Space Art (the French duo of Dominique Perrier and Roger Rizzitelli who played with Jean-Michel Jarre and released three albums from 1977-1981) was a great influence too.

What were your first analog instruments?
Kyle – I typed "analog synth" on Bay and ended up with Univox MiniKorg, it's a kind of preset synth but with a weird filter, then after that my first real variable synth was a Korg Monopoly which we still use.

Michael – I bought a lot of keyboards but wasn't really getting along with them, then I got a Roland SH101 and I though oh, this analog sound is what I like – then I started building up a modular system, reading books like yours to find out what I needed, at the time when the 5U DIY format was popular. Then I thought that I should be able to buy one synth that did all that, but I'm glad I had my synth binge before they reached today's prices because finding some of the stuff I have now would be ridiculous.

Kyle Dixon and Michael Stein on stage in London on the surrealistic *Upside Down* set, performing music from *Stranger Things* with the DSI, Oberheim, and ARP2600 synths.

Kyle – We paid $600 for the MonoPoly and now they're $2000, and we ask for an ARP2600 to be hired for each live show because we use ours a lot, but to buy one now is impossible. The interesting thing is that some of them sound better than others, the filters are different and some days we're lucky and get a particularly good one.

The band S U R V I V E started in Austin, Texas where Kyle and Michael were in school together, and later found they'd both start creating music independently.
Kyle – The first time we got together as a band we ended up making the first S U R V I V E song in a day, put it up for download, and when we had

four songs some friends had a label and wanted to put it out as an EP. We were using the arpeggiator on the MonoPoly a lot, we had an Akai MPC but we didn't really have any way to sync it all up. We were sending audio out of the MPC as clicks. Then we got this FutureRetro Mobius sequencer so we could send out S-trigs and V-Trigs to the MPC and program gates, but certainly for a long time it was a case of "don't touch the rate on the arpeggiator", and it all sounded very human. But what we wanted was for it to sound very tight! So when we could sync it up using the Mobius we really liked that effect.

Michael – Later on I had ProTools and has it set up that I could select any MIDI-CV convertor I liked so I could pick each instrument as if they were soft synths, a way of work that was pretty efficient and I'm coming back to that now. We all switched to MPC's at the same time but there's a pretty steep learning curve on them, it took about 8 months because essentially it's a weird old computer.

So now in TV work, is the sync to picture question much more demanding?

Kyle – A little bit, each cue is different. If we need to have beats that hit specific points you either don't pay attention to tempo, or you try to dial it in so they're right, making sequences on the MPC then adjusting timing so it fits, we're still working very much in an audio mode though we have a lot of MIDI going on now – if you are asked to make last minute changes it's really nice to have the MIDI going on within in the computer so you can see it all happening.

And in terms of sound creation, what are your favourite instruments?

Michael – We get a lot of FM type sounds out of analog synths, the ARP2600 is the only one where you get a super complex crossmod to do that.

Kyle – My favourite is the ARP Solus though, you get some really gnarly sounds out of it, some of the ARP models have a strange ring modulator which gives very aggressive behaviour which sounds really cool. Now we have the most modern versions of the Prophet and the Oberheim because you can depend on them, and they sound good. In S U R V I V E, I still play the MonoPoly and Michael plays the Roland SH101, there was one point where all four of us had Sequential Six-Traks, the Korg PolySix became too unreliable so we stopped using that.

Michael – Now that Korg had re-made the ARP Odyssey they were kind enough to give us the whole set with the three different filters. Playing bass I may want to play the Mk1 and playing leads I can switch to the Mk3, the companies are doing a good job with these re-issue instruments.

Kyle – We use laptops but having the hardware is much more fun. On the laptops we have the basic sequencing of tunes going on, and we use PPG Phonem (vocal synthesizing software) a lot, going through UAD plug-ins which are a big part of the reason we use plug-ins, and Omnisphere for some atmospheric sounds. Also Eventide Octavox which is an 8-channel pitch/delay with panning, feedback and harmonies, which thickens things up a lot, we use it a lot on the TV show. You can make a sound cluster creating atonal stuff or make a sequence out of just one note, it has a very specific timbre which is not just spinning out a long delay, it emulates the sound-on-sound effect from a tape echo, with a feedback effect which has a bit of a wobble too it.

What's lined up for the future?

Kyle – Apart from Series 3 of *Stranger Things* we've been working on a VR project for the Oculus headset, it's a three-part series of short films called Spheres with Darren Aronofksy (*Pi, The Fountain, Black Swan*) producing, and on the first one Jessica Chastain narrating. It's about the universe, the first is about black holes – there's a very interactive section towards the end with different layers of sound going on. That stuff is just under our two names but the other S U R V I V E members are getting more involved again now, we're trying to figure out now how to make it all work as a team again.

Dixon and Stein are great examples of musicians whose influences from the earlier synth pioneers has helped them ride the wave of new interest in analog sound.

https://survive.bandcamp.com/

ADAM LASTIWKA ON *TRAVELERS*

From 2016, Netflix and the Canadian channel Showcase aired *Travelers*, created by Brad Wright of *Stargate* renown and having perhaps an unlikely star and co-producer in Eric McCormack, best known for his light comedy role in *Will & Grace*. To be fair, McCormack turns in a terrific performance as one of an unconnected group of people saved from death by having their personality erased and replaced by one from a dystopian future. Unsurprisingly, being for one reason or another at the point of death and having to retain a present day persona while fighting to avoid disastrous changes to the future timeline gives all these characters tremendous difficulties. Soundtrack composer Adam Lastiwka responded to the challenges of these characters by creating different musical and sonic textures for each one; moody analog textures sweep through tonal changes, fast sequences run across the stereo field, and a rich electro-acoustic acoustic element is layered on by instruments such as guitar, cello, and koto.

Adam Lastiwka working on the *Travelers* soundtrack with Arturia, DSI, and other synths.

What's your music education background – classical?
No, I suppose it would be considered fairly non-traditional. I'm a bit of a musical "mutt", but I suspect that is more of an asset than a liability when it comes to working in film and TV. I grew up listening to classical music at a young age thanks to my mom, and 70s prog rock thanks to dad, but started to experiment with sequencing software around the age of twelve, despite lacking any actual music background or comprehension (or training, it was all just trial and error). That drew me down the path of production and composition well before I was interested in playing any instruments, which I suppose in hindsight is a bit unusual and certainly a motivator to focus on film and TV. I picked up guitar around sixteen and was really, really captivated by that, which enabled me to catch on quickly enough to be teaching full-time when I was eighteen. Later on I ended up apprenticing with a really brilliant Berklee trained composer named Shawn Pierce. We worked and studied together for almost seven years. So I have no formal music education, but in this field I strongly believe the apprenticeship system lends itself to an accelerated maturity, and I'm very grateful for the opportunities I got to work with Shawn as well as other great composers.

What was your first exposure to hearing and then playing electronic instruments?
Software sequencers really made sense to me, something about seeing the sequences and patterns in that way just really fit the way my brain worked. I got started around the cusp of software instruments and DAWs really starting to become powerful and useful, but my ears never seemed to be captured by the sound of soft-synths. I didn't have the finances to get hardware synths so early on I relied on sampling and field/audio recording to extract interesting sounds. It's still a massive part of my production style. Eventually I would get in to the analog world, and I'm happy to utilise anything in the software world that helps me effectively get the job done.

How did you move into creating music commercially and for soundtracks?
I think it's kind of a strange interest for a seventeen year old that should be dreaming about girls and being in a rock band, but I just always really wanted to be working in film. I didn't want to play in bands or rely on other people for collaboration. My goal was to be totally self contained and self-sufficient, and the music I was creating just had a naturally filmic sound. I made demos and sent them around and eventually got a record contract with an indie label to create music geared at licensing and trailers. The first album I made had a track on it that eventually got placed in the trailer for the Ridley Scott film *Body of Lies*. It was an exclusive license and was for more money than I ever could have dreamed of – which came in right around the time I was really facing some serious defeats. It reinvigorated the possibilities for me, and within a couple years I had met and was working full time with Shawn, who definitely steered me down the path of television, which I previously had not really been aware of as an option.

What was your earliest equipment setup and instruments? Did you start very early on to mix analog, digital and acoustic sounds?

I was never happy in just the software world, though my rig early on was definitely bound to it. It encouraged me to record as much audio as I could and learn to layer the sounds in a more abstract way. I was learning as much as I could about classical orchestration and developing a good degree of knowledge in the electronic world, while simultaneously learning and recording as many different instruments as I could. I began to apply the very well formed principles of orchestration to the electronic setting which enabled me to create more unusual and complex arrangements, without being totally bound stylistically or by convention. I looked at sound in a more abstract way, and the principled classical experience enabled me to transcend and that and then evolve away from it. What I mean is that if I record something unusual like stripping velcro back, or flipping the pages of a book, I'll replace whatever conventional element that sound shares the same sonic niche with, and then edit the performance to create a new part. Sometimes I'll layer the parts and sometimes I'll replace them. In the case of say a snare drum part, the additional layers of velcro or paper bag crumpling, tin foil scraping etc. add life and space and air that is sometimes absent in electronic music.

Travelers has some terrific textural passages and themes for each character. What was your approach to creating these? Did you find the Roli Seaboard and software useful for new textures?
The Roli was really helpful in finding new degrees of expression from synths. I love the naturalness of the movement over the keyboard and how akin it is to the mechanics of playing a stringed instrument. I find it very useful for doubling parts of my performed instruments as you can very easily match them to create additional reinforcement layers. I am about to receive my order of the Deckard's Dream synth [a rackmount Yamaha CS80 clone] which is one of the few MPE [MIDI Polyphonic Expression]-enabled analog synths out there which can take full advantage of what the Roli does. Unfortunately it took them a bit longer to build it so I think I'll only be able to use it on the final episode of season three. But I'm really excited to see how that will integrate!

My approach on *Travelers* was to really utilise my ability to create a fusion of analog-digital-orchestral-traditional folk music in to the sound of the universe. The idea I pursued was that in the future the entirety of the remaining human race lives crammed in to these survival domes. I created my own sort of narrative for that and thought, with all these cultures coming together and sort of being forced to homogenise in a way, what would it be like if they brought instruments with them that everyone played, but they weren't aware of the proper technique or sound the instrument was produce. So I basically spent the entire budget of the show on instruments, some of which I truly had no idea what they were, but I applied what technique I had to try and elicit interested sounds and emotions from them, and it became the foundation of the sound. The goal was to assimilate them in to the electronic world, which I just found easier with analog synths, which tend to occupy a distinct sonic niche, and lend more musical character more efficiently. Layering just a couple of analog synth tracks starts to create a world that I could then use the instrumental performances to meld with. I didn't want it to be distracting or too obvious, but there are some fun moments in the show where I highlight it.

What's your feeling now about the balance between analog, digital/ software and orchestral/sampled sounds in compositions?
Each of these worlds has so much to offer and there is so much learn from each discipline that can transcend to the others. I am very much a non-purist and I firmly believe that acquiring knowledge in different domains and applying it to others is where some great innovations can occur. I absolutely love studying music from all around the world and use that to further my understanding and appreciation of not only music but of the people and cultures as well.

I think utilising the strengths of each field is what it takes to effectively create a new voice in music. The goal at the end of the day with all of it is to move people, and I'm eager to utilise whatever I can to accomplish that.

What's your current studio instrument lineup and favourite analog instruments?
My first synth was a Moog Voyager RME, and since being bitten by the analog bug my collection has grown quite rapidly. I own all the analog Dave Smith keyboards but my two favourites are by far the Pro6 and the OB6. I love the Elektron Analog Keys as well. For the price it's the most diverse and sonically capable synth out there. The interface is rather frustrating compared to the ease and intuition of the Dave Smith stuff, but the capability of it is insane. I've gotten in to the modular world as well and even though it's a very time consuming exploratory thing (the antithesis of what TV scheduling allows) I will treat it sort of like a fixed architecture synth, and will use a patch as a theme for that specific episode I'm on. The sound is just so distinct and unique it would be a shame not to utilise it.

At the time of writing *Travelers* Season 3 is just being completed, and no doubt will be available through Netflix pretty much indefinitely. It's a great example of the current trend for commissioning one-man electronic-based scores, provided they can find a way to match some of the tone and richness of classical and acoustic instruments.

www.adamlastiwka.com

Suzanne Ciani on TV advertising music and the Buchla Modular

The last of our interviewees new to this 2nd Edition has experienced a resurgence of interest in her pioneering electronic music work, though she's equally at home with classically styled piano and with movie and TV advertising music. Suzanne Ciani, born in 1946, graduated in 1970 with an MA in composition from the University of California Berkeley, where she met synthesizer designer Don Buchla. By 1974 she'd moved to New York and was creating sound effects for TV advertising using the Buchla modular and other instruments, and mostly notably created a "pop and pour" sound for Coca-Cola. Performing up to 50 sessions a week, she was interviewed on TV shows such as David Letterman, giving many people their first view of an electronic music system in action.

Suzanne Ciani and the Buchla 200e, performing live in London 2018.

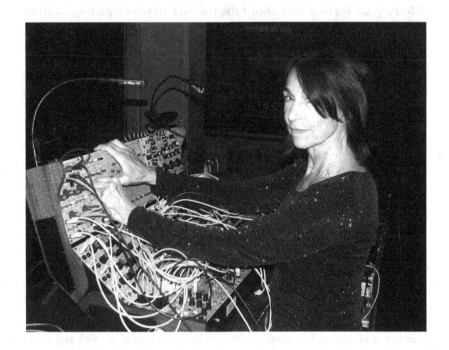

She started to produce instrumental albums using a Roland MC8 sequencer, Sequential Prophet 5 synth, and other instruments, first for Atlantic and later for Peter Baumann's Private Music label. Through her own label, Seventh Wave, she released albums in an orchestral style but her early Buchla concerts were re-discovered by the UK label, Finders Keepers, and in 2016 she gave her first Buchla concert in 40 years, using a modern 200e system and a Buchla 227 Quadrophonic Output module to make a quadrophonic vinyl release possible. We met at her London performance at the Jazz Cafe.

When I started my "comeback" it was Andy Votel from Finders Keepers Records based in Manchester who was really responsible, he released two Buchla concerts from 1975, we hadn't recorded in quad but we had some kind of reference recording and that's what he put out. I had met Don Buchla in 1969 when I was at UC Berkeley and went to work for him, by that time it was 5 or 6 years after he started and he had evolved his concepts about what electronic instruments could be. The original conception of the analog modular system was not a performance instrument, it was a recording instrument, the idea was to record sounds and assemble them into a composition, but by the time I met Buchla he had this vision that it was an instrument that could be performed in real time live. I didn't know just how hard that was, but I did it.

That was in marked contrast to the Moog system?
Yes, it's the same distinction with what we're seeing now in Eurorack modular – early on, the presence of the keyboard short-circuited all that

original thinking, and people were using the machines to play existing compositions like *Switched-On Bach* or Tomita, so nobody was exploring the new music that could be performed. The Buchla was designed to be portable, it was compact and that's how I made my splash in New York, most synthesists were confined to the studio with these huge systems, I was able to take my system out and do work where they wanted it to be done.

The Buchla was ahead of its time and it still is, the Eurorack idea has started to catch up, but when Don was designing he started with the human body. Electronics has become pretty settled in a way, the functions of oscillator and envelope have become distilled, but the interfaces – Don said, "how big are human hands, how are they shaped?", and people are only now starting to say that again. There's the Roli touch keyboard which has some nice features though again it's a traditional keyboard layout, but to Don a keyboard is just an access point, it's not fully predictable – it's a command system, from the Buchla keyboard I can start something, stop something, transpose something . . .

But with a classical background too, weren't you set in those ways?
I was proselytised by Don Buchla, and I was aware of the danger of perception, I was on a mission for Don and I wanted to communicate his vision, I didn't want to confuse anybody, so I didn't touch a piano in ten years, there's no picture of me playing a keyboard. That was the critical point, once people saw a keyboard their thinking went backwards.

When I moved to New York from California, I arrived with just my Buchla, I was doing concerts in art galleries and very quickly faced hunger and extinction, so I found a way to make money in advertising, and that's what I did with the Buchla. Simultaneously from around 1974 electronics shifted and after Bob Moog and Don Buchla came the Japanese instruments in larger numbers, and also my Buchla broke, so that was pivotal – I couldn't get it fixed, so I moved to the more conventional electronic music instruments. On my first album *Seven Waves* which was all electronic but orchestrated, I used the Prophet 5 which had just come out, later a Prophet T8 and a Yamaha DX7, I thought of all the instruments as the individual players so I listed absolutely every instrument I used on all the albums.

Later on I also had the Synclavier, at the time I had a deal with RCA Red Seal which was a classical label, and when I was leaving New York I wanted to sell it but the thing was over $200,000. Once you've used a Synclavier you find it's so fast that using something like an Akai sampler seems very different. Since selling that I've never sampled, I only use sampling to design. Can you quickly sample something yourself, something like a champagne bottle – now you can do it, but using something like Spectrasonics there are so many stages, sample it, import it, export it . . .

Recently I was at Dave Smith's studio and he was showing me the Prophet X which handles samples as well as analog sounds, it's not just a sampler, the thing about Dave's instruments is that I never used presets – you had all the controls you needed, and on the X you can do that to sampled sounds and modify them, it's really powerful. I can't stand samples, to me they're dead, although I know there are wonderful ones which we hear on TV all the time, but unless it has some movement in the sound they're dead.

How did you come back to live performance with the Buchla?

I was happily living at the beach doing my studio albums, although my last one was in 2005 called *Silver Ship* and after that in 2008 I did a DVD project about the Galapagos but that didn't have much of a life, then we went through that horrible period of everyone stealing music so I just had no energy to record. But I always had an independent mentality, I had my own label and everything was just licensed out. I've been indie since 1994, but then I was contacted by Finders Keepers about the old Buchla tapes and they're putting out pieces I would never have had time to put out.

So each live performance now is based on ideas, I have maybe four or five that I'm going to include, and I don't over-rehearse because I think part of the interest of live performance is that you're in the moment, creating and finding your way, you know what you want to do but you have to be flexible, things change – for example today I set up and everything was tuned up a half step and then my keyboard froze, so you have to be flexible. But I have starting points. I always start with the ocean because that's always been my muse, waves are my comfort zone so I start there and I feel relaxed. Then I start to explore, there is a memory in the Buchla and when I started using the 200e which is quite different from the old 200 I found there's even more, but I can't use them, in fact I use fewer and fewer. My approach is that I have two basic patches, and everything I do is based on the same sequences I used in the 1970s. I'd written a 40-page paper about playing the Buchla, and when I came back to playing the new instrument I looked back at that paper and I learned a lot, those materials cover four different 16-stage sequences and those are the starting points, then it's an improvisation on these sequences.

Are you completely in command of the Buchla when you perform?

Because I'm classically based I do use pitches, and the 200e was never designed to use pitches, so it's been a process and a real challenge. When I talked to Don about this I had a non-working Multiple Arbitrary Function Generator which couldn't be fixed, and Don had a new version called the Dual Function Generator, and it just would not tune. I'd take it over to him and I'd say: "Don, it just doesn't tune, what am I going to do?" And he thought about it, and he said: "Do something else!" But I really wasn't ready to do that, so I just added a clone of the module, made by a Russian guy who now lives in Japan.

But still the heat wreaks havoc on the system, it's not happy in either high humidity or heat, so I also use Animoog on the iPad, you have some dependability from something like that, it's wonderful and I've thought a lot about the instruments of the future now that Don has passed away, and I do think this sort of graphic interface is the way forward. The trick is how you integrate it with the patching and the tweaking and all the stuff we love. The whole module and Eurorack idea is an interesting field because it's always very customised, people pick and choose what they want and you won't find any two systems alike. The starting point is that's it's all very individual, but because the Buchla is not from the same family it's hard to integrate with Eurorack modules. The Eurorack to me is a little too compact, I have small hands but it's hard to get in there, the Buchla is designed more based on ergonomics. Like on the Teenage Engineering products, they're so small you can lose them in your pocket – though I do have a little Makenoise system Eurorack now which I love.

Suzanne's performance in London made the most of the Buchla and the quadrophonic sound system, with textures, sequences, and individual notes spinning around in dramatic fashion. It's great to hear electronic music created live and spontaneously in this way, and the Finders Keepers Records releases on vinyl as well as on CD are cataloguing the fact that this was done as far back as the 1970s but can remain equally compelling today. See the Discography, Appendix E, for more, and Suzanne's quick video tour around her Buchla 200e system on the book's website: www. routledge.com/cw/jenkins.

www.sevwave.com

Incidentally, the legacy of Suzanne Ciani and the Buchla remains strong elsewhere in California. Ciani has recently collaborated with Kaitlyn Aurelia Smith, a classical musician like Ciani given an early introduction to the Buchla Modular System.

Kaitlyn Aurelia Smith and her Buchla Modular.

She's now built up an impressive electronic music studio and performs in Europe as well as the USA (www.kaitlynaureliasmith.com). Working in San Francisco, Carolyn Fok is a painter and musician whose *Memoir of Sound* project now archives hundreds of cassette, DAT, and digital projects recorded using early modular synths, Moog and other keyboards, many under the name Cyrnai and often alongside collaborators like keyboardist Tim Story and guitarist Elliot Sharp (www.carolynfok.com).

To come almost full circle in this history of analog synthesizer music, 2018 saw Jean-Michel Jarre celebrating 50 years in music with a massive compilation album, *Planet Jarre*, followed by a new studio album and tour. Jarre (interviewed in Chapter 5) started in beat groups before the synthesizer was invented, created his early music with tapes and oscillators, and quickly picked up on the possibilities of the EMS VCS3 synth as soon as it was launched.

Jean-Michel Jarre celebrated
50 years in music in 2018.

Following his massively successful *Oxygene* and *Equinoxe* albums,
which benefitted from the abilities of custom sequencers designed by
Michel Geiss, he mastered sound sampling using the Fairlight CMI, later
went into areas of techno and dance music, and returned to his analog
roots with a series of sequels to *Oxygene* as well as collaborations with
other rock, pop, and electronic artists on the *Electronica* albums in 2015
and 2016. The new *Planet Jarre* is a massive 41-track compilation spanning from early "lost" works to new unreleased tracks, including *Coach-
ella Opening*, and in itself stands as a compelling history of the fascinating
development of electronic music.

https://jeanmicheljarre.com

Talking to the musicians in this final chapter shows that it's once again
a terrific time for analog style music, which is benefitting on a worldwide
basis from a resurgence of interest particularly among the TV and movie
community. Designers and manufacturers are taking advantage of the latest technology to re-create the best of the original analog designs, and to
integrate them (particularly within Eurorack and other modular instruments) into digital and software systems so that analog can take its place
alongside acoustic, orchestral, and other sound sources when required, in
order to make the most of its striking sonic possibilities.

In the appendices we look at how values are holding up for the original classic analog instruments and how to choose, and if necessary service them, for second-user purchase, at what newer designs are currently on the market, and at some contacts and some example music (including new vinyl!) and books for further study.

Finally, there is a listing of all the audio examples and some of the video examples on the book's website, www.routledge.com/cw/jenkins. These include demonstrations of basic sound types and techniques, sound samples from a huge range of analog synths, which can readily be turned by any hardware or software sampler user into playable sound sets, and

Mark Jenkins, 2018 studio setup.
Keyboards 1: Korg Microstation, Yamaha AN1X with SU200 sampler, Yamaha CS2X with QY10 sequencer, Technics WSA1, Korg KARMA, Roland JD800, Moog Sonic 6.
Rhythms & arpeggiators: Yamaha DJX2B, Zaquencer, Novation Nova, Korg Micro, Roland EF303 time synchronised effects, Roland A300 Pro MIDI controller, and Korg TinyPiano.
Modular 1: Blue Snowball microphone, Zoom FX, Fostex monitor, oscilloscope, MFB Eurorack modular, Korg MS20i controller, MIDI bass pedals, Korg MS04 modulation pedals.
Modular 2: briefcase Eurorack system with TipTop power, HexInverter sequencer and various modules, snake to custom built joystick controller.
Display for Mac Mini running Logic Pro X, Korg Triton Le 76, M-Audio Keystation Pro88, guitar effects including Electrix Repeater, Korg AX1500G and Wem Copicat, Alesis DM Pro drum module with Roland Handy Pad.
Behringer mixer for Moog Theremini and MIDI modules: Alesis MIDIverb, Korg NS5R, Yamaha VL70-M for WX7 wind controller, two E-mu Proteus 2000s, Alesis S4, Korg N1R, Solton TS4 accompaniment module, Roland JV1080, Kawai mixer.
Keyboards 2: Korg Prophecy, Roland SK88, Novation Xiosynth, E-Mu Vintage Keys.
Guitars, rack including Hohner XE9 accompaniment module and Novation A-Station.

video demos which go into much more depth on the facilities of many of the instruments discussed, particularly the Eurorack and other modules newly included in Chapter 8. Edits from some of the artist interviews are also on the website in video form. It is worth checking the website every few months for new material as it is added.

Happy Synthesizing!

www.markjenkinsmusic.com

Appendix A

Classic instruments: specifications and values

The table here covers classic analog instruments which are now up to 50 years old or even slightly more, the majority of them pre-MIDI. It doesn't attempt to include most generally reliable later midrange MIDI/digital instruments, such as the Korg MS2000/R, Access Virus B, Novation Nova and Xiosynths, Yamaha CS2X and Roland XP10/early Fantoms, which are all readily available in relatively large numbers on the second user market, usually reliable, and an unproblematic purchase, or modern instruments still in production, such as those from Arturia, Roland Boutique and Korg re-issue keyboards, Novation's Peak module, Dave Smith Instruments Evolver and Prophet series, Studio Electronics or recent Behringer models, which haven't yet come onto the second-user market in any significant numbers.

Eurorack and other modules for which there is a busy second-user market are also not included; they tend to remain in good condition so invariably change hands on second-user markets at about two-thirds of the retail price for those still available, maybe a little more if discontinued, for example in the case of Doepfer's handy all-in-one single voice module the A-111–5, which has become sought after.

At the time of writing, prices in GBP, USD and Euro are all roughly equivalent. P = number of notes polyphony (P = fully polyphonic); O = number of oscillators per note; K = number of keyboard octaves; M = number of memories or presets; FX = effects (1 = chorus, 2 = delay, 3 = digital multi-effects including reverb, or a spring reverb); V = value (k = 1000).

Model	P	O	K	M	FX	V	Information
Akai							
AX60	6	6	5	64	1	200	MIDI, split, arp, sampler in; two versions, second has more controls.
AX80	8	16	5	96	–	300	MIDI, 32 memories preset. Large graphic editing display.
AX73	6	6	6	100	1	200	MIDI, split, sampler in, master functions like Akai MX73.
VX90	6	6	–	100	1	150	MIDI rack-mount version of AX73, sampler in.
VX600	6	12	3	90		600	For wind controllers. Multi, matrix modulation, audio in, MIDI.

(continued)

Model	P	O	K	M	FX	V	Information
ARP							
2500*	2	5	5	–	–	15k	Large modular, systems varied, usually with 3 × ten-step seq.
2600*	2	3	4	–	3	4k	Flexible, usually duo variable synth, spring reverb.
Explorer 1	1	1	3	8	–	150	Variable, semi-preset small mono synth.
Little Brother	1	1	–	8	–	150	Keyboardless, preset expander version of Explorer 1.
Pro Soloist	1	1	3	30	–	250	Preset solo synth with aft.
Pro DGX	1	1	3	30	–	250	Update of Pro Soloist with digital selector buttons.
Odyssey*	2	2	3	–	–	950	Versatile duophonic. Three versions, all now reproduced by Korg.
Axxe	1	1	3	–	–	250	Simplified Odyssey with one VCO. Three versions, last adding PPC.
Solus	1	2	3	–	–	350	Two-VCO portable synth built into flight case.
Quartet	P	1	4	10	–	150	String/piano/organ poly presets; made in Italy by SIEL.
Avatar	6	2	–	–	–	300	Keyboardless Odyssey with six-channel guitar synth input.
Sequencer	–	–	–	–	–	450	2 × eight-step analog sequencer, various revisions.
Omni/Omni II	P	1	4	4	1	250	String synth with filter, phaser, chorus, split bass.
Quadra	P	1	5	16	1	1.5k	Strings, limited prog poly, lead and bass synths, phaser.
Solina	P	1	4	4	1	450	String synth, can trigger external monophonic synth.
Bit							
One	6	12	5	64	–	250	Prog poly with velocity, split, MIDI.
99	6	12	5	99	–	250	Improved prog poly with velocity, split, MIDI.
01	6	12	–	99	–	250	Rack-mount of Bit 99, MIDI.
MKB	–	–	–	99	–	150	Three-zone MIDI master keyboard with sequencer.
Casio							
CZ101*	8	8	4	32	–	150	Four-part multi, mini-key MIDI Phase Distortion synth.
CZ1000	8	8	4	32	–	150	Four-part multi, MIDI Phase Distortion synth.
CZ3000	16	16	5	64	–	150	Eight-part multi, MIDI Phase Distortion synth.
Cheetah							
MS6	6	12	–	96	–	350	Six-part multi-rack, MIDI, only one audio out, 320 ROM memories.

340

Model	P	O	K	M	FX	V	Information
Crumar							
DS1	1	2	3.5	-	-	350	Duophonic DCO synth, Gate in/out but no CV in/out.
DS2	1	2	3.5	-	-	475	DS1 with simple poly section added.
Spirit*	1	2	3	-	-	750	Bob Moog design with arp, sync, three wheels, two filters, audio in.
Trilogy	P	2	4	7	-	150	Strings, brass and organ, simple polyphonic.
Stratus	P	2	4	-	-	150	Simplified Trilogy.
Composer	P	1	4	14		175	Strings/brass/organ/poly and mono, breath control, multiple outs.
EDP							
Wasp	1	2	2	-	-	650	Powerful mini touch keyboard synth, speaker, battery power.
Wasp Deluxe	1	2	3	-	-	750	Wasp with mechanical keyboard and oscillator balance.
Gnat	1	1	2	-	-	300	Simplified Wasp with strong pulse width mod setting.
Gnat Deluxe	1	1	3	-	-	350	Same as Gnat with mechanical keyboard.
Spider	-	-	-	-	-	300	Powerful digital sequencer with Gate/CV outputs.
Caterpillar			3			150	Four-note poly keyboard for Wasps and Gnats.
E-H							
MiniSynth	1	1	2	-	-	150	Tiny two-octave touch keyboard vel micro synth.
Elka							
Rhapsody 490	P	1	4	4	-	150	Strings, rich sound.
Rhapsody 610	P	1	5	4	-	275	Strings, piano, harpsi, strong layers, splits.
Synthex*	8	16	5	80	1	2k	Powerful prog polysynth, four-channel seq, split, later with MIDI.
EK22/EM22	6	12	5/-	96		175	5/1 split/layer, vel, aft, MIDI; and non-rack expander version.
EMS							
VCS3*	1	3	-	-	3	3.5k	Pin bay, spring reverb. Keyboard optional
DK1/2	-	-	4	-	-	250	Mono- and duophonic optional keyboards for VCS3/Synthi A.
Synthi A	1	3	-	-	3	2.5k	Portable version of VCS3, spring reverb.
Synthi Aks*	1	3	-	-	3	3.5k	Synthi A plus 256-note touch keyboard sequencer.
Synthi E	1	1	-	-	-	600	Simplified single-oscillator Synthi A for schools.
Synthi Logik	1	1	-	-	-	600	Improved German-designed Synthi E with small keyboard.
Polysynthi	P	1	4	-	2	450	Single filter poly, analog delay, aft.

(continued)

Model	P	O	K	M	FX	V	Information
Synthi Hi-Fli						600	Guitar processor with phasing, ring modulation, etc.
Ensoniq							
ESQ1/ESQM	8	24	5/-	100	-	350	Strong multi digital synth, res filters, seq on ESQ1 keyboard.
SQ80*	8	24	5	100		350	ESQ1 with more waveforms and seq memory, individual key aft.
Farfisa							
Syntorchestra	P	1	3	13	-	400	Compact piano/brass/string synth with mono section.
Jen							
SX1000	1	1	3	-	-	150	Simple fully variable mono, no CV/Gate inputs.
SX2000	1	1	3	7	-	130	Organ-top preset/variable version of SX1000.
Kawai							
S100P	1	1	3	32	3	200	Mono preset/variable, spring reverb and flanger, aft – see Teisco.
SX210	8	8	5	32	1	250	Prog polysynth with large LED display, no MIDI.
SX240	8	16	5	48	-	300	Prog polysynth, 1500-note sequencer, portamento, MIDI.
K3, K3M	8	16	5/-	64	1	150	DCO poly, vel, aft, MIDI, chorus; and 2U rack-mount version.
K4, K4R*	8	16	5/-	64	1	200	Digital waves, resonant VCFs, MIDI; multiple outs on K4R rack.
Korg							
Mini 700	1	1	3	-	-	250	Basic single-oscillator monosynth.
700S	1	2	3	-	-	350	Dual VCO version of 700.
800DV	2	2	3.5	-	-	450	Duophonic, dual layer version of 700.
770/S	1	2	2.5	-	-	300	Two LFOs, two VCFs, audio input.
SB100 Bass	1	1	2	-	-	300	Small bass version of 800DV.
900PS	1	1	3	24	-	250	Preset mono with touch modulation bar.
PE1000	P	1	5	7	1	250	In fact, one key short of five octaves; two filters though.
PE2000	P	1	4	8	1	275	One key short of four octaves; phaser, mixable sounds.
M500 Micro/SP	1	1	2.5	30	-	200	Small preset/variable mono, noise; SP version has speaker.
MS10	1	1		-	-	250	Useful semi-patchable mono, audio in.
MS20*	1	2		-	-	550	Powerful semi-patchable mono pitch-to-voltage input, now replicated by Korg.
SQ10 Sequencer	-	-	-	-	-	350	3 × 12-step analog sequencer.
MS50	1	1	-	-	-	650	Keyboardless expander, like MS10 with added functions.
VC10 Vocoder	P	1	2.5	-	1	350	Keyboard vocoder with ensemble, white noise, microphone.

Model	P	O	K	M	FX	V	Information
MP4 MonoPoly	4	4	3	–	1	450	Powerful unison mode, thin poly mode with single filter.
PS3100	P	1	4	–	1	550	Semi-patchable, like large polyphonic MS20, with chorus.
PS3200	P	2	4	16	1	3k	Semi-patchable, remote keyboard, chorus, digital memories.
PS3300	P	3	4	–	–	4k	Three independent patchable poly synth sections, remote keyboard.
Lambda	P	3	4	8		150	String synth with phaser and chorus speed controlled by joystick.
Sigma	1	3	3	19	–	150	Preset/variable mono with aft, ring mod, two joysticks.
Delta	P	1	4	–	–	100	String synth with filter, white noise, joystick.
X911	1	1	–	8	–	250	Mono preset/variable guitar synth, pitch and CV/Gate ins, fuzz.
Trident	8	16	5	16	1	450	String, brass, polysynth, joystick, flanger, split.
Trident 2	8	16	5	32	1	650	Improved Trident with more memories and better editing.
Polysix	6	6	5	32	1	350	Arp, unison, chorus, phaser. Later with MIDI and 120 memories.
Poly 61/61M	6	12	5	64		250	Arp, no unison; MIDI on 61M.
Poly 800	8	8	4	64	1	200	Poly, seq, one filter, MIDI, chorus; some with inverse key colours.
EX800	8	8	–	64	1	150	Rackmount of Poly 800 retaining sequencer, MIDI.
Poly 800 Mk2	8	8	4	64		200	Delay replaces chorus, improved MIDI.
DW6000	6	12	5	64	1	150	Digital synth, 8 waves, res filters, MIDI.
DW8000	8	16	5	64	2	250	16 waveforms, delay, vel, aft, arp, res filters, MIDI.
EX8000*	8	16	–	64	2	250	Rack-mount of DW8000, no arp, MIDI zoning.
DSS1	8	16	5	64	3	250	12-bit sampler using floppy disk, delay, res filters.
DSM1	8	16		64		200	Rack-mount of DSS1, no delay, larger memory, multiple audio outs.
Moog							
I/IC/IP	1	2	5	–	–	4k	Two VCOs, 12 modules (16 on IC), two panels; P versions were portable.
II/IIC/IIP	1	5	5	–	–	8k	5 VCOs, 16 modules (28 on IIC), two panels.
III/IIIC/IIIP	1	13	5	–	–	15k	13 VCOs, 25 modules (39 on IIIC), three panels, optional 3 × 8 seqs.
10	1	2	5			4k	Simplified 1C with one VCA, five-octave keyboard.

(continued)

Classic instruments: specifications and values

Model	P	O	K	M	FX	V	Information
12	1	2	4			5k	Like 10 with four-octave duo keyboard, better VCOs, two VCAs.
15/15A	1	2	4	–	–	5k	Small integrated system; 15A version lacked filter banks.
35/35A	1	5	5	–	–	7k	Mid-sized integrated system; 35A version lacked filter banks.
55/55A	1	7	5	–	–	8k	Large integrated system; 55A version lacked sequencer.
Sonic 6	2	2	4	–	–	950	Duophonic synth in suitcase with speaker, ring mod, two LFOs.
Satellite	1	1	3	13	–	200	Preset organ-top design, also in Thomas and Cordovox organs.
Taurus	1	2	1	4	–	700	Powerful pedal bass synth, portamento and release buttons.
Taurus 2	1	2	1.5	–	–	500	Revised Taurus, a Rogue on a chrome pole; bend, mod wheels.
Liberation	P	2	3.5	–	–	650	Strap-on mono, simple poly section, aft, ring mod, remote PSU.
Minimoog*	1	3	3.5	–	–	2k	Classic portable mono performance synth, VCO3 acts as LFO, re-issued 2017.
Minitmoog	1	2	3	13	–	450	Rare improved Satellite with two VCOs, sync, touch sensitivity.
Opus 3	P	1	4		1	350	Strings/brass/organ, chorus, single filter, pitch bender.
Micromoog	1	1	2.5	–	–	400	Budget monosynth with sub-osc, pitch bend ribbon.
Multimoog	1	2	3.5	–	–	450	Two-VCO model, aft, sync, pitch bend ribbon, sync, S&H.
Source	1	2	3	16	–	550	Programmable, sync, arp, 2 × 88-note sequencer.
Prodigy	1	2	2.5	–	–	450	Budget mono, sync; most with CV/Gate and audio in.
Polymoog	P	2	6.5	8	–	900	Poly, split, vel, single VCF, preset/ variable, three-band filter, S&H.
Polymoog Kybd	P	2	6.5	14	–	700	Preset, most with the Vox Humana sound.
Memorymoog-	6	18	5	100	–	2k	Three VCOs per voice, sync, arp, unison; MIDI and seq on "+" models.
Rogue	1	2	2.5	–	–	400	Budget mono; sync, wheels on panel, noise, external PSU.
Realistic MG1	P	2	2.5			400	Budget mono with simple poly section, S&H.
Oberheim							
SEM	1	2	–	–	–	600	Keyboardless two-VCO expander with good interfacing, now rebooted.
2-Voice	2	4	3	–	–	750	Comprised two SEMs, often with dual eight-step analog sequencer.

Model	P	O	K	M	FX	V	Information
4-Voice	4	8	4	–	–	950	Comprised four independent SEMs, often with programmer.
8-Voice	8	16	4	–	–	2k	Some units with 2 × 4 or 2 × 5 octave keyboards and programmer.
6-/12-Voice	12	24	4	–	–	4k	Very few made, various keyboard configurations.
OB1	1	2	3	8	–	650	Powerful early programmable mono, CV/Gate, sync, S&H.
OBX	8	16	5	32	–	650	Stereo output, poly portamento, S&H, no MIDI.
OBsx	6	12	4	24	–	600	Preset poly, no MIDI, some with fewer voices or more memories.
OBXa	8	16	5	32	–	500	Updated OBX; different chips, split/layer, sequencer interface.
OB8	8	16	5	120	–	750	Usually MIDI; arp, split/layer, many portamento facilities.
Matrix 6/6R	6	12	5/-	100	–	550	Powerful MIDI programmable poly, and 19-inch rack-mount version.
Xk	–	–	5	–	–	150	Versatile compact MIDI master keyboard with arpeggiator.
Xpander*	6	12	–	100	–	2k	Keyboardless multi MIDI module (not 19-inch) with CV/Gate ins.
Matrix 12	12	24	5	100	–	3k	Doubled Xpander, MIDI, no CV/Gate, vel and (usually) aft.
Matrix 1000*	6	12	–	1k	–	400	1000-preset 1U 19-inch MIDI, edit with Matrix 6/6R or software.
OBMX	12	24	–	256	–	1k	Powerful prog poly MIDI multi-rack, some with fewer voices.

Octave

Model	P	O	K	M	FX	V	Information
Cat/SRM2	2	2	3	–	–	650	Duo like Odyssey, sub-octaves, noise, sync; SRM2 has more in/outs.
Kitten/Mk 2	1	1	3	–	–	350	Mono like ARP Axxe, sub-octaves, S&H; Mk 2 kybd digitally scanned.
Voyetra 8	8	16	5	100		800	Early MIDI 19-inch rack-mount, remote vel/aft keyboard, seq, arp.

OSC

Model	P	O	K	M	FX	V	Information
OSCar	2	2	3	36	–	2k	Duo, prog, arp, seq, MIDI (usually), various software versions.

PPG

Model	P	O	K	M	FX	V	Information
Wave 2	8	8	5	100	–	1k	Wavetable digital synth, resonant analog filters, seq.
Wave 2.2	8	16	5	200	–	1.5k	Adds MIDI, second bank of digital oscillators.
Wave 2.3	8	16	5	200	–	1.6k	Adds more memories and sequencer functions.

(continued)

Model	P	O	K	M	FX	V	Information
Roland							
SH1000	1	2	3	10	–	200	Preset/variable organ-top synth.
SH2000	1	1	3	30	–	350	Preset/semi-variable organ-top synth, aft.
SH1	1	1	2.5	–	–	350	Small variable synth, audio in with envelope follower.
SH2	1	2	3	–	–	350	Like SH1 with more keys, second VCO, only one envelope.
SH3/a*	1	1	3.5	–	–	375	Sub-oscillators, S&H, three LFOs, no Gate/CV ins.
SH09	1	1	2.5	–	–	350	Sub-osc, S&H, CV/Gate inputs.
SH5	1	2	3.5	–	–	750	Much extended SH3 adding CV and Gate ins.
SH7	2	2	3.5	–	–	950	Extended SH5, ring mod, CV/Gate in, S&H.
Saturn SA09	P	1	3.5	–	1	175	Compact piano/organ, chorus/vibrato, accent for key click.
RS101	P	1	5	3	1	150	Brass/string keyboard, split, chorus.
RS202	P	1	5	3	1	150	Like RS101, adds trigger out for synths, vibrato delay.
RS09	P	1	3.5	4	1	175	Compact organ and string synth, stereo outs, audio in for chorus.
101 Synth	1	1	3	–	–	750	Single VCO synth, part of semi-modular system.
102 Expander	1	1	–	–	–	750	Matching expander with S&H, ring mod.
104 Sequencer	–	–	–	–	–	550	Matching 2 × 12-step sequencer; speakers and mixer also made.
100M 190 rack	1	1	–	–	–	450	Three-mod rack, various combinations of VCOs, mono and poly kybds.
100M191/191J	1	2	–	–	–	650	Five-mod rack, "J" model had multiple socket section.
MRS2 Pro Mars	1	2	3	18	–	275	Programmable mono, CV/Gate in, sub-oscillator.
RS505	P	1	4	–	1	375	String/poly/bass synth, chorus, audio input.
SPV355	1	2	–	–	–	400	Rack-mount single VCO synth with pitch-to-voltage input.
VP330	P	1	4	–	1	850	Keyboard vocoder, strings and choir, chorus.
System 700	1	9	5	–	–	10k	Large modular system, usually with seq.
Jupiter 4	4	4	4	18	1	450	Four-voice polysynth, arpeggiator.
Jupiter 8	8	16	5	64	–	900	Poly, split, solo, unison, arp, porta, sync, most DCB, some MIDI.
TB303 Bassline	1	1	1	–	–	750	Bass synth/seq with CV/Gate out, glide, micro keyboard.
Jupiter 6*	6	12	5	48	1	650	MIDI split prog poly.
SH101*	1	1	2.5	–	–	450	Portable mono, battery power, sub-octaves, arp, seq.
MC202	1	1	2.5	–	–	350	Two-channel CV/Gate seq with single analog voice like 101.

Model	P	O	K	M	FX	V	Information
Juno 6/60	6	1	5	56	1	350	Six-voice poly, DCB except on early 6s; 60s programmable.
Juno 106/HS60	6	1	5	128	–	350	Six-voice poly, MIDI, portamento; speakers on HS60.
JX3P	6	12	5	64	1	350	Six-voice poly, MIDI, seq, PG200 editor option.
MKS30	6	12	–	64	1	350	Vel rack-mount of JX-3P, no seq, PG200 editor option.
JX8P	6	12	5	64	1	350	Vel/aft upgrade of JX-3P, patches nameable, PG800 editor option.
Super JX10*	12	24	6	64	1	550	Stereo outs, seq, sync, chorus, PG800 editor option.
MKS70	12	24	–	64	–	500	2U rack of JX-10, PG800 editor option.
MKS7 Super Qrt	6	1	–	100	–	250	2U MIDI rack, preset Juno bass/chords/melody, TR707 drums.
MKS80	8	16	–	128	–	500	2U MIDI rack-mount, vel and aft, crossmod, sync, two LFOs.
MPG80	–	–	–	–	–	300	Optional editor for MKS80, sends control changes by MIDI.
Alpha Juno 1	6	6	4	64	1	250	Compact prog DCO poly, MIDI.
HS10	6	6	4	64	1	250	Domestic version of Alpha Juno 1 with speakers.
Alpha Juno 2	6	6	5	64	1	250	MIDI prog poly, vel, aft, PG300 editor option.
HS80	6	6	5	64	1	250	Domestic version of Alpha Juno 2 with speakers.
MKS50	6	6	–	64	1	250	Rack-mount of Alpha Juno 2, PG300 editor option.
EM101	8	1	–	16	–	100	Piano top module with preset Juno poly and lead/bass sounds.
D50	8	16	5	64	3	250	Prog MIDI digital waveform synth, res filters, PG1000 editor.
D550	8	16	–	64	3	350	Rack-mount of D50, PG1000 editor option.
D110	8	16	–	128	3	150	Prog multi digital wave multi-out MIDI module, res filters.
MT32	8	16	–	128	3	120	Piano-top multi module, simplified D110.
JD800	8	16	5	64	3	850	Best all-round keyboard synth ever.
JD990	8	16	–	128	3	650	Module of JD800, less distinctive.
Sequential							
700 Programmer		–	–	64	–	150	Prog, dual LFO, dual ADSR memory unit for modulars.
800 Sequencer	–	–	–	–	–	150	Digital sequencer with CV/Gate outs.
Prophet 5	5	10	5	40	–	1.5k	Prog poly, many revisions, last with MIDI; polymod, unison.
Prophet Remote	–	–	4	–	–	250	Four-octave remote keyboard for Prophets

(continued)

Classic instruments: specifications and values

Model	P	O	K	M	FX	V	Information
Prophet 10	10	20	2x5	80	–	2.5k	Dual keyboard programmable poly, most with seq.
Pro One	1	2	3	–	–	650	Variable mono, two channels, 40-note seq, arp, many inputs.
Prophet 600*	6	12	5	100	–	650	Prog poly, MIDI, seq, arp, unison, polyglide, polymod.
Six-Trak	6	6	4	100	–	350	Prog multi poly, seq, arp, stack mode, MIDI.
Multi-Trak	6	6	5	100	1	400	Prog multi poly, split, vel, seq, arp, chorus, MIDI.
Max	6	6	4	100	–	250	Simplified Six-Trak, no wheels or stack mode, external PSU.
Split 8/Pro 8	8	8	5	64	1	450	US/Japanese revisions of Multi-Trak, split not multi, no seq.
Prelude	P	1	4	8	–	150	Mixable brass/piano/organ/strings from SIEL, like Orchestra 2.
Fugue	P	1	4	10	–	150	Renamed SIEL Cruise, semi-variable mono with simple poly.
Prophet T8	8	2	6.25	64	–	2k	Wooden weighted kybd vel/individual after, MIDI, seq.
Prophet VS	8	32	5	100	–	1.5k	MIDI digital poly, res filters, 128 waves, vel/aft, poly arp.
Prophet VS Rack	8	32		100		800	Rack-mount of VS "vector synthesis" instrument.
SIEL							
Orchestra	P	1	4	8	1	150	Simple brass/strings/reeds/piano, like ARP Quartet.
Cruise	P	1	4	24	1	200	Preset/variable mono synth and brass/reed/string/piano poly.
Mono	1	1	3	10	1	150	Preset, semi-variable mono with white noise.
Opera 6/DK600	6	12	5	99	–	250	Prog poly, most MIDI, vel, split; revised in Germany as "Kiwi".
Expander 6	6	12	–	99	–	200	Non-rack-mount MIDI expander version of Opera 6.
DK70	8	8	4	50	–	200	Seq, layers, shoulder strap, battery power, MIDI.
DK80/EX80	12	12	5	50	–	150	Seq, split, only two filters, MIDI; expander non-split, no seq.
Synton							
Syrinx*	1	2	3.5	–	–	1.2k	Variable mono with two filters, no progs or MIDI.
Teisco							
S60F	1	1	2.5	–	–	250	HPF, vibrato delay, rubber pad for pitch bend, CV/Gate in.
S60P	1	1	3	16	–	250	Preset model with aftertouch.
S100F	1	2	3	–	–	250	ADSR repeat, CV/Gate in, pitch mod by ADSR, VCF mod by VCO.

Model	P	O	K	M	FX	V	Information
S100P	1	2	3	32	3	200	Mono preset/variable, spring reverb and flanger, aft – see Kawai.
S110F	2	2	3	–	–	350	Expanded S60F, duo, fixed filter bank, three pads, CV/Gate in.
SX400	4	8	4	16	1	375	Prog poly, aft, mono CV/Gate in, chorus, no MIDI.
Yamaha							
SY1	1	2	3	27	–	250	Preset/variable organ top with aftertouch.
SY2	1	2	3	27	–	350	Slightly modified SY-1 in flight case.
CS50	4	4	4	13	1	800	Preset/variable poly, aft, chorus, res HPF, ring mod, audio in.
CS60	8	8	5	13	1	2k	Longer keyboard, pitch ribbon, one preset programmable.
CS80*	8	16	5	26	1	4k	Weighted key individual aft/vel, four presets prog, weight 100 kg.
CS70M	6	12	5	30	1	450	Prog poly, seq, unison, chorus, CV/ Gate out.
GX1/EX1	8	16	2x5	40	1	12k	Third mono kybd; pedals, drum unit, speakers; weight 387 kg.
CS5	1	1	3	–	–	150	Very simple one-VCO variable mono, pitch bend slider.
CS10	1	1	3	–	–	150	Similar to CS5 but with two envelopes.
CS30*/CS30L	1	2	3.5	–	–	450	Variable mono, eight-step seq, ring mod, ext audio in; L version omits seq.
CS15	2	2	3	–	–	250	Fully variable mono like two CS5s, two CV/Gate ins for duo.
CS15D	1	2	3	29	–	250	Layerable preset performance mono, not very similar to CS15.
CS20M	1	2	3	8	–	250	Large programmable mono, CV and Gate in.
CS40M	2	4	3.5	20	–	350	Large prog duo, ring mod.
SK10	P	1	4	–	1	150	Very limited organ/string/brass with chorus.
SK15	P	1	4	–	1	175	Strings/organ/poly.
SK20	P	1	5	–	1	200	Strings/organ/poly, split, tremolo, no memories.
SK30	P	2	5	–	1	250	Large strings/organ/poly/solo, split, no memories.
SK50D	P	2	2x5		1	650	Very large dual kybd strings/organ/bass/poly/solo, pedals option.
CS01/CS01MkII	1	1	2.5			150	Mini portable variable mono, speaker, battery power, noise, BC.
AN1X	8	2	5	128	3	550	Flexible VA synth capable of CS80 style sounds.
AN200	8	2	–	128	3	350	Stripped down module version of AN1X.

*A highly recommended purchase at the price shown or below!

Abbreviations

ADSR = attack/decay/sustain/release envelope; aft = aftertouch (pressure) sensitive; arp = arpeggiator; BC = breath control; CV = control voltage; DCB = digital communication buss (on Roland synths); DCO = digitally controlled oscillator; in = input; mod = modulation; mono = monophonic; kybd = keyboard; multi = multi-timbral; out = output; poly = polyphonic; PPC = proportional pitch control (on ARP synths); prog = programmable; PSU = power supply unit; res = resonant; seq = sequencer; S&H = sample-and-hold; sync = oscillator synchronisation; VCF = voltage-controlled filter; VCO = voltage-controlled oscillator; vel = velocity sensitive.

A note about values

Values given are for instruments in reasonable condition and represent a compromise between private sale prices (for example on eBay) and dealer prices, which are invariably higher. Extended numbers of voices or memories, software updates, or MIDI where not always fitted can add to these values. Collectors will sometimes pay even more for certain items regardless of their actual musical usefulness. The real usefulness of any polyphonic synthesizer, which lacks MIDI, or any monophonic synthesizer without CV and Gate inputs for control by a MIDI-to-CV converter, should be carefully considered before purchase. Re-issues and re-boots of instruments, such as the Minimoog, Korg MS20 (Mini), (Korg) ARP Odyssey, and so on, may have slightly driven down second user prices for the originals. These tend to recover as the genuine originals become increasingly difficult to find in good working order.

Not covered

The following instruments are not included as they are unlikely to be seen, except through very specialised outlets, or for other reasons indicated:

- *Very rare older modular systems*: Aries, Boehm, Delta Music Research, E-Mu, PPG, Polyfusion, Serge, and Synton.
- *Rare, but not much sought after*: Firstman, Hillwood, Pulser, Multivox, Solton, Baldwin, Eko, and Hammond synthesizers.
- *Kits*: Digisound, Dewtron, Elektor/Formant, ETI/Transcendent, Maplin, PE/Practical Electronics, and (usually) PAiA.
- *Very rare*: older Buchla, EML, Gleeman, Musonics, RSF, Steiner-Parker, Syntec, Davoli, Rhodes Chroma, and Polaris.
- *Not really analog synthesizers*: Mellotrons, all electromechanical and electronic pianos, organs, many string and brass synthesizers, most digital synthesizers and sampling keyboards even if provided with filters for example the Ensoniq Mirage, most sequencers, most remote and master keyboards, and drum machines including sampling drum machines.

Appendix B

Analog and virtual analog instruments: currently or recently in production

Current or recently discontinued, relatively mass-market true analog and virtual analog instruments likely to be seen in stores or prominently in second-hand listings. Many excellent product lines, such as Trax Controls and Harpur Instruments in the UK, Modal with their 002 synth, the John Bowen synth in the USA, the Schmidt in Germany, and so on, are not included because they are more or less boutique productions or sold direct to users. Some pioneering companies, such as Quasimidi and Spectral Audio, are definitively closed; some such as Waldorf appeared to close for a while but came back. No attempt has been made to include manufacturers solely of modular systems as there are more than 300 in Eurorack alone, plus many more working in other module formats, but there is an extensive list of Eurorack manufacturers on the book website: www.routledge.com/cw/jenkins

ACCESS

- **Virus c/Virus kc** Twenty-four-voice, 16-part desktop/rack-mount virtual analog synth, effects, audio inputs, vocoder and five-octave keyboard version.
- **Virus Classic** More affordable desktop/rack-mount model roughly equivalent to the older Virus b.
- **Virus Rack** One-unit, 19-inch module version with limited controllers. Still in production.
- **Virus Indigo** Three-octave keyboard version of Virus c.
- **Virus TI Keyboard** Total Integration version with software version running in parallel.
- **Virus TI Rack** Desktop/rack-mount version of TI.
- **Virus TI Polar** Three-octave version of TI with white styling.

ALESIS

- **A6 Andromeda** Five-octave, 16-voice, 16-part multi-timbral analog synthesizer; two oscillators with sub-oscillators and five waveforms, three LFOs and three seven-stage envelopes, and two multi-mode filters per voice.

- **Ion** Compact 16-voice virtual analog synth with dual filters and effects, powerful and flexible, so a good introduction to virtual analog sound.
- **Micron** 3-octave, compact version of Ion with arpeggiator and effects but limited controllers.

None now in production. Alesis now markets home keyboards, pianos and piano workstations.

ARTURIA

- **Microbrute, Minibrute 2 and desktop version** Compact mono synths with some modular possibilities, either one a great introduction to analog synthesis.
- **Matrixbrute** Large modular-like mono synth for the advanced user.
- **Rackbrute** Powered cabinet in two sizes to mount Eurorack modules above a keyboard.

BEHRINGER

- **Deepmind 12 and desktop, Deepmind 6** Behringer's first entries into the synth field, along the lines of classic Roland Juno instruments.
- **Model D** Paradigm shifting Eurorack compatible clone of the Minimoog at a bargain price.
- **Neutron** similarly price-busting semi-modular desktop Roland System 100-style modules, Oberheim style keyboard polysynth, EMS Synthi A clone and others imminent?

CLAVIA

- **Nord Lead III/Nord Rack** Twelve-voice virtual analog synth and rack-mount version.
- **Nord Modular/Rack** Eight-voice PC- or Mac-programmable modular synthesizer emulator and rack-mount version.
- **Micro Modular** Four-voice desktop version of Nord Modular. Discontinued – Mac OSX version of editing software is a beta test version but works well.
- **G3 Modular** Current models of the Nord Modular, keyboard and rack versions.
- **G3 Modular Engine** Nineteen-inch mount module without controllers, for use with computer.

All now discontinued

- **Nord Lead Mk 4** Current model still in production.
- **Nord Electro** Keyboard with sounds of electromechanical instruments.

- **Nord Stage** Sounds of electromechanical instruments plus simplified synth engine.

DAVE SMITH INSTRUMENTS (SEQUENTIAL)

- **Prophet** and **Evolver** synth lines, Prophet X which has a huge sample memory is the latest, also Oberheim's **OB6** and at the time of writing three **DS Modular** Eurorack modules.

ELEKTRON

- **Analog Keys** currently shows as discontinued though still in many stores. The **Analog Four Mk2** desktop synth is still in production though, at around 1200 pounds.

FUTURERETRO

- **FutureRetro** 777 Desktop/rack mono analog two-VCO synth with sequencer.
- **Mobius** Two-unit rack sequencer based on the 777's sequencer.
- **Revolution** Radical design circular bass synth/sequencer.

All discontinued. **Zillion** and **Orb** sequencers running out at the time of writing, 512 Touch Keyboard available. The company appears to be transitioning to Eurorack designs.

JOMOX

- **AirBase99** Rack-mount 1U MIDI analog percussion sound module.
- **SunSyn** Desktop eight-voice, eight-part virtual analog synthesizer. Not in production. Jomox now manufactures analog drum machines and some Eurorack modules.
- **T-Rackonizer** (£273.00) Eurorack filter with delay and reverb.

KORG

- **Electribe EA1** Simple desktop monophonic virtual analog synth with sequencer. Various Electribe synth, sampler and sequencer models appearing subsequently.

- **MS2000/R** and **MS2000B/BR** Three-and-a-half-octave, four-voice virtual analog polyphonic keyboard, sequencer, arpeggiator, vocoder and its Mk2 and rackmount versions.
- **Legacy** Initially a 5/8ths sized MS20-style USB controller keyboard with emulation software. Now various software and app packages including MS20, Polysix, Wavestation and Korg M1.
- **RADIAS** Superseding the MS2000BR, a physically similar rack module with virtual analog sounds and vocoder, slotting into an optional keyboard.
- **MicroX/X50** From the Trinity/Triton series, so including many analog-style sounds plus physical controllers for internal sounds or soft synths, two and five octaves respectively.
- **RK100S** keytar virtual analog synth in black, white or red now discontinued and instantly became collectable.

All now out of production.

- **MicroKorg** Three-octave mini-key virtual analog with battery power options, effects, vocoder, same sound engine as MS2000. Still in production in 2018.
- **Volca** Small desktop mono synth, drum machine, sampler and mixer.
- **Monotron, 2-Voice, Monotron delay** Small battery operated pocket size synths.
- **MiniKorg** 5/8ths size reproduction of MS20 synth.
- **ARP Odyssey** 5/8ths or full size reproductions of the Odyssey with a choice of 3 filter types.
- **MonoLogue. MiniLogue, ProLogue** Battery operable monosynth, small four-voice synth and full size upmarket eight- or 16-voice analog polysynth.

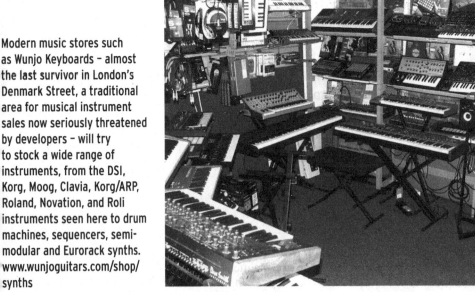

Modern music stores such as Wunjo Keyboards – almost the last survivor in London's Denmark Street, a traditional area for musical instrument sales now seriously threatened by developers – will try to stock a wide range of instruments, from the DSI, Korg, Moog, Clavia, Korg/ARP, Roland, Novation, and Roli instruments seen here to drum machines, sequencers, semi-modular and Eurorack synths. www.wunjoguitars.com/shop/synths

MAM

- **DRM1** Eight-channel analog percusssion generator.
- **DRM2** Single-channel analog percussion generator.
- **ACM2** Handclap analog percussion generator.
- **SQ16** Step-time MIDI sequencer for TB303 and TR808/909 type patterns.
- **Warp 9 MIDI Space Filter** One-unit, 32-memory filter

All discontinued.

- **MAM MB33** Re-issued in 2015 in desktop form rather than the original rack, a simple TB303 Bassline emulator. Designer Stefan Schmidt went on to design the considerably more complex Schmidt Synthesizer.

MOOG

- **Moogerfooger, Minifooger Series**. Low pass filter, ring modulator, analog delay, phaser and MURF stepping filter pedals, a CP251 Control Processor to interface these and add LFO, sample-and-hold and more, and an input interface for the Voyager.
- **Voyager V3 and rack version** Monophonic three-and-a-half-octave, three-oscillator programmable MIDI synth with X-Y controller pad. XL version incorporated the CP251.
- **Little Phatty, Stage Phatty, Sub Phatty** Simplified keyboards (alongside a **Slim Phatty** module) with fewer front-panel controls, no X-Y pad and a shorter keyboard, but a typical fat Moog sound.
- **Sub37, Subsequent 37** Two-voice paraphonic programmable synths.
- **Minimoog, Modular System 15/35/55/Sequencer** Re-issues – very limited numbers.
- **Mother 32, DFAM, Minitaur** Synth and percussion modules and desktop bass module.
- **Grandmother** Semi-modular keyboard synth with springline reverb.
- **Moog One** Polyphonic synthesizer with eight or 16 voices.

In time for NAMM 2019, Moog previewed the Moog One, and impressive eight- or 16-voice polyphonic synth with clear influences both from the Voyager and the Memorymoog. With retail prices of 6000 and 7500 dollars or pounds respectively, clearly an upmarket instrument boasting a Fatar velocity and afterttouch sensitive 61-note keyboard, three-part playing, dual filters and Eventide effects.

NOVATION

- **Super Bass Station** Monophonic two-DCO, 1U programmable rack-mount synth, arpeggiator.
- **Supernova/Pro** Rack-mount virtual analog synth, effects.
- **Nova** Twelve-voice desktop/rack-mount virtual analog synth, effects, vocoder.
- **Supernova II/Pro/Pro X** Five-octave virtual analog keyboard, effects, arpeggiator.
- **Nova II/X/XL** Four-octave, 12-voice virtual analog keyboard, effects.
- **KS5/4, K-Station, KS Rack, A-Station** Five-, four-, two-octave, desktop/rack-mount and 1U rack VA synths with effects and arpeggiator.
- **ReMote 25 (updated to X-Station)** Two-octave VA synth with USB audio interface and MIDI control functions.
- **XioSynth** Two- and four-octave compact virtual analog synths with USB interface, audio inputs and MIDI controller functions.

All discontinued

- **Bass Station 2** Compact keyboard mono synth.
- **Peak** Powerful desktop polyphonic synth.

Many groovebox and controller products including **Mono Station** with a built-in synth.

OBERHEIM

- **OB12** Twelve-voice, two oscillator per voice, four-octave, four-part multi-timbral synth with S/PDIF digital audio out produced in collaboration with Viscount in Italy. Striking blue panel finish. A good second-hand purchase at around £600, though the LCD backlight regularly fails.
- **TVS, SEM** The two-voice keyboard and desktop module newly re-issued in various versions.
- **OB6** Powerful modern polysynth created in collaboration with Dave Smith at DSI.

QUASIMIDI (NOW CLOSED)

- **Raven/Raven Max** Virtual analog techno synth with front-panel controllers and sequencer based on that of the Cyber 6 master keyboard.
- **Sirius** Polyphonic virtual analog keyboard workstation, sequencer, effects, drum sounds, vocoder.
- **Rave-O-Lution 309/309 Groove-X** Monophonic rack-mount virtual analog synth with sequencer, drum sounds and a black limited edition version signed by Klaus Schulze with expansions fitted.

- **Polymorph** Sixteen-voice, four-part rack-mount virtual analog synth, sequencer, effects.

RADIKAL TECHNOLOGIES

Spectralis and **Accelerator** digital polyphonic synths currently in production. **Delta CEP A** which is a digital desktop or rack synth with analog style controls just about to appear. Three Eurorack modules also in production.

Joerg Schaaf of Radikal Technologies in his studio.

ROLAND

- **JP8000** Four-octave splittable layerable keyboard with RPS, effects, arpeggiator.
- **JP8080** Six-unit rack-mount splittable, layerable, ten-voice virtual analog synth with arpeggiator, effects, vocoder.
- **Juno-D** Lightweight entry-level samploid synth, digital resonant filter, effects, arpeggiator, D-Beam controller.
- **Juno-G** Styled after the Juno 106 but based on the FANTOM samploid workstation.
- **VP770** Polyphonic vocoder using timestretch/formant shift technology.
- **SH201** VA synth with basic display and effects.

All discontinued.

- **Boutique Series** desktop modules with an optional small keyboard replicating the Vocoder Plus, Bassline, JX3P, Jupiter 8, TR909 drum machine and more, and a Minimoog clone the SE02 co-designed with Studio Electronics.
- **System 500** Eurorack system of 5 modules descended from System 100, 100m and 700.

- **Aira** System including a "plug-out" reprogrammable mono synth and module version, Touch Bassline, and more. Related to **System 8** large polyphonic "plug-out" keyboard synth.
- **JDXi, JDXa** Mini and full size flexible analog/digital poly synths with sequencers.

SPECTRAL AUDIO (NOW CLOSED)

- **Pro Tone** Rack-mount monophonic non-programmable MIDI two-VCO synth.
- **Neptune/Neptune 2** Rackmount mono non-programmable MIDI 2-VCO synth with fuzzer.
- **Cyclus 3 Sequencer** Released early 2006, apparently the Swiss company's final product.

STUDIO ELECTRONICS

- **19" Rackmount** Omega 4–8, C.O.D.E. 4–8
- **Desktop** Boomstar 4075, 5089, SEM, SE80, SE-02.
- **Eurorack** Tonestar2600, Tonestar8106 complete synth voices.
- **Boomstar Modular Eurorack** 3003, 4075, 5089, 8106 (filters), Amp, BBox; Charcot
- Circles, Grainy Clampit, Levels, LFO 2, MIDI 3, Multiple, Oscillation, Outs, Quadnic, Sci Fi, SE88, SEM, Shapers, Slimo, STE. 16, Utility Modules: Attenulag, Mix4, Router, VCA2.
- **Boomstar Modular Eurorack Systems** Modstar Sensei Hybrid,
- Modstar Sensei Analog, Modstar Seito Rising.

STUDIOLOGIC

- **Sledge V2** Keyboard synth in orange or black with black keys finishes. Sound engine from Waldorf and a terrifically versatile and great value for money synth.

WALDORF

- **Pulse** Three-unit mono rack-mount programmable synth, arpeggiator; Pulse + version with CV/Gate and Audio Input. Now on **Pulse 2** version.
- **Q-Synth** Five- and three-octave version VA and wavetable synth with effects.

- **Micro Q/Micro Q Omega** Two-unit rack version of Q in blue or yellow styling.

All now discontinued

- **Blofeld/Blofeld Desktop** Simple, powerful analog imitative synth in black or silver, keyboard or desktop versions.
- **KB37** Keyboard for mounting Eurorack module.
- **Waldorf Eurorack** Complete range of modules.
- **Streichfett, Rocket, 2-Pole** Desktop string synth, mono synth and filter.
- **Quantum** Upmarket 5-octave keyboard synth.
- **STVC** Keyboard reproduction of Roland VP330 Vocoder Plus.
- **Nave, PPGWave, Lector, etc** Software and app packages.

YAMAHA

- **ReFace CS** A CS80 synth replicant in mini 3 octave format. Other ReFace models are for DX7, YC organs, CP electric pianos.

Yamaha's ReFace CS – the huge CS80 synth replicated in a compact new format.

CURRENT MODULAR SYSTEMS 2018

The first edition of this book was able to list a couple of dozen active manufacturers of analog modules. Now at the time of writing there are more than 300 manufacturers in the Eurorack field alone, with around 6,500 products on offer between them and a few dozen more making Moog, Modcan, Serge, Blacet, MOTM, FracRac, and other formats.

This makes modular synthesis one of the fastest growing areas of sound creation, though it remains one for the more committed user, given the necessity of choosing the components of a system, arranging cabinets and power, patching modules together, and so on.

Many users of modular systems create their own cabinets and install ready-made power systems, while others interface rackmount modules to pedals and stand-alone instruments, such as the Korg Volca and Roland Boutique series. It's a field full of possibilities and one which will no doubt expand in unpredictable directions in coming years.

Since a full list of modular manufacturers with contact details would take up around eight pages of this book, it has been transferred to the book's website www.routledge.com/cw/jenkins and will be updated occasionally.

Appendix C

Purchasing guide for analog instruments

Purchasing classic analog instruments can be a difficult business, there are several important factors to be taken into account. Failure to attend to these can result in the purchaser ending up with an expensive instrument that does not meet his or her requirements, may not be working well enough to re-sell, and is difficult, costly, or impossible to service.

The first important decision is whether the instrument is being purchased for serious musical use or simply as a collector's item or museum piece, or maybe for some combination of the two, say, to look great on stage, even though it's not really doing very much. If the instrument is filling a serious musical need on stage or in the studio, and particularly if it is going to be used in a studio, hire company or other commercial situation by third parties, then it obviously needs to be in the best possible condition. This may not be the same as the best cosmetic condition, obviously, an instrument can play very well despite having a scratched or battered cabinet, although you would have to wonder what effect the scratching or battering process was having on the internal electronics. But there are ways of checking out the probable performance and reliability of an instrument before purchase, which we'll look at later.

Some classic analog instruments – such as this never released prototype Model B Minimoog – demand preservation in a museum rather than risking exposure to stage conditions.

If, like some collectors, you simply love (for instance) Roland keyboards, already have an SH1, SH2, SH3, SH3a, SH5, and SH7 that you never play, and simply want an SH09 to complete the collection, then you may or may not be very concerned as to whether every switch, knob, and slider works perfectly. You may be more concerned about the cosmetic appearance of the cabinet and willing to pay more for an instrument which looks unused but does not actually work than for one which looks extremely battered but which plays perfectly.

Often, analog synthesizers are used either for processing external sounds through their filters, or for control from an external MIDI-to-CV converter. If you are sure that your applications for an instrument are limited to these, then you could get a bargain price on an instrument on which the oscillators were not working, or on which the oscillators were working but the keyboard was dead, or had a few dead keys.

It's important to realise that some older instruments are relatively easy to service and modify because of the simplicity of their design, while others of the same period that appear equally accessible can actually be a bad bet, because they use custom chips and other electronic or mechanical components that are no longer available. Instruments that mix analog and digital technology can be particularly frustrating because faults that appear straightforward can actually turn out to require the replacement of a digital processor, which in many cases will not be available. Even simple cosmetic repairs can sometimes be difficult; finding the exact style of slider cap to replace missing ones can often be impossible, for example.

CHECKING TECHNIQUES

Even if the possible purchase is at a very low price, it is usually best to check various aspects of any instrument you are offered in an extremely systematic way. The first step, of course, is to switch on the power; just because the power light (which may be built into the power switch) does not come on does not mean the instrument is not working because the small bulbs used for power lights early on were rather unreliable. If an instrument does appear completely dead there may be a fault in its external power supply if it has one, or it may simply have a cracked or burned-out internal fuse and will work perfectly once this is replaced. There is nothing more satisfying than purchasing an instrument at the price of one that is not working and finding that a new fuse costing a few pennies is all that's needed to bring it back to perfect working order. However, there's nothing more frustrating than assuming that this is going to be the case, then finding that an entire new power supply or processor board is required.

Once power is applied, on keyboard instruments, start by checking the keyboard itself. Most keyboards are designed using one of two systems: either key contacts or more commonly later on, conductive rubber cells. Key contacts are flexible lengths of wire that are moved from one conductive rail to another by the movement of each key; the completion

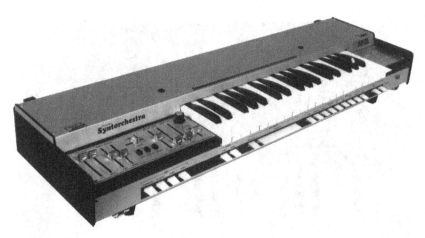

Farfisa's Syntorchestra – an Italian-built typical second-hand find that could reward careful examination.

of the circuit on the destination conductive rail plays the appropriate note. If the key contact is bent or absent it can fail to complete the circuit and that note will fail to sound; others on either side of it may continue to work perfectly well. Being dropped or knocked can misalign a few key contacts on an instrument; on old or very frequently played instruments they can simply wear out, becoming either too soft or too brittle and snapping. Dirt on a key contact can also prevent the circuit from being completed and the result will be the same even if the contact is more or less in place; sometimes the note will work intermittently, depending on how firmly or from what angle the key is played.

In more complex instruments there can be two or more sets of key contacts for each note. These either trigger off different layers of sound (such as harpsichord and strings in the Elka Rhapsody) or two independent monophonic voices (as in some duophonic synthesizer designs). It's quite possible for one set to be misaligned while the other set works well. So, for instance, you could get an instrument that plays strings from every key but has several keys on which the harpsichord layer does not sound.

Rubber cells in the form of a small "bubble" containing a conductive pad, which is pushed down onto two contacts on a circuit board to complete a circuit between them, were introduced in an attempt to solve the problems of wire key contacts but have different problems of their own. Firstly, the rubber cells will never be obviously bent or completely missing, so it's not so easy to see which keys may be problematic. Secondly, they are subject to simply wearing out with age and repeated use, and to the entry of dirt and grease. Thirdly, they are extremely dependent on precise positioning, so a tiny shift of the rubber cell, which cannot even be seen by the person trying to analyse a fault, can cause it to stop working. Fourthly, they are usually moulded in long strips, so to replace just one or two can be impossible, while replacement strips of cells can often be unobtainable. Instruments using this system include the Roland SH101 and the Casio CZ101.

A simple monophonic synth such as the ETI Transcendent 2000 (shown here and opposite) can have a large and fragile circuit board.

To test the keyboard, select a short, simple sound and play every single key. After doing this with an average touch, try it again with a much softer touch; sometimes keys that are working only intermittently will then show up.

If the keyboard is aftertouch sensitive, select a sound on which this is obvious and play a few notes. Aftertouch can be implemented in various ways: on the EMS Polysynthi and PPG Wave, the whole keyboard bends downwards, so there is relatively little to go wrong. On other designs such as the Oberheim Xk MIDI controller, the keyboard sits on a long dual conductive strip, and a very small downward movement applies pressure to this and generates a voltage for creating vibrato, filter opening or other parameter changes. The dual conductive strip is often covered with felt or another fabric; this can become permanently compressed or the dual conductive strip can move slightly. In either case the aftertouch facility can be lost. On many older instruments, a great deal of pressure is needed to have any noticeable effect, certainly more than the average player could comfortably apply during the course of a performance. Unfortunately, while this can be due to one of the problems previously mentioned it can also be due to poor design from the outset. Keep in mind that on most polyphonic instruments with aftertouch, applying pressure on just one note will affect all the notes currently sounding. On only a few instruments, such as the Yamaha CS80, Prophet T8, and Ensoniq SQ80, the aftertouch facility operates independently on every key, so it should be possible, for instance, to induce vibrato on a high melody note without adding any vibrato to the notes of a chord played simultaneously with the other hand.

Velocity sensitivity is also implemented in various ways, some involving a second set of key contacts, some based on a digital analysis of the speed of movement of the first set. In the former case, individual keys can experience problems that can often be fixed easily enough; in the latter case, if there is no velocity response at all on the instrument, this is usually the sign of a more serious problem.

Having checked the keyboard, it's then best to check the front-panel controls. On preset synths, select every sound and play, if briefly. On programmable synths, call up a few different sounds and, if there is a keypad, check that every digit is working. It is possible to find that some memories are fine while others contain garbage or no sound at all; usually, this is just a matter of corrupted data rather than of memories actually being defective, and this can be solved by reloading from cassette, floppy disk or MIDI. The internet is a good source of original and additional sound banks for loading via MIDI, but data cassettes for instruments that can only be reloaded in this way can be very hard to find, the end result usually being an appeal to an existing user of the instrument to make one up specially.

The question of problems with memory back-up batteries for programmable synthesizers has become a serious one. These batteries (usually of the lithium type) were designed to last five or perhaps ten years, so almost all synthesizers designed in the 1980s and 1990s are now experiencing problems. Unfortunately, these batteries have not always been made easy to replace (perhaps the manufacturers thought their products would be out of fashion in less time than the batteries took to run down) and without a working back-up battery the synthesizer has to be reprogrammed every time it is switched off – an impossibly impractical task.

Suffice it to say that the memory back-up battery is usually easy enough to identify once an instrument has its cabinet opened, and is usually easy enough to remove, although this sometimes involves a little de-soldering. It is seldom too specialised to replace, since many types of battery designed for cameras and hearing aids are widely available. On some instruments, such as some of the early PPG Wave models, the memory back-up battery is rechargeable; it very rarely needs to be replaced but if the instrument is not powered up every few weeks it will go flat and the instrument will lose its memories and will need to be reloaded. These rechargeable memory back-up batteries should not be replaced with non-rechargeable lithium types without a corresponding change to the circuit because they are under constant charge whenever the instrument is switched on, and a non-rechargeable replacement may explode.

There is one further group of problems with memories that can become extremely confusing. Some manufacturers of portable instruments that could use an external power transformer or battery power provided programmable memories, but did not see fit to build in a lithium or other battery to back them up. In these cases (on Korg Poly 800s and some Casios, for instance) the instruments had to contain batteries even when operated from an external power transformer because these batteries would back up the memories. Allow them to go flat, or remove them because an external power transformer was in use, and the result would usually be a completely mute instrument, although the memories were generally retained for the few minutes it took to change a set of batteries. To be frank, there may be opportunities for a bargain to be found here from sellers who do not realise that a set of batteries is all that is needed to fix an apparently insoluble memory problem.

On programmable synthesizers (and certainly on modular systems) there may be as many as 80, 90, or more switches and sliding or rotary controls that need to be checked. The problems that arise with these are obvious enough; they can wear out and become intermittent with age, become affected by dust or dirt or, particularly in the case of sliders, be physically snapped off. It has to be up to the individual purchaser to decide how forgiving to be in these situations. Obviously, a slightly intermittent switch to select an inverted mode for the second envelope is going to be much less significant than a completely non-functional oscillator level control, which seriously limits the functionality of the instrument.

As far as possible, check every slider, rotary, and switch, paying particular attention to the highest and lowest setting of each one, which, particularly on sliders, are the most likely positions to be affected by a build-up of dust or dirt. On output volume controls in particular, a buildup of dirt is likely to lead to crackling when changing the volume, which may be acceptable in a studio situation, but can be very disturbing on stage. Where sliders are completely snapped off, it's often possible to continue to use the control with a fingertip, and while not ideal, it's a good reason for negotiating a lower price while not actually limiting very much the enjoyment to be gained from the instrument.

Having checked all the keys and controls of an instrument, it would be helpful to be able to check all the outputs and inputs, including control voltage and gate inputs. Unfortunately, this is not always practical, because the seller may not have any suitable source of control voltages available for testing. But it's fair to say that if the keyboard works on an instrument, it's unlikely that the external control voltage and gate inputs will be problematic, and other inputs such as audio inputs, foot pedal inputs and footswitch connections rarely exhibit faults.

Looking for loose connections: the author's early equipment setup in 1980 as heard on the *Analog Archives* CD. TASCAM 4-track, Powertran Transcendent 2000 synthesizer, Electro-Harmonix Vocoder, Clef Master Rhythm drum machine, Practical Electronics analog sequencer, Park reverb mixer, Moog Sonic 6, Welson string ensemble, EDP Wasp, EDP Spider sequencer, custom footswitch transposer, Oberheim Mini Sequencer, guitar and pedals.

REPAIR TECHNIQUES

It's extremely satisfying to find that many apparent defects, particularly on early synthesizers, can be repaired quickly and easily. Key contacts are often accessible through a removable bottom panel; an apparently dead key can sometimes be repaired in seconds simply by bending a key contact back into place. Intermittent keys are often improved by cleaning the key contacts. As they are too delicate to actually wipe this is usually done by spraying on a dirt-dissolving aerosol switch cleaner, such as EML, available from all electrical stores. Some recommend WD40 but this is really more suitable as a lubricant for moving mechanical parts, as it is flammable and has a rather distinctive odour.

When key contacts are actually broken or missing, the situation becomes much more difficult. It is possible to buy conductive wire, again from electrical stores, but cutting and fitting a piece that exactly matches the other key contacts can be difficult. When multiple key contacts are involved the difficulties are increased; sometimes key contacts are simply soldered to a contact strip, but when they are individually supported by small plastic blocks, these are hard to replace if cracked or broken. Rubber cell-based key contacts, as mentioned earlier, are hard to replace; conductive rubber is readily available, but cutting, shaping, and fixing it down in order to dependably replace defective rubber cells is extremely difficult.

Front-panel keypads sometimes use conductive rubber as well. These are hard to replace if cracked or worn. Often, they are simply dirty but are still difficult to remove, clean, and replace with any degree of success. Membrane switches or touch keyboards that have worn down with long use will now be impossible to replace (for example, the touch keyboard on an EDP Wasp or EMS Synthi Aks, or the membrane switches on the

Moog Source or Sequential Prophet 600, though modern replacements for the latter pair have been known to appear on eBay).

Rotary controls, switches, and sliders suffering from dirt can usually be improved using aerosol switch cleaner without having to take the instrument apart; simply spray switch cleaner into the affected controls (with the power off) and move them repeatedly to spread it and to clean up the internal contacts. If rotary controls are snapped off they are usually easy to replace; so-called rotary "potentiometers" (variable resistors) have not changed much in construction, size, or available values since the earliest synthesizer designs. Rotary switches, such as those to select oscillator octaves, can be slightly more problematic as can some small types of two- or three-position switches.

Snapped sliders are a much more annoying problem. For some reason, small sliders, which have been widely used particularly on Japanese instruments, now seem impossible to obtain consistently in Europe. Roland, ARP, and many other instrument designs are frequently found suffering from broken sliders that cannot be easily replaced. Attempting to glue new shafts onto slider potentiometers is seldom successful; the area available to apply adhesive is simply too small to make a join strong enough to withstand everyday use. In addition, the type of slider caps used by companies such as ARP and Roland in the 1970s now seems impossible to exactly match. On some synthesizers, such as the Roland SH101, some values of sliders are still available through Roland service centres, while others of different values are not.

To be fair, some companies, such as Roland, have tried as hard as possible to keep even the older components available. Items such as envelope chips for early models, for example the Roland Juno 6, can still be obtained from major service centres at very modest cost. But carrying out more major repairs to circuitry is beyond the scope of this book; on early instruments, such as the Minimoog, not only are all the components discrete (that is, comprising individual transistors, resistors and capacitors), they are in many cases extremely common and should be very easy to replace. The problem is that circuit diagrams are often not available. Even good service engineers can be reluctant to take on a repair that involves identifying a defective component without reference to a circuit diagram. Also, many early instruments (notably those from ARP) concealed part of their circuitry (usually the filter) within blocks of resin, making them difficult or impossible to service.

Later instruments that used integrated circuits (chips) simplified this situation; if a synthesizer has gone silent and there is one IC near the audio output, it's a good bet that this is part of the voltage-controlled amplifier that controls the output and may need replacement. With luck, it will be in an IC socket and can simply be pulled out; if not, a special soldering iron head can help in de-soldering all the eight or 16 pins of the IC simultaneously prior to removal. However, some replacement ICs can be simply impossible to replace; the Curtis and SSC oscillator, amplifier, and filter chips used in many Sequential products are now very rare and expensive. Regrettably, a fine instrument that could be repaired in a few minutes given the right chip can become an impractical job simply because the right chip is no longer obtainable.

This situation also applies in great measure to Yamaha instruments. Yamaha very early on became involved in manufacturing their own ICs, which were not available elsewhere. Problems with these on instruments, such as those from the CS range, are often totally impractical to repair. In addition, the amount of hand-wired circuitry involved in some instruments, such as the Yamaha CS80 and YC45D organ, made them extremely demanding to repair even when they were current, technicians being referred to special sets of service manuals and even to special training courses in order to master the complexities of the instruments. The chance of finding an appropriately trained technician now is extremely low.

MODIFICATIONS

For the purposes of modernising classic instruments to integrate them into current studio set-ups, they can be roughly divided into four groups: polyphonic instruments with MIDI, polyphonics with no external control inputs or outmoded interfaces, monophonics with control voltage and gate inputs, and monophonics with no external control inputs.

Polyphonic instruments with MIDI are seldom problematic. Some, such as the Roland JX3P and Sequential Prophet 600, had rather simple implementations of MIDI at first, with software updates sometimes becoming available later. Instruments fitted with early software versions are generally difficult to improve any further now, but any problems (such as only responding to a limited choice of MIDI channels) can sometimes be overcome using an external MIDI processor, or a multioutput MIDI interface from your computer with one output assigned only to the problematic synth. Sometimes third parties made software revisions available, such as the GIGLI chip for the Sequential Prophet 600.

Some instruments, such as the Elka Synthex, OSC OSCar, and Korg Poly 61, had MIDI added during their lifetime (the Poly 61 becoming the Poly 61M), while some Roland instruments, such as the Juno 6 and 60 and the Jupiter 8, were initially usually fitted with DCB (Digital Communication Bus). Roland made a (now rare) DCB-to-MIDI converter unit, Kenton in the UK made a DCB interface, and later Jupiter 8s, for instance, were MIDI equipped instead, so DCB instruments are relatively easy to equip with MIDI. There are several specialist companies that can carry this out and can also add MIDI (with a varying number of features) to many other processor-controlled instruments. Ultimately, almost any keyboard instrument (including everything from Yamaha CS80s to acoustic pianos) can be MIDI-equipped, if necessary by adding a whole new set of digitally scanned key contacts, but this can be a very expensive process. Adding Moog's PianoBar was also a simple external (though quite costly) method of adding MIDI to an acoustic piano or other keyboard.

The release of several extremely powerful, flexible programmable polyphonic synthesizers just predated the introduction of MIDI; the Oberheim OBX, Sequential Prophet 5, Korg Polysix, and several others

Moog's Taurus 2 remains highly sought after. Expect to pay even more for the original floor standing model or for the recent modernised Taurus 3.

come to mind. Before taking on any of these instruments, the cost of adding MIDI rather than looking for a later version or model with MIDI already fitted should be carefully considered. Adding MIDI could, for instance, easily double the cost of a Korg Polysix bought on the second-hand market. Some instruments do not, in any case, benefit fully from being MIDI equipped: a Yamaha CS80 can have MIDI Out added fairly easily, but adding MIDI In is difficult and would ignore all the benefits of playing the CS80's responsive, weighted, individual-note aftertouch keyboard.

There are very few classic MIDI monophonic instruments (later OSC OSCars being the most obvious example), but earlier monophonics, whether programmable or not, were usually equipped with control voltage and gate inputs. These are pleasingly easy to control through MIDI; several manufacturers make affordable one-, two-, four- or even eight-channel

MIDI-to-Control Voltage interfaces (Roland's MPU101 was an early and very flexible four-channel model, but is now very rare). Often, these will not have the ability to handle pitch bend or modulation, but at least the basic sounds of the instrument can be controlled from a MIDI set-up. Moog instruments often had an "S-trigger" (switch-to-ground trigger) input rather than the more common gate inputs, but these are available direct from several MIDI interfaces, or the input can be easily modified by any engineer to the Gate standard found on Roland and other instruments. As is well known, Korg and some Yamaha instruments often used a hertz/volt keyboard scale rather than the one volt per octave scale used by other manufacturers. Again, many MIDI interfaces have an output or an option to cope with this, or the voltage input can be adjusted (for instance, on the Korg MS20) and a position found that gives a correct scale response, or a suitable interface can sometimes be found or built (Korg's MS03 signal processor carried out this task very satisfactorily, but is now very rare). Modular systems by their very nature are, of course, very easy to control from MIDI-to-control voltage interfaces, and in these cases the user can have fun playing notes through one channel of MIDI while "playing" a filter or other parameter through another. A small Eurorack system including a MIDI-to-CV interface from any one of a number of manufacturers is worth building just to control any other CV equipped monophonic synths in the studio.

One group of instruments that can be problematic is the monophonics without control voltage and gate inputs. The potential purchaser has to wonder why they were not so equipped in the first place; often, the question was simply one of economy, and inputs can be very satisfactorily added at very modest cost. But sometimes there are other factors involved, for instance, on instruments which appear to have analog oscillators, but in which they are digitally controlled and played from a digitally scanned keyboard. In these cases, the designers would have had to fit a complex analog-to-digital conversion circuit in order to provide control voltage inputs, and this would not usually have been seen as economical. Good examples are: the EDP Wasp (hard to add control voltage inputs, and almost as hard to control by MIDI, except from the Kenton MIDI interface, which offers a Wasp option); the Eko P15 and Jen SX2000, which are analog preset synths but which have a digitally scanned keyboard; the ARP Pro DGX, and so on.

Some monophonics, such as early models of the Moog Prodigy, which had no external input sockets, are very easy to modify. Simply drill two holes for new sockets in the rear panel, locate the points at which the control voltage and gate leave the keyboard, add wires to these points, and external control becomes instantly possible (adding an audio input to the synthesizer's filter is equally straightforward on many instruments). But others present unexpected problems; apparently, early Roland instruments such as the SH1000 and SH2000 are not easily modified for control voltage input, others turn out to use subtly different control standards internally (such as the EMS synths, which used one-third of a volt per octave).

In conclusion, purchasing any instrument with a view to having it repaired, retrofitted, or modified, can be extremely unpredictable. The real

Some modifications are simpler than others: in this analog studio the Moog Prodigy (right) has been separated from its keyboard (centre).

usefulness of any monophonic instrument with no control voltage and gate inputs, or any polyphonic instrument without MIDI, has to be carefully questioned, and if a purchase is made on the assumption that either of these can be fitted, it's as well to be sure of the facts, or to be willing to make a further investment that may be as much as if not more than the purchase price of the instrument itself.

THE MARKET

There is a tendency these days to see advertised as "classic analog synthesizers" or "valuable collector's items" some instruments that have little value to collectors and, more to the point, are of almost no use in any practical stage or studio setting. One recent example has been the ARP Pro DGX; it is extremely heavy for its size, it has no control voltage and gate inputs and these are not easy to add, and it has little flexibility in its sounds. Thirty-five years ago it was an expressive lead line instrument, but with no MIDI fitted and with incredibly flexible, MIDI-equipped instruments, such as the Korg Prophecy, themselves now selling for ridiculously low prices secondhand, it is difficult to see any reason to pay big money for a Pro DGX. An original ARP Odyssey or Minimoog are, of course, different matters, despite both instruments now having modern re-issues or reboots.

Here are six purchasing categories for the more commonly available classic instruments, for consideration alongside the information in Appendix A. The dictates of fashion, the laws of supply and demand, the availability of replacement parts or modifications, the release of new versions and reboots of old products, and many other factors may move these instruments from one category to another, as may the offer of an unusually low or high price for any of them.

Firstly, some exceedingly overpriced instruments have been known to sell for much more than their actual usefulness would dictate (the Roland

TB303 Bassline being the most obvious example), and there are others which maintain more cachet than their real musical usefulness will justify.

Other instruments placed in the second category have become expensive, but justifiably so, as they have classic characteristics that were hard to reproduce until the recent spate of reboot instruments; the Minimoog, the ARP Odyssey, and so on.

A third group offers some excellent characteristics of the better-known models but have never become collectable or too expensive, and so represent the best value here.

A fourth group provides reasonable value for money only, either because they are flexible but remain quite costly, such as the original Korg MS20, or are unfashionable so very inexpensive but may only have a few useful facilities. Of the MIDI instruments, the Roland D110 module and Kawai K4 keyboard or K4R module come to mind. A very few small analog keyboards, such as the Korg Micro Preset, exist, which while not very powerful, can often be found so cheaply that they do become a worthwhile purchase.

The fifth group comprises instruments that really should be turned down these days, even if offered at next to nothing, and especially if advertised as "classic collector's items". They incorporate serious design compromises, lack control inputs or MIDI or memories, sound thin, or are extremely large, heavy, underspecified, unreliable, overpriced, or sometimes a combination of many of these. There is always a more useful instrument than these just around the corner, usually at no greater price and with better facilities, sound, or construction quality. Take on any of these instruments at your own risk, though be aware that some players will still love them and that a very low price can make even the most seriously compromised instrument design attractive.

A final group of (sometimes excellent) instruments vary widely in price, but have to be considered very carefully before purchase at any price, simply because they have developed a reputation for unreliability and are now extremely difficult and expensive to service. Examples include the Memorymoog and the Yamaha CS80, and some other instruments that can be easily damaged, even during the course of delivery to the new owner.

The Polymoog and Polymoog Keyboard now suffer from serious servicing problems.

The SynthaSound from the Baldwin organ company – rare and potentially hard to service.

1. *Often very highly priced*. Roland TB303 Bassline; EDP Wasp; Moog Taurus 1 and 2, Micromoog, Multimoog, Moog Source, Minitmoog, Prodigy; E-H MiniSynth; Korg PS3100, PS3200, PS3300; Oberheim OBX, OBsx, OBXa, OB8; Roland VP330, Juno 106, Jupiter 4, MC202, Juno 60, 101 Synth/102 Expander/104 Sequencer; Korg Mini 700/700S, MS50, MP4 MonoPoly, VC10 Vocoder; Yamaha CS30; RSF models.

2. *Expensive but worthwhile*. ARP 2600, original Odyssey; Minimoog; Crumar Spirit; Synton Syrinx; Elka Synthex; EMS VCS3/Synthi A/Synthi Aks; OSC OSCar; Sequential Prophet 5, 10, Prophet VS/Rack, T8; Oberheim original SEM, OB1, 2/4/6/8/12-Voice, Xpander, Matrix 12, OBMX; Moog modular systems; Roland 700 system, 100M system, Jupiter 8, Jupiter 6, SH5/7.

3. *Excellent value*. Sequential Pro One, Prophet 600; Ensoniq ESQ1, SQ80; Kawai SX240, K4/K4M; Oberheim Matrix 1000, 6, 6R; Roland D110 (if completely re-programmed), D50/550, JX3P, Super JX10, SH101 (maybe); Korg DSS1, DW8000/EX8000, Poly 800 Mk II, M500 Micro Preset/M500SP.

4. *Good value*. Bit One, 01, 99; Octave Cat (though prices have risen lately); Moog Rogue, Realistic MG1; Roland SH1/2/3/3a; Yamaha CS30L, CS50, CS60 (though now becoming very expensive); Korg original MS20, MS10, Polysix, Poly 61/61M; Sequential Six-Trak; ARP Axxe; Casio CZ101.

5. *Not so desirable*. Sequential Prelude, Fugue, Max, Roland SH1000/2000, Juno 6; ARP Pro Soloist/DGX, Quartet; Crumar Stratus, Trilogy, Composer, string synths; SIEL Orchestra, Cruise, Mono, Opera 6/DK600,

Expander 6, DK70, DK80/EX80; Kawai SX210; Moog Satellite, Opus 3, Yamaha SK10/15/20/30/50D, CS70M; Korg Trident/Trident 2, Lambda/ Sigma/Delta; most Multivox, Firstman, Pulser, Solton models, Russian-built Polyvox and other instruments.

6. *Serious service problems*. ARP Avatar, ARP Quadra; Polymoog Synthe-sizer, Polymoog Keyboard; Memorymoog; PPG Wave 2/2.2/2.3; Yamaha CS80.

Appendix D

Bibliography

Out-of-print items are often available through second-hand book dealers on Amazon.com.

Bergman, Billy and Horn, Richard. *Experimental Pop; Frontiers of the Rock Era.* Blandford Press, ISBN 0-7137-1550-2. Actually featuring very little about pop or techno-pop acts, for instance dismissing Kraftwerk in a few words, but very good on the minimalists Reich, Riley and Glass, and the early electronic music studios.

Bussy, Pascal. *Kraftwerk – Man, Machine and Music.* SAF Publishing, ISBN 0946719-38-1. Useful info on the German techno-pop band, though compiled without access to the band itself and perhaps superseded by Wolfgang Fluer's insider autobiography.

Bussy, Pascal. *The Can Book.* SAF Publishing, ISBN 0946719055 (1989). Fascinating history of the German avant-garde rock band, featuring Irmin Schmidt's innovative use of analog keyboards and processing.

Colbeck, Julian. *Keyfax,* omnibus edition. Hal Leonard/Mix Books, ISBN 0918371082 (1998). A buyers' guide to over 1000 keyboards (not only analog) – 193-page version compiled from all earlier editions of the book.

Crombie, David. *The Synthesizer and Electronic Keyboard Handbook.* Doring Kindersley, London, ISBN 0-86318-076-0 (1984). Now less than up to date, but some excellent black-and-white and colour photos and diagrams of various keyboard performers, instruments and systems.

Emerson, Keith. *Pictures of an Exhibitionist.* Blake Publishing, ISBN 9781844540532 (2003). The ultimate insider autobiography, spanning Emerson's very earliest years and influences to the career peaks of The Nice and ELP. Lots of detail, of course, on Emerson's thoughts while developing these bands – not so much on the technology, but fascinating nevertheless.

Fluer, Wolfgang. *I Was A Robot,* 2nd edition. Sanctuary Publishing, ISBN 1-86074-417-6 (2003). Fascinating insight into the history of the German techno band Kraftwerk from the perspective of their sometime electronic drummer, including some very rare gear photos of items such as the "percussion cage".

Forrest, Peter. *The A–Z of Analogue Synthesizers* (out of print), Part One, A–M revised, ISBN 09524377-2-4; Part Two, N–Z revised, ISBN 09524377-4-0. Guide to analog instrument specifications and values.

Forrester, George et al. *Emerson, Lake & Palmer – The Show That Never Ends*. Helter Skelter, ISBN 19009-2417X. Comprehensive history of the band, superseded in 2007 with a new edition featuring more photos (ISBN 1900924714).

Griffin, Mark. *Vangelis – The Unknown Man*. Albedo, ISBN 0-9523187-25. Probably the only detailed biography of the Greek keyboard prodigy. Available from Albedo, Smithy Croft, Ythanbank, Ellon, Aberdeenshire, AB41 7UA, UK.

Hanson, Martyn. *Hang On To A Dream*. Helter Skelter, ISBN 1-90092443-9 (2002). Excellent history of The Nice, including no contributions from Keith Emerson himself but some fine detail from other members, up to and including the 2002 reunion concerts.

Holmes, Thomas. *Electronic and Experimental Music*. Charles Scribner, New York, MacMillan, London, ISBN 0-684-18395-1 (1986). Dry, but with useful chapters on the earliest electronic instruments, such as the telharmonium, a good illustrated discography of albums that are mostly no longer available, and some fascinating photos.

Mason, Nick. *Inside Out*. Weidenfeld & Nicholson, ISBN 0-279-84387-7 (2004). Fascinating and sumptuously illustrated insider's history of the progressive rock band Pink Floyd, with some excellent gear photos.

Moraghan, Sean. *Mike Oldfield – A Man and His Music*. Britannia Press, ISBN 0-9519937-5-5 (1993). Very interesting history covering Oldfield's early bands, the launch of *Tubular Bells*, his experience with Exegesis and revitalized work up to *Tubular Bells 2*.

Newcomb, Martin. *The Museum of Synthesizer Technology*. No ISBN number (1994). Guide to the UK's analog synthesizer museum as it stood at its opening by Bob Moog in 1994 (though it closed shortly after). Many excellent colour pictures.

Pinch, Trevor and Trocco, Frank. *Analog Days*. Harvard University Press, ISBN 0-674-01617-3 (2002). Very detailed look at the early history of Robert Moog's company on the East Coast of the USA, and the parallel work of Don Buchla and the use of his instruments in Ken Kesey's "Acid Tests" on the West Coast.

Remilleux, Jean-Louis. *Jean-Michel Jarre*. Editions Olivier Orban France, McDonald Futura UK, ISBN 0-7088-4263-1 (1988). Large-format paperback with excellent photos of Jarre's studio equipment and live concert set-ups for the early Paris, China, Houston and Lyon concerts.

Russ, Martin. *Sound Synthesis and Sampling*, 2nd edition. Focal Press, ISBN 0-240-51692-3 (2005). Very detailed insight into the physics of analog sound, and of FM and other types of digital synthesis, MIDI applications, and much more.

Samagio, Frank. *The Mellotron Book*. Pro Music Press, ISBN 1-93114014-6. Comprehensive history of the tape-driven electromechanical monster that was an indispensable partner to the analog synths of Tangerine Dream, Klaus Schulze, Kraftwerk, and many other bands.

Stump, Paul. *The Music's All That Matters*. Quartet Books, ISBN 0-70438036-6 (1997). Interesting history of progressive rock, including much on Emerson, Lake & Palmer, Pink Floyd, Yes and Rick Wakeman, King Crimson, and the Canterbury bands.

Stump, Paul. *Digital Gothic.* SAF Publishing, ISBN 0-946719-18-7 (1997). Subtitled "A Critical History of Tangerine Dream", examining the band largely through their recorded output and including many extracts from contemporary reviews and interviews, plus some interesting photos and a discography, though now outdated in its references to fan magazines and other sources of information.

Vail, Mark. *Vintage Synthesizers.* GPI Books, ISBN 0-87930-275-5 (1993). Expanded from articles in *Keyboard* magazine, featuring some excellent photos of classic synthesizers and some less clear but interesting black-and-white photos of rare instruments. Covers in detail the rise and fall of Moog Music, the history of ARP, the European design companies, individual instruments, such as the Yamaha CS80, various sequencers and samplers, and a US-oriented guide to purchasing, prices, and servicing.

Wooding, Dan. *Rick Wakeman - The Caped Crusader.* Robert Hale, London, ISBN 0-7091-6487-4 (1978). Biography of Wakeman's early career, including information on his analog keyboards.

Appendix E

Discography

There have been many thousands of LP and CD releases that are worth hearing through the history of analog sound synthesis. Here are just a few suggestions of titles that are more readily available, to represent each musical category generally associated with the technology.

PURE SYNTHESIZER

KITARO

Silk Road 1, Domo, 71050-2 (1980)
Typical romantic, Eastern-tinged cosmic synthesizer album from one of Japan's most popular artists, particularly featuring lead melodies on the early Korg synthesizers.

JEAN-MICHEL JARRE

Oxygene, Polydor, 800-015-2 (1976)
Oxygene 7-13, 486984-2 (1997)
Planet Jarre (2018)

Jean-Michel Jarre celebrated fifty years in music in 2018.

Of the output of France's most popular synthesist, *Equinoxe*, *Chronologies*, and various live albums including *The China Concerts* are also worth hearing, while *Magnetic Fields* and *Zoolook* are more dominated by sound sampling and digital instruments. The "50 years of music" compilation, *Planet Jarre*, also has some great previously unheard old and new pieces in a wide variety of styles.

ASH RA

New Age of Earth, Virgin Records, CDV2080 (1977)
A classic of German analog synthesis. *Blackouts* of the following year was much better recorded, the lush sound of *New Age* being sadly rather muddy but was much more dominated by electric guitar. The later *E2-E4*, a near-live demo made in a much more minimalist systems music style for Klaus Schulze's label, later became a favourite with ambient techno fans.

MICHAEL GARRISON

In the Regions of Sunreturn, Windspell, RE79CD
Just one of a whole catalog of almost interchangeable CDs from one of the USA's most distinctive early masters of analog synthesis. Garrison was one of few players of the Syntar, seen here with designer George Mattson who went on to market a line of mini-modular systems.

George Mattson with his Syntar, which featured in the work of Michael Garrison.

TIM BLAKE

Crystal Machine, Mantra, 067 (1977) later re-issued with bonus tracks
New Jerusalem, Mantra, 068 (1978) later re-issued with bonus tracks
A master of sequencing, soloing, and abstract sound, Blake played a double
EMS Synthi A, Elka Rhapsody, a Minimoog and, live, a Moog modular and
Yamaha CS80. There are great photos on both CDs. The double EMS gets
rack-mounted some time between albums into a striking blue case.

PETER BAUMANN

Trans Harmonic Nights, Virgin, CDV2124 (1978)
The ex-Tangerine Dream member's solo album includes the four-minute
Chasing the Dream, one of the most brilliant demonstrations of multi-
layered analog sequencing ever recorded. The earlier album *Romance '76*
is worth hearing; the later and more song-oriented *Repeat Repeat* less so.

MICHAEL HOENIG

Departure from the Northern Wasteland, Kuckuck, CD079 (1978)
A masterpiece of multiple analog sequencing in odd time signatures, layered melodies, and abstract sounds, featuring ARP sequencer, Elka Rhapsody, and EMS synths.

Michael Hoenig's 1980s move
to the USA saw a Synclavier
added to the analog synth
setup.

HELDON

Stand By, Spalax, 14233 (1978)
Bolero Pts 1–8 is an exercise in fast Moog modular sequencing, abstract
effects, guitars, voices, Minimoog lead lines, and very heavy bass drones
over jazzy drums. All the other Heldon albums, the earliest ones using

an EMS VCS3 and the later ones a Moog modular, are worth hearing too, *Electronique Guerilla, Un Reve Sans Consequence Speciale, It's Always Rock'n'Roll, Agneta Nilsson* and *Interface* included.

RICHARD PINHAS

Chronolyse, Cuneiform, RUNE 30CDX (1976)
Iceland, Cuneiform, RUNE 44X (1979)

A synth and a smoke. Richard Pinhas and Heldon recorded some of the heaviest electronic music ever heard, here with EMS Synthi A and Moog Modular synthesizers and guitar.

Variations I–VII Sur Le Theme De Bene Gesserit and *Duncan Idaho* on the former album were recorded live on a Moog modular system. Pinhas is the founder of Heldon and later recorded several more solo albums, including *Rhizosphere, East West, L'Ethique, DWW, Cyborg Sally, De L'Un et Le Multiple* and *Le Plan*, releasing *Trax* with Yoshida Tatsuya on the Bam Balam label in 2018. The first three have plenty of big Moog sequencing, the next three add more band, jazz fusion, and baroque-oriented material, the most recent tend almost entirely towards electronically processed guitar loops.

KLAUS SCHULZE

Timewind, Caroline, CDCA2006 (1975)
Live, Manikin, MRCD7008 (1979)
Timewind, based on overdubbed sequencer improvisations, is a well-loved example of the epic cosmic synthesizer style. *Live* shows how Schulze could achieve almost the same levels of complexity, and perhaps even greater power, in a stage setting, Arthur Brown guesting on vocals.

TANGERINE DREAM

Rubycon, Virgin, CDV2025 (1975)
Ricochet, Virgin, CDV2044 (1976)
Like the two Klaus Schulze albums listed, these two albums show how to manipulate, process, and modify analog sounds and sequences, first in the studio and then live. Chris Franke's constant variation of the live sequencing on *Ricochet Pt. 2* remains unparalleled to this day.

Chris Franke around the Ricochet era with Tangerine Dream – Moog modular synthesizer and sequencers, mixers, EMS SynthiA, Elka Rhapsody, Mellotron.

WHITE NOISE

White Noise, *An Electric Storm*, Island, 3D CID 1001 (1968)
David Vorhaus, White Noise, *2 – Concerto for Synthesizer*, Virgin, CDV2032 (1974)
White Noise, *III – Re-Entry*, AMP-CD 031 (1980)
The first White Noise album dates from the late 1960s and doesn't even include synthesizers as such, it's more about tape cut-ups, effects, and analog processing. But *Concerto* is all analog, mainly showing the strengths as well as the weaknesses of the early EMS synths, while *Re-Entry* has some great multichannel analog sequencing, expressive melody lines, and abstract sounds. *White Noise 4* uses more sound sampling, while *White Noise 5* creates analog-style loops and sequences using current software (3–5 available through www.markjenkinsmusic.com).

VANGELIS

Heaven and Hell, RCA, ND71148 (1975)
Blade Runner, East West, 4509–96574–2 (1982)

The former album defined Vangelis's lush, romantic sound, while the latter, not on CD until 1994, particularly showcases the strong brass, string, and abstract sounds of the Yamaha CS80 in an atmospheric film music setting.

Vangelis in an early multi-keyboard session. Organs, electric pianos, Korg Micro Preset, Roland SH2000, Korg 700, Elka Rhapsody, Farfisa Syntorchestra.

ROLF TROSTEL

Recall Level (double CD), Manikin, MRCD7035
Three albums (*Inselmusik*, *Two Faces* and *Der Prophet*) on two CDs, which effectively comprise highly imaginative demo discs for the sequencing and synthesis capabilities of the PPG Wave synths.

ROBERT SCHROEDER

D.Mo Vol. 1, CUECD-113 (Germany) Vols 2–4 also available on Spheric Music
Fascinating document of the early days of an analog master who built much of his own equipment. Epic, repetitive sequencer parts and expressive melodies on the Multimoog and other synths.

T.O.N.T.O.'S EXPANDING HEADBAND

Zero Time, Real Gone Music RGM0174 USA 2013 (CD)
It's About Time, Polydor 2363 308 (1974) LP
T.O.N.T.O. Rides Again, Viceroy Vintage VIN6036–2 1996 (CD)
The first pair are long deleted LPs (occasionally on CD) showcasing the expanded Series III Moog synthesizer T.O.N.T.O. in a melodic, innovative fashion; the later CD release *T.O.N.T.O. Rides Again*, apparently quickly deleted, compiles both albums, renaming the tracks from the second (so these aren't new bonus tracks, as some listeners think) and unaccountably omitting the best track *Beautiful You*. T.O.N.T.O. was recently due to go on exhibition at the National Music Center in Calgary (www.nmc.ca), which in 2017 ran a stunning exhibition "A Synth Odyssey" featuring many rare analog instruments such as Patrick Moraz's double Minimoog.

Malcolm Cecil with T.O.N.T.O. – distinctive curved shape of synth cabinets not due to use of a fish-eye lens. www.tontostudio.com

BAFFO BANFI

The Sound of Southern Sunsets, Innovative Communications IC710.065
Frontera, AMS AMS248 CD (2015)
Giuseppe "Baffo" ("Hairy") Banfi played with Italian prog rock group Biglietto Per L'Inferno before getting into solo synth music, collaborating

with Klaus Schulze and appearing on his IC label. *The Sound of Southern Sunsets* is a compilation featuring the whole of his album *Ma Dolce Vita*, two tracks from *Hearth* and an unreleased piece. His earlier *Galaxy My Dear*, which is less smoothly produced is a nice example of early abstract synth music, using multiple echoes on the accompaniment section of organ-style instruments, underneath many analog layers.

Baffo Banfi in the studio with Yamaha CS60 and Minimoog.

AMBIENT

VARIOUS

Artificial Intelligence, Warp, CD6 (1992)
"For long journeys, quiet nights, and club drowsy dawns", featuring Alex Paterson, Richie Hawtin, Autechre, etc.

APHEX TWIN

Selected Ambient Works 85–92, R&S Records, AMB3922 (1992)
More extremely ambient analog music.

PETE NAMLOOK

The Definitive Ambient Collection, Rising High Records, RSNCD11 (1993)
Lots of EMS Synthi and Oberheim 4-Voice over drum loops, deep bass drones, and highly resonant sweeps. Also a lot of Roland JD800 preset sounds.

KLAUS SCHULZE/PETE NAMLOOK

Dark Side of the Moog IV, Fax, PK08/108

Collaborative album – one of at least nine – alternating Schulze's later style of fast techno synthesis with the late Pete Namlook's more abstract passages.

DERRICK MAY

Innovator, Transmat, TMT-2 CD (1997)
Experimental dance-oriented material from 1987 to 1997.

ROCK/PROGRESSIVE ROCK

PINK FLOYD

Dark Side of the Moon, Harvest, CDP7-46001-2 (1973)
Synthesizer and sequencer parts mostly courtesy of EMS gear. Later albums, such as *Wish You Were Here* and *Animals*, were more dominated by organ and Minimoog solos, created by the late Richard Wright.

RICK WAKEMAN

Journey to the Centre of the Earth, A&M, CD3156 (1974)
Six Wives of Henry VIII, A&M, D32Y3115 (1972)
Journey remains probably the best combination ever of live analog synth soloing with an orchestral backing, a tremendously powerful pairing. *Six Wives* shows similar techniques, particularly on the Minimoog, in a more controlled studio setting.

EMERSON, LAKE & PALMER

Emerson, Lake & Palmer, Atlantic, 7567-81519-2
Brain Salad Surgery, Atlantic, 19124-2 (1973)
Welcome Back My Friends to the Show That Never Ends, Atlantic, AMCY 215-6
The band's first album doesn't include much synthesizer, but the gliding Moog solo on *Lucky Man* was disproportionately influential. *Brain Salad Surgery* includes *Karn Evil 9* with its famous random sample-and-hold interlude, electronic percussion, and rare recording of the early Moog Constellation polyphonic ensemble prototype. *Welcome Back . . .*, one of the most powerful live albums ever released, features much use of Minimoog as well as modular Moog and Clavinet. As for later albums, *Works Vol. 1* (Atlantic, 7000-2, 1977) features Yamaha's GX1 on *Fanfare for the Common Man*; contemporary live recordings also feature it on performances of *Pirates*. The albums *Trilogy, Tarkus, Pictures at an Exhibition*, and *Works Live* are worth hearing too, *Works Vol. 2* and *Love Beach* much less so.

A rarely seen photo of Keith Emerson with ELP and the Moog Constellation prototype.

CAMEL

The Snow Goose, Deram, 800–080–2 (1975)
Compellingly melodic progressive rock featuring the work of the late Peter Bardens on highly expressive Minimoog.

MINIMALIST/SYSTEMS/AVANT-GARDE

TERRY RILEY

A Rainbow in Curved Air, CBS, MK7315 (1969)
There are no actual analog synthesizers on this album, just electric organs, electric harpsichord, rocksichord, and percussion. But the multi-layering and echo techniques remain astonishingly imaginative, the album influencing the opening of Mike Oldfield's *Tubular Bells* and The Who's *Baba O'Reilly* and giving the name to the band Curved Air.

GEORGE HARRISON

Electronic Sounds, EMI, 7243-8-55239-2-2
Two long tracks, recorded in 1969 in the UK and in 1968 in the USA with the assistance of Bernard Krause, with some strong Moog modular sounds, but mostly extremely abstract and tuneless in what sounds like more or less a live performance on a Moog modular, featuring sporadic filtered white noise and other sounds, plus a lot of tape and spring reverb.

MORTON SUBOTNICK

Silver Apples of the Moon, Elektra Nonesuch 71144, 1967

Subotnick's influential early compositions performed on Buchla instruments. At the time of writing in 2018, Subotnick is still performing live around the USA and Europe at the age of 85.

Morton Subotnick in performance with Buchla music systems.

MOTHER MALLARD'S PORTABLE MASTERPIECE COMPANY

Like A Duck To Water, Earthquack, 1976
Led by David Borden, the Mother Mallard ensemble was an early user of multiple Moog and other synths, performing original music and pieces by Steve Reich, Philip Glass, and others.

SUZANNE CIANI

Buchla Concerts 1975, Finders Keepers Records FKR082 (LP) (2018)
A long-lost gem re-discovered by Andy Votel of Finders Keepers Records, the album spins quadrophonic panning sequences between wave and wind sounds and abstract analog textures.

TECHNO-POP

GOKCEN KAYNATAN

Anatolian Invasion 4, Finders Keepers Records FKR019 (LP) (2017)
A truly bizarre find from Andy Votel and Doug Skipton at Finders Keepers Records, this vinyl release compiles EPs from a Turkish electro-pop pioneer who went on to compose for the national TV station. Kaynatan discovered the EMS Synthi in Germany and went on to build an eye-catching

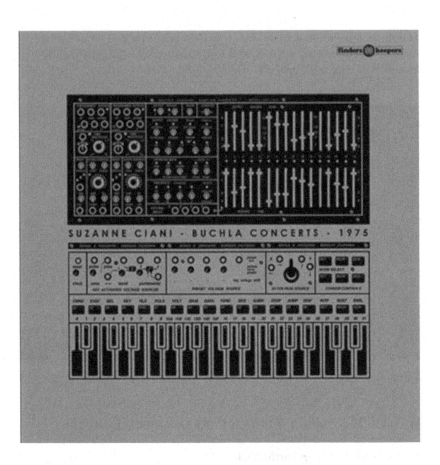

Suzanne Ciani's early Buchla Modular system heard on the *Buchla Concerts* 1975 LP. www.finderskeepersrecords.com

studio and live performance setup, apparently using old arcade game cabinets. The music? A cross between Joe Meek, guitar-led surf music, and early White Noise or Radiophonic Workshop styles, all with a striking Turkish element. Kaynatan is still performing, these days with a whole band, female backing singers and keyboardists and a laser harp.

Gokcen Kaynatan brought electro-pop music to Turkey.

THE HUMAN LEAGUE

Reproduction, Virgin, CDV2133 (1979)
Travelogue, Virgin, CDV2160 (1980)
Analog synths fuzzed, sequenced, pitch shifted, echoed, and reverberated, with some brilliant substitutes for horns, guitars and voices.

ORCHESTRAL MANOUEVRES IN THE DARK

Orchestral Manouevres in the Dark, Virgin, Dindisc DIDCD2 (1980)
Including the singles *Red Frame White Light* and *Electricity*, good examples of simple analog melody sounds used in catchy pop compositions. Korg's Micro Preset featured heavily.

YELLOW MAGIC ORCHESTRA

Yellow Magic Orchestra, Restless Records, LS9156-Z (1979)

Hits such as *Firecracker* and *Yellow Magic* from Ryuichi Sakamoto, Yuki-hiro Takahashi, Haruomi Hosono, and Hideki Matsutake (who went on to work as Logic System), analog with an Oriental edge, and an early inspiration for the *Stranger Things* composers Dixon & Stein.

Yellow Magic Orchestra on stage – a truly frightening amount of analog equipment.

KRAFTWERK

Autobahn, EMI, CDP 564-7-46153-2 (1974)
Trans Europe Express, EMI, CDP 564-7-46133-2 (1977)
Defining the techno-pop style, although with the more romantic sounds of electric guitars and flutes still included on the former, while the latter becomes completely electronic, showcasing sequencing, percussion, and vocoding techniques. *Radio Activity*, *The Man Machine* and *Computer World* are also well worth hearing. *The Mix*, which features re-recorded tracks from the previous, less so, while *Electric Café* was more concerned with sampling technology.

GARY NUMAN

Replicas (various editions)
The Pleasure Principle (various editions)
Featuring Numan's hit singles *Are Friends Electric?* and *Cars*, these albums show off strong, basic sounds on the Minimoog, Polymoog, and ARP Odyssey in a stripped-down techno-pop setting.

Kraftwerk's secret weapon, the optical disk playing Optigan by toymakers Mattel, never the Mellotron. Choirs and other Optigan sounds on *Radio Activity* and *Trans Europe Express* counterpointed analog Minimoog basses, ARP Odyssey leads and custom percussion sounds.

CLASSICAL

Isao Tomita

Snowflakes Are Dancing, RCA, RD84587 (1974)
Rich string sounds from the outset, a much better use of dynamics than most, many ring-modulated sounds, and the late Isao Tomita's distinctive portamento whistling and human voice tones on *Arabesque No. 1*.

Isao Tomita with the Moog Modular system.

WALTER CARLOS

Switched On Bach, CBS, MK63501
A Clockwork Orange, Warner, 927-256-2

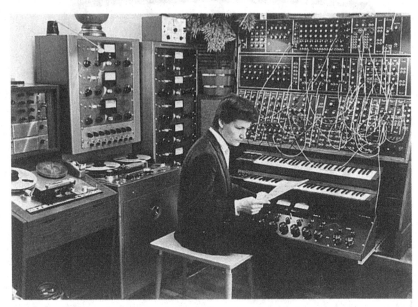

Walter Carlos at the Moog Modular System working on *Switched-On Bach*.

Sounding perhaps a little sterile now, *S.O.B.* nevertheless remains a landmark album. *A Clockwork Orange*, from Stanley Kubrick's movie, includes many classical arrangements plus a 4m13s version of the much more experimental *Timesteps*, and some later versions include a much longer edit.

SYNERGY (LARRY FAST)

Electronic Realizations for Rock Orchestra, Passport, PBCD6001
A progressive rock release that verges on classical music territory in its highly arranged, multilayered compositions, mainly achieved just using one Minimoog.

CRAIG LEON

Bach To Moog, Sony Classical 88875 052612 (2015)
Producer Craig Leon fuses the sound of a newly manufactured Moog Modular System 55 almost inseparably into the sound of the Sinfonietta Cracovia. Includes a set of *14 Canons on the Goldberg Ground* on the Moog alone, plus some nice sleeve pictures.

Craig Leon's "Bach to Moog".

JAZZ ROCK FUSION

JAN HAMMER

Early Days, Nemperor, 2K40382 (CBS)
Featuring tracks from *The First Seven Days* (1975), *Oh Yeah?* (1976), *Like Children* (1974) and *Melodies* (1977), coloured by Hammer's expressive, often guitar-like, Minimoog soloing. Hammer's *Miami Vice* soundtrack material is also worth hearing, combining analog, digital, and sampled sounds.

GEORGE DUKE

Faces In Reflection, MPS Records 21 22018-4
Feel, MPS Records MC25355
All of Duke's 1970s albums, also including *The Aura Will Prevail*, *Liberated Fantasies*, *Reach For It*, and many more, offer blistering jazz/rock fusion analog synth solos alongside strong funky rhythm playing and many more experimental sounds. Duke was an early master of the portable analog

synth controller, including the Davis Clavitar from around 1978 (probably after trying having the keyboard separated from his Minimoog), which he used alongside drummer Billy Cobham, bassist Stanley Clarke, and others in many excellent collaborative projects.

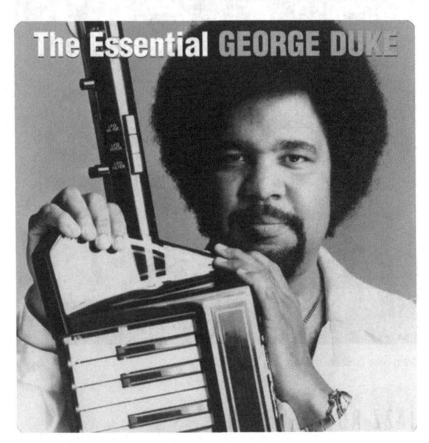

George Duke was a prolific jazz and fusion keyboardist.

TV MUSIC

DIXON & STEIN _____

Stranger Things, Vol. 1, Lakeshore Records
31 cues from the TV series showing just how to create atmospheric, ambient music for TV

JOHN CARPENTER _____

Lost Themes, Sacred Bones SBR123 (LP) 2015
Carpenter composing for a change without a movie to soundtrack, a well-received album.

Stranger Things co-composer Kyle Dixon in his studio in Austin, Texas.

COMPILATIONS

VARIOUS ARTISTS

Switched On Bob – A Tribute to Bob Moog, Cherry Red CASA11CD (2009) Italian compilation featuring Funki Porcini, Dan Lacksman (Telex/Deep Forest), Dana Countryman, Jean-Jacques Perrey, who writes some of the sleeve notes, and many others. Moog and other synths played helpfully identified for each track

Electronic Music – It Started Here, Not Now Music NOT2CD627 (2016) Very good value double CD compilation of electronic music pioneers, including Tom Dissevelt, Vladimir Ussachevsky, Kid Baltan, Karlheinz Stockhausen, Pierre Henry, Daphne Oram, Max Mathews, Otto Leuning, Edgard Varese, John Cage, and many others. Gleefully skips from avant-garde music to TV advertising and pop, great presentation as a mocked up double vinyl disk.

Electricity – A Brief History of Future Sound, MOJO (2012) Compiled by Daniel Miller of Mute Records for Mojo magazine as a cover CD, this is available very inexpensively through eBay and elsewhere despite a "not to be sold separately" legend. Hits the mark as regards early electro-pop, Fad Gadget's *Back To Nature*, and tracks by DAF, Throbbing Gristle and others, plus the brilliant but very much not electronic *Oh Yeah* by Can.

Music for the 3rd Millennium, AMP Music, AMP-CD039 (1999) Showcase for all the techniques of progressive rock, orchestral synthesis, ambient dance, and experimental electronics. Features rare and unreleased tracks contributed by Keith Emerson (ELP), Rick Wakeman (Yes),

One of many CDs released in tribute to Bob Moog.

One of several recent compilations of pioneering electronic music.

Patrick Moraz (Yes/Moody Blues), Larry Fast (Synergy/Peter Gabriel Band), Karl Bartos (Kraftwerk), White Noise, Isao Tomita, Richard Pinhas (Heldon), Steve Jolliffe (Tangerine Dream), MC Russell, Alquimia, Steve Baltes (Ash Ra Tempel), Neil Ardley (*Kaleidoscope of Rainbows*), and the author.

ELECTRICITY

A BRIEF HISTORY OF FUTURE SOUND
COMPILED FOR MOJO BY MUTE'S DANIEL MILLER
Throbbing Gristle, Can, Plastikman, Moby, Nitzer Ebb,
Cabaret Voltaire, LCD Soundsystem, Daft Punk and more...

A good introduction to the electro-pop field.

Music for the 3rd Millennium Vol. 2, AMP Music, AMP-CD041 (2000)
Features rare and unreleased tracks contributed by Michael Rother (Neu!)
with Dieter Moebius (Cluster), Nick Rhodes (Duran Duran), Deep Forest,
Logic System (YMO), Paul Haslinger (Tangerine Dream), Clearlight, Dave
Greenslade (Greenslade), Bernard Xolotl, Ryo Okumoto (Spock's Beard),
WaveWorld, Nash The Slash, Spiral Of Silence (Mark Jenkins and Alquimia),
Bernd Kistenmacher, Modulus, Michael Stearns, and Nick Magnus (Steve
Hackett Band).

Music for the 3rd Millennium Vol. 3, AMP Music, AMP-CD042 (2001)
Features rare and unreleased tracks contributed by Robert Schroeder,
Vincent Clarke (Erasure) and Martyn Ware (Heaven 17), Damo Suzuki
(Can), Geoff Downes (Asia), Warren Cuccurullo (Duran Duran), Alqui-
mia Vs. Steve Baltes (Ash Ra Tempel), Johannes Schmoelling (Tangerine
Dream), Vidna Obmana, Rudiger Gleisberg, Mario Schoenwalder and Har-
ald Grosskopf.
All three volumes available through www.markjenkinsmusic.com

The three volumes in the "Music for the 3rd Millennium" CD series.

MATTHIAS BECKER/KLAUS STUEHLEN

Synthesizer Von Gestern, Vols 1-3, BMG Ariola, OW19-876292-907 (1990, 1992, 1994)
Sonic demo CD featuring pieces composed entirely using one synthesizer on each track. Volume 1 features the ARP Odyssey, Yamaha CS60 and CS15, Oberheim SEM, Korg MonoPoly, Poly 800, MS20 and PS3100, Minimoog, Memorymoog and System 55, Roland Jupiter 8, Juno 60, System 100 and SH5, Sequential Pro One, Rhodes Chroma, Mellotron, EMS Synthi, and PPG Wave.

Volume 2 features the ARP 2600, EDP Wasp, EMS Synthi100, Mixtur-Trautonium, Oberheim OBXa, OSC OSCar, Roland Jupiter 6 and System 100M, RSF Kobol, SCI Prophet 5, and DKI Synergy.

Volume 3 features the Buchla System 100, E-Mu Modular, Gleeman Penta-phonic, Korg PS3300, Mixtur-Trautonium, Oberheim Xpander, Sequential Prophet VS and T8, Steiner-Parker Synthacon, Variophon, Yamaha GS2, and TX816.
There is also an accompanying book in two volumes, but all are hard to find now.

Appendix F

Contacts

Access – www.virus.info/home
Akai – www.akai.com
AMP Music – www.markjenkinsmusic.com; www.youtube.com/markjen kinsmusic
Analog Zone – www.synthzone.com/analogue.htm
Analogue Solutions – www.analoguesolutions.co.uk
Apple – www.apple.com
Cakewalk – www.cakewalk.com
Creamware – www.creamware.de
Cycling 74 – www.cycling74.com
Dave Smith Instruments – www.davesmithinstruments.com
Devil Fish (Roland TB303 and TR606 modifications) – www.firstpr.com. au/rwi/dfish/
Doepfer – www.doepfer.de
EMS (modifications) – www.hinton-instruments.co.uk
E-Mu/Ensoniq – www.emu.com
FutureRetro – www.future-retro.com
JOMOX – www.jomox.de
Keith Emerson archive – www.keithemerson.com
Kenton Electronics – www.kentonuk.com
Kitaro – www.kitaro.live/
Korg – www.korg.de; www.korg.co.uk
Kraftwerk – www.kraftwerk.com
Mark Jenkins – www.markjenkinsmusic.com; www.youtube.com/mark jenkinsmusic
Metasonix – www.metasonix.com
Moog (Big Briar) – www.moogmusic.com
Novation – https://global.novationmusic.com/
Oberheim – http://tomoberheim.com/
Patrick Moraz – www.patrickmoraz.net
Rick Wakeman – www.RWCC.com
Roland – www.roland.us.com; www.roland.co.uk
Sequential Circuits (servicing) – www.winecountrysequential.com
Serge – www.serge.synth.net
Sherman – www.sherman.be
Spectral Audio – www.spectralaudio.ch (closed)
Steinberg – www.steinberg.net
Synth Fool (Kevin Lightner's synthesizer database) – www.synthfool.com
Synth Museum (synthesizer database) – www.synthmuseum.com

Tangerine Dream – www.tangerinedream-music.com
Terry Riley – http://terryriley.net
Yamaha – https://usa.yamaha.com; https://uk.yamaha.com

Many other website contacts are listed under individual entries through-out the book.

WEB AND OTHER RESOURCES

www.analogindustries.com – Discussions, book lists and more.
http://analogsynthmuseum.free.fr – Some useful photos and links in French and English.
www.analoguehaven.com – Store on Broadway in Santa Monica, CA, which handles many lines including Anyware, Cwejman, Moog, Doepfer, Macbeth and MOTM.
https://www.audities.org – Foundation about 30km from Calgary, Canada including a collection of over 150 instruments.
www.harmony-central.com – Enormous musical instrument database and discussion group.
www.keyboardmuseum.org – 1300 keyboards collected in Winter Park Florida, this now appears to be closed though.
www.kvraudio.com – Huge resource for demo, shareware and freeware virtual instruments.
https://www.matrixsynth.com – News and discussion of all things synthesizer.
www.modulargrid.net – Invaluable regularly updated resource for modular synths.
www.ruskeys.net/eng/synths.php – Handy database of Russian synths, in English.
www.synrise.de – Huge database of information and photos of synth products; some in English but mostly in German, no longer being updated but with some interesting photos of rare instruments.
http://www.synthark.org – Handy database with some great instrument photos.
http://www.synthesizerstudio.com/info – Recording studio in Hamburg, Germany offering rental of 60 different synths and drum machine.
www.synthmuseum.com – List of manufacturers and books.
www.synthzone.com – Synthesizer database.
http://www.vintagesynth.com – Useful database.
www.youtube.com – Search for video clips including much on Kraftwerk, Can, Delia Derbyshire, Klaus Schulze, Gary Numan, Tangerine Dream, Bob Moog, BBC Radiophonic Workshop, scores of modular synth jams and instrument demos, etc.

The **Vintage Synthesizer Museum**, http://vintagesynthmuseum.com, in Emeryville, California has a huge collection of classic and modern instruments and can be hired for recording sessions. Just part of the collection is listed on the next page.

Contacts

Delia Derbyshire with Desmond Briscoe in the very early days of the BBC Radiophonic Workshop, around the time of her realisation of Ron Grainer's theme for *Dr Who*. There's a complete documentary *The Delian Mode* by Kara Blake on YouTube, as well as material on Daphne Oram, Morton Subotnick, and many other UK and US pioneers of electronic music.

Moog – Minimoog Model D/Taurus
EMS – Synthi AKS/Vocoder 2000
ARP – 2600 w/3620 / 1613 Sequencer/Odyssey MkII/Rhodes Chroma
Sequential Circuits – Prophet 5 / Prophet T8 / Pro-One
Oberheim – 4-Voice/OB-Xa/Xpander
Roland – Jupiter 8 / 6 / 4 / Juno 6 / Juno 106 / SH-101 / TR-808 / TR-909 / RS-505
Korg – MS-50 / MS-20 / MS-10 / SQ-10 / VC-10 / Mono/Poly/Univox MaxiKorg/PS-3100
Yamaha – CS-70m/CS-60 / CS-20m
EML – Electro-Comp 101 / Electro-Comp 200
Steiner Parker – Synthacon
Gleeman – Pentaphonic Clear
Crumar – Spirit
Octave – Cat MkI
Oxford – OSCar
PAiA – Proteus
Fairlight – CMI
DKI – Synergy
Buchla – Music Easel
Kilpatrick – Phenol
Eurorack – 18U by 90hp Modular System
MOTM: 25 Module 5U System

Just a handful of the instruments at the Vintage Synthesizer Museum in Emeryville CA.

Appendix G

Website content: www. routledge.com/cw/jenkins

AUDIO MATERIAL

All sounds and music prepared by Mark Jenkins at Unicorn Studio, London.

1. Demo track, *Analogical!* Performed on various analog and virtual analog instruments. © Mark Jenkins 2006

ANALOG MODULAR SYSTEM SOUNDS

2. Various wave shapes
3. Frequency modulation
4. Multiple oscillators/beat frequencies
5. Four filter types
6. Filter modulation
7. Complex filter modulation from several LFOs
8. Envelopes

ANALOG MODULAR SYSTEM SEQUENCES

9. Voice-like fast modulation
10. Filtering sequences
11. Overdriven amplifiers
12. Different envelope lengths
13. Velocity-sensitive sequences
14. Mixing waveforms, footages and modulation
15. Complex, fast, voice-like modulation
16. White noise, resonant filters
17. Oscillator glide
18. Multi-source filtering

KORG MS10

19. Bendy (C × 3), Wobbly (C × 3), Stringy (C × 3), Growly (C × 3)
20. Power Drill, White Noise Whoosh Up, White Noise Whoosh Down, Resonant White Noise Weebles × 4, White Noise Growl × 3

ELKA SYNTHEX

21. Zap (C × 4), Swoosh (C × 4)
22. Filtersquidge (C × 4), Big Synth Ensemble (C × 4)
23. Jarresync (C × 4), Jarrewhoosh (C × 4)

ARP 2823 ODYSSEY (ORIGINAL)

24. Dual Osc Slow Attack Lead (C × 4), Sync Lead (C × 4)
25. Filter Wobbles (×5)

EDP WASP

26. Detuned Bass (C × 4), Detuned Twang (C × 4)
27. Squarewave Mod (C × 4), Sample-and-Hold Drone (C), Inverted Filter (C × 4)

ROLAND JUPITER 4

28. Hard Bass (C × 4), Cosmic (C × 4), String Ensemble (C × 4)

MOOG 204D MINIMOOG (ORIGINAL)

29. Deep Res Twang (C × 3), Short Res Twang (C × 3)
30. Wah Res Twang (C × 3), Deep Held Fuzz (C × 3), Slow Attack Beat Fuzz (C × 3)
31. Smooth Flute (C × 3), Square Vibrato Lead (C × 3), Acid Blap (C × 3)

ROLAND TR808 (ORIGINAL)

32. Kick 1, 2, 3 with/without accent, Cowbell, Open High-Hat 1, 2, 3, Closed High-Hat, Maraca, Rimshot
33. Clap × 2, Cymbal 1, 2, 3, Tom 1, 2, 3, Conga 1, 2, 3, Clave, Snare 1, 2, 3, 4

ROLAND MC202

34. Bass A (3Cs × 4), Wow (C × 4)
35. Bass B (3Cs × 4), Synthi (C × 4)

ROLAND TB303 BASSLINE (ORIGINAL)

36. Acid A (C × 4), Acid B (C × 4)
37. Acid C (C × 4), Acid D (C × 4), Acid E (C × 4)

OBERHEIM OB1

38. Resonant Twang (C × 3), Smooth Twang (C × 3), Sync Sweep (C × 3)
39. Sample-and-Hold Drone (C × 3), Filter Mod Delay (C × 3), Detuned Drone (C × 3), Acid Short Blap (C × 3)

KORG M500 MICRO PRESET

40. Synthe 1 16' (C × 3), Synthe 2 32' (C × 3), Trumpet 8' (C × 3), Dbl. Bass 32' (C × 3), Bassoon (C × 2)
41. Clarinet (C × 2), Oboe (C × 2), Flute (C × 2), Recorder (C × 2)
42. Bass Voice (C × 2), Tenor Voice (C × 2), Alto Voice (C × 2), Soprano Voice (C × 2), Whistle (C × 2), Noise Wobble, Bass Wobble

MOOG 5330 SATELLITE

43. Lunar (C × 2), Bell (C × 2), String Pick (C × 2), String Strike (C × 2), String Pluck (C × 2), String Blow (C × 2)
44. Reed Bright (C × 2), Reed Full (C × 2), Reed Hollow (C × 2), Reed Thin (C × 2), Brass Open (C × 2), Brass Mute (C × 2)
45. Siren, Up From The Depths

ARP 2731 EXPLORER 1

46. All Footages Riff, All Footages (C × 3), Brass (C × 2), Hollow (C × 2)
47. Reed (C × 2), Fuzz (C × 2), Descending Wobble (C)
48. Resonant Noises (×6)

ARP PRO SOLOIST DGX

49. Bassoon (C × 2), English Horn (C × 2), Oboe (C × 2)

ARP 2351 SOLUS

50. Space Weebles (×3)

OSC OSCAR

51. Mid Bloop (C × 3), Wah Sync Bass (C × 3), Synthorgan (C × 3)
52. Wah Sync Bass 2 (C × 3), Digital Syn (C × 3), Delay Digibass (C × 3), Swoop Lead (C × 3)
53. Acid Riffs, Spacey Arpeggio

MOOG MULTIMOOG

54. Filter Fuzz Drone (C × 3), Filter Fuzz Twang (C × 3)
55. Light Acid Twang (C × 3), Sharp Acid Twang (C × 3), Percussive Fuzz (C × 3)

MOOG SONIC 6

56. Resonant Filter Epic Weebles
57. Moog Drone, High Bubbling Drone

YAMAHA CS80

58. Schulzian Epic Ringmod Blasts
59. Resonant Funk/String (C × 3), 8 Octave Swoop
60. Resonant Wah/String (C × 3), Portamento Riff

ROLAND VP330 VOCODER PLUS (ORIGINAL)

61. Male Voice (C × 3), Slow Attack String (C × 3)

62. Male Voice and String (C × 3), Glide Up Voice and String (C)

SEQUENTIAL PROPHET 600

63. Bass Twang (C × 3), Rich String (C × 3), Deep String (C × 3)

64. Resonant Wah (C × 3), Steel Drum (C × 3), Filter Crossmod (C × 3)

65. Deep Detuned Bass (C × 3), Deep Twang (C × 3), Minimoog-like Riff

DSTEC OS1 ORIGINAL SYN

66. Rez Bass, Wah, Rez Wah, Res Sweep Up/Down, Random Filter, Random Glide

SPECTRAL AUDIO PRO TONE

67. Plonk 1/2, RezTwang 1–4, Filter Sequence

WALDORF PULSE V1

68. Bass 1–4, Sync Sweep, Random Sequence

ROLAND SH5

69. Plonk 1–4, Noise/VCO Sequence

SHERMAN FILTER BANK

70. Drum Loop

71. Roland SH101 Sequence

METASONIX PT-1 PHATTYTRON

72. Swept Filter Resonance

RED SOUND SYSTEMS DARKSTAR

73. 6 Sounds

ACCESS VIRUS B

74. 17 Sounds and Loops

NOVATION NOVA

75. 11 Sounds and Loops

KORG 800DV

76. Dual Filter Melody

CASIO CZ101

77. Eyeamamohg

KORG PROPHECY

78. Prophetic Steps, Studio Moog, The Big One

ROLAND JD800

79. Deep Seamphony, Sync Lead, Harsh Lead

STUDIOLOGIC SLEDGE V2

80. Bass and lead sounds
81. Pad and textural sounds
82. Arpeggio and patterned sounds

STUDIO ELECTRONICS BOOMSTAR SEM

83. Sounds 1
84. Sound 2
85. Sounds 3

KORG MS20MINI

86. Monophonic sounds
87. Abstract sounds
88. Sequences and patterns

BEHRINGER MODEL D

89. Leads
90. Basses
91. FX sounds

BEHRINGER NEUTRON

92. Monophonic sounds
93. FX sounds
94. Textures and patterns

JHS PRO RHYTHM

95. Analog percussion sounds 1
96. Analog percussion sounds 2
97. Analog percussion sounds 3

The JHS Pro-Rhythm drum synth offered some terrific analog percussion sounds.

YAMAHA EX-1

98. Synths
99. Pedals
100. Drum machine

VIDEO MATERIAL

Also on the website are many video clips featuring artists and instruments covered in the book. These will be updated regularly and some will also appear on a related YouTube channel.

Eurorack modules:

- York Modular, Expert Sleepers, Division 6, Instruo etc. in-depth video demos
- Wunjo Keyboards London, store overview
- Roli Seabord Rise with Equator software video demo by Heen-Wah Wai
- Dixon & Stein "Stranger Things" video interview with soundcheck
- Steve Jolliffe Tangerine Dream *Cyclone* interview, website only
- Suzanne Ciani Buchla 200e video overview
- Yamaha EX-1 video demo
- Moog Sub 37 video demo
- MFOS semi-modular system video demo
- Studio Electronics Boomstar SEM video demo
- Studiologic Sledge video demo
- JHS Pro Rhythm video demo
- Animoog app video demo
- Sunrizer (Horizon) app video demo

...and more

Keep checking the website www.routledge.com/cw/jenkins for new material as it is added.

Index

nted in the United States
Baker & Taylor Publisher Services